Been-Chian Chien and Tzung-Pei Hong (Eds.)

Opportunities and Challenges for Next-Generation Applied Intelligence

Studies in Computational Intelligence, Volume 214

Editor-in-Chief
Prof. Janusz Kacprzyk
Systems Research Institute
Polish Academy of Sciences
ul. Newelska 6
01-447 Warsaw
Poland
E-mail: kacprzyk@ibspan.waw.pl

Further volumes of this series can be found on our homepage: springer.com

Vol. 194. Ian Dempsey, Michael O'Neill and Anthony Brabazon (Eds.)
Foundations in Grammatical Evolution for Dynamic Environments, 2009
ISBN 978-3-642-00313-4

Vol. 195. Vivek Bannore and Leszek Swierkowski
Iterative-Interpolation Super-Resolution Image Reconstruction: A Computationally Efficient Technique, 2009
ISBN 978-3-642-00384-4

Vol. 196. Valentina Emilia Balas, János Fodor and Annamária R. Várkonyi-Kóczy (Eds.)
Soft Computing Based Modeling in Intelligent Systems, 2009
ISBN 978-3-642-00447-6

Vol. 197. Mauro Birattari
Tuning Metaheuristics, 2009
ISBN 978-3-642-00482-7

Vol. 198. Efrén Mezura-Montes (Ed.)
Constraint-Handling in Evolutionary Optimization, 2009
ISBN 978-3-642-00618-0

Vol. 199. Kazumi Nakamatsu, Gloria Phillips-Wren, Lakhmi C. Jain, and Robert J. Howlett (Eds.)
New Advances in Intelligent Decision Technologies, 2009
ISBN 978-3-642-00908-2

Vol. 200. Dimitri Plemenos and Georgios Miaoulis *Visual Complexity and Intelligent Computer Graphics Techniques Enhancements*, 2009
ISBN 978-3-642-01258-7

Vol. 201. Aboul-Ella Hassanien, Ajith Abraham, Athanasios V. Vasilakos, and Witold Pedrycz (Eds.)
Foundations of Computational Intelligence Volume 1, 2009
ISBN 978-3-642-01081-1

Vol. 202. Aboul-Ella Hassanien, Ajith Abraham, and Francisco Herrera (Eds.)
Foundations of Computational Intelligence Volume 2, 2009
ISBN 978-3-642-01532-8

Vol. 203. Ajith Abraham, Aboul-Ella Hassanien, Patrick Siarry, and Andries Engelbrecht (Eds.)
Foundations of Computational Intelligence Volume 3, 2009
ISBN 978-3-642-01084-2

Vol. 204. Ajith Abraham, Aboul-Ella Hassanien, and André Ponce de Leon F. de Carvalho (Eds.)
Foundations of Computational Intelligence Volume 4, 2009
ISBN 978-3-642-01087-3

Vol. 205. Ajith Abraham, Aboul-Ella Hassanien, and Václav Snášel (Eds.)
Foundations of Computational Intelligence Volume 5, 2009
ISBN 978-3-642-01535-9

Vol. 206. Ajith Abraham, Aboul-Ella Hassanien, André Ponce de Leon F. de Carvalho, and Václav Snášel (Eds.)
Foundations of Computational Intelligence Volume 6, 2009
ISBN 978-3-642-01090-3

Vol. 207. Santo Fortunato, Giuseppe Mangioni, Ronaldo Menezes, and Vincenzo Nicosia (Eds.)
Complex Networks, 2009
ISBN 978-3-642-01205-1

Vol. 208. Roger Lee, Gongzu Hu, and Huaikou Miao (Eds.)
Computer and Information Science 2009, 2009
ISBN 978-3-642-01208-2

Vol. 209. Roger Lee and Naohiro Ishii (Eds.)
Software Engineering, Artificial Intelligence, Networking and Parallel/Distributed Computing, 2009
ISBN 978-3-642-01202-0

Vol. 210. Andrew Lewis, Sanaz Mostaghim, and Marcus Randall (Eds.)
Biologically-Inspired Optimisation Methods, 2009
ISBN 978-3-642-01261-7

Vol. 211. Godfrey C. Onwubolu (Ed.)
Hybrid Self-Organizing Modeling Systems, 2009
ISBN 978-3-642-01529-8

Vol. 212. Viktor M. Kureychik, Sergey P. Malyukov, Vladimir V. Kureychik, and Alexander S. Malyoukov
Genetic Algorithms for Applied CAD Problems, 2009
ISBN 978-3-540-85280-3

Vol. 213. Stefano Cagnoni (Ed.)
Evolutionary Image Analysis and Signal Processing, 2009
ISBN 978-3-642-01635-6

Vol. 214. Been-Chian Chien and Tzung-Pei Hong (Eds.)
Opportunities and Challenges for Next-Generation Applied Intelligence, 2009
ISBN 978-3-540-92813-3

Been-Chian Chien and Tzung-Pei Hong (Eds.)

Opportunities and Challenges for Next-Generation Applied Intelligence

Springer

Prof. Been-Chian Chien
Department of Computer Science and Information on Engineering
National University of Tainan
33, Sec. 2, Su-Lin St.
Tainan City 700
Taiwan
E-mail: bcchien@mail.nutn.edu.tw

Prof. Tzung-Pei Hong
Department of Computer Science and Information Engineering
National University of Kaohsiung
700, Kaohsiung University Rd.
Nanzih District, Kaohsiung 811
Taiwan
E-mail: tphong@nuk.edu.tw

ISBN 978-3-540-92813-3 e-ISBN 978-3-540-92814-0

DOI 10.1007/978-3-540-92814-0

Studies in Computational Intelligence ISSN 1860949X

Library of Congress Control Number: Applied for

Typeset & Cover Design: Scientific Publishing Services Pvt. Ltd., Chennai, India.

Printed in acid-free paper

9 8 7 6 5 4 3 2 1

springer.com

Preface

The term "Artificial Intelligence" has been used since 1956 and has become a very popular research field. Generally, it is the study of the computations that enable a system to perceive, reason and act. In the early days, it was expected to achieve the same intelligent behavior as a human, but found impossible at last. Its goal was thus revised to design and use of intelligent methods to make systems more efficient at solving problems. The term "Applied Intelligence" was thus created to represent its practicality. It emphasizes applications of applied intelligent systems to solve real-life problems in all areas including engineering, science, industry, automation, robotics, business, finance, medicine, bio-medicine, bio-informatics, cyberspace, and man-machine interactions.

To endow the intelligent behavior of a system, many useful and interesting techniques have been developed. Some of them are even borrowed from the natural observation and biological phenomenon. Neural networks and evolutionary computation are two examples of them. Besides, some other heuristic approaches like data mining, adaptive control, intelligent manufacturing, autonomous agents, bio-informatics, reasoning, computer vision, decision support systems, expert systems, fuzzy logic, robots, intelligent interfaces, internet technology, planning and scheduling, are also commonly used in applied intelligence.

The book belongs to the "Studies in Computational Intelligence" series published by Springer Verlag. It discusses both the chances and the difficulties of applied intelligent systems in all aspects. It consists of 52 chapters authored by participants to the 22nd International Conference on Industrial, Engineering & Other Applications of Applied Intelligent Systems (IEA-AIE 2009). Each chapter was reviewed by at least two anonymous referees to assure the high quality. The book is divided into 12 parts including Bio-informatics, Computer Vision, Data Mining and Knowledge Discovery, Decision Support Systems, Genetic Algorithms, Intelligent Design & Manufacturing, Intelligent Systems, Integration Systems for Real Life Applications, Machine Learning, Multi-Agent Systems, Planning and Scheduling, and Mining Interesting Knowledge. The material of each part of this book is self-contained. Many approaches, applications, restrictions, and discussions are presented in the chapters. Readers can take any individual part according to their research interests. We hope that this book can provide some useful reference values to the researchers and the students in the field of applied intelligence. We also hope the readers can find opportunities and recognize challenges in the development and applications of intelligent systems.

We would like to thank Springer Verlag, Professor Janusz Kacprzyk, the editor of this series, for their help in publishing the book. We would also like to thank Professor Ngoc Thanh Nguyen for his advice and the reviewers for their hard work in assuring the high quality of the book. Besides, a grateful appreciation is expressed to Professors Chih-Chin Lai, Wen-Yang Lin, Chao-Lin Liu, Hsin-Chang Yang, Rong-Ming Chen, Tsung-Yen Chuang, Jiann-Shu Lee, Ming-Yi Ju and Mang-Chun Wang for their help that made the conference and the book possible. Finally, we cordially thank all the authors, who did important contributions to the book. Without all the efforts, this book can not be successfully carried out.

March 2009

Been-Chian Chien
Tzung-Pei Hong

Contents

Bio-informatics

An Intelligent System for Analyzing Gene Networks from Microarray Data 1
Austin H. Chen, Ching-Heng Lin

Predicting Residue-Wise Contact Orders in Proteins by Support Vector Regression with Parametric-Insensitive Model 7
Pei-Yi Hao, Lung-Biao Tsai

SNP-Flankplus: SNP ID-Centric Retrieval of Flanking Sequences 13
Cheng-Hong Yang, Yu-Huei Cheng, Li-Yeh Chuang, Hsueh-Wei Chang

Tumor Classification on Mammographies Based on BPNN and Sobel Filter 19
Enrique Calot, Hernan Merlino, Claudio Rancan, Ramon Garcia-Martinez

Computer Vision

An Adaptive Biometric System Based on Palm Texture Feature and LVQ Neural Network 25
Chen-Sen Ouyang, Ming-Yi Ju, Han-Lin Yang

A Binarization Approach for Wafer ID Based on Asterisk-Shape Filter 31
Hsu Wei-Chih, Yu Tsan-Ying, Chen Kuan-Liang

Feature Particles Tracking for the Moving Object 39
Tao Gao, Zheng-guang Liu, Jun Zhang

Break-Segment Detection and Recognition in Broadcasting Video/Audio Based on C/S Architecture 45
Xiangdong Wang, Xinhui Li, Yuelian Qian, Ying Yang, Shouxun Lin

Data Mining and Knowledge Discovery

A Time Series Case-Based Predicting Model for Reservation Forecasting 53
Tsung-Hsien Tsai, Sheryl E. Kimes

A Filtering Approach for Mining Frequent Itemsets 59
Jen-Peng Huang, Huang-Cheng Kuo

Mining Data Streams with Skewed Distribution by Static Classifier Ensemble 65
Yi Wang, Yang Zhang, Yong Wang

Accumulative Influence Weight Collaborative Filtering Recommendation Approach.................................. 73
Nan Li, Chunping Li

RAD: A Radar-Alike Data-Clustering Algorithm for Large Databases 79
Cheng-Fa Tsai, Jiun-Huang Ju

Decision Support Systems

Applying ELECTRE and Maximizing Deviation Method for Stock Portfolio Selection under Fuzzy Environment 85
Chen-Tung Chen, Wei-Zhan Hung

Constraint Programming Approach for a University Timetabling Decision Support System with Hard and Soft Constraints 93
Li-Yen Shue, Pei-Chun Lin, Chia-Yin Tsai

Behavioral Scoring Model for Bank Customers Using Data Envelopment Analysis 99
I-Fei Chen, Chi-Jie Lu, Tian-Shyug Lee, Chung-Ta Lee

Decision Support for Combinatorial Reverse Auction with Multiple Buyers and Sellers 105
Fu-Shiung Hsieh, Cheng Chung Hua

Visualization-Based Approaches to Support Context Sharing towards Public Involvement Support System 111
Shun Shiramatsu, Yuji Kubota, Kazunori Komatani, Tetsuya Ogata, Toru Takahashi, Hiroshi G. Okuno

Genetic Algorithms

A Case Study of Genetic Algorithms for Quay Crane Scheduling 119
Yi Wang, Yun Chen, Kesheng Wang

Genetic Algorithm and Time-Scale Separation on the Reduced-Order Observer-Based Control Design for a Class of Discrete Systems 127
Shing-Tai Pan, Ching-Fa Chen

AKDB–Tree: An Adjustable KDB-Tree for Efficiently Supporting Nearest Neighbor Queries in P2P Systems 133
Ye-In Chang, Lee-Wen Huang, Hung-Ze Liu

Optimal Lot Sizing and Algorithm for Dynamic Pricing under Random Yield and Demand 139
Guo Li, Shihua Ma, Wenya Chou

Intelligent Design and Manufacturing

Collaborative Production Machine Choice System Based on Hierarchical Case Based Reasoning 147
Shian-Shyong Tseng, Fengming M. Chang, Huan-Yu Lin, Chi-Chun Hsu, Chung-Chao Ku, Jui-Pin Tsai, Jun-Ming Su

Passive Analog Filter Design Using GP Population Control Strategies 153
Mariano Chouza, Claudio Rancan, Osvaldo Clua, Ramón García-Martínez

Thruster Fault-Tolerant for UUVs Based on Quantum-Behaved Particle Swarm Optimization 159
Jing Liu, Qi Wu, Daqi Zhu

Intelligent System for Locating, Labeling, and Logging (ISL^3) 167
Yan Huang, Terry Griffin, Salomon Lompo

Probabilistic Analysis of Information Center Insecurity 175
Shyue-Liang Wang, Jyun-Da Chen, Tzung-Pei Hong, Paul A. Stirpe

Intelligent Systems

Application of Combined Forecasting Models to Intelligent Transportation Systems 181
Yang Zhang, Yuncai Liu

Fuzzy Performance Analysis Model Based on Grid Environment 187
Huey-Ming Lee, Chia-Hsien Chung, Tsang-Yean Lee, Jin-Shieh Su

Emotional Processes in Computer Games 193
Khalil Shihab, Nida Chalabi

Probability Reasoning Internet Assessment System with Rule Usage Structures and Similarity Comparisons 199
Yuan-Horng Lin

Integration Systems for Real Life Applications

Combining Genetic Algorithm and Simulated Annealing to Design $H_2/H_{i}nfty$ Deconvolution Filter with Missing Observations 207
Jui-Chung Hung, Wei-Chi Lee

The Hybrid Model Development of Clustering and Back Propagation Network in Printed Circuit Board Sales Forecasting 213
Yen-Wen Wang, Chen-Hao Liu, Chin-Yuan Fan

Toward the Knowledge Circulation between Institutions and Public Audiences with Virtual Agents 219
Hung-Hsuan Huang, Hidekazu Kubota, Toyoaki Nishida

Muon Tomography Algorithms for Nuclear Threat Detection 225
Richard Hoch, Debasis Mitra, Kondo Gnanvo, Marcus Hohlmann

Machine Learning

A Model-Based Software Reasoning Approach to Software Debugging 233
Rui Abreu, Peter Zoeteweij, Arjan J.C. van Gemund

Learning Rules from Multiple Criteria Decision Tables 241
Chien-Chung Chan

A Mobile Location Algorithm with Least Range and Clustering Techniques for NLoS Environments 247
Cha-Hwa Lin, Chien-Chih Wang, Chih-Hung Tsai

Learning Interaction Structure Using a Hierarchy of Dynamical Systems 253
Yasser Mohammad, Toyoaki Nishida

Fuzzy Assessment for Survey Defuzzified by Signed Distance Method 259
Lily Lin, Huey-Ming Lee

Multi-Agent Systems

Multi-Mobile Agent Based Virtual Alliance Formation (MAVAF) Platform 265
Chang Yu Chiang, Chuan Jun Su

On Supporting Cross-Platform Statistical Data Analysis Using JADE 271
Chien-Ho Wu, Yuehjen E. Shao, Jeng-Fu Liu, Tsair-Yuan Chang

Building Agents That Learn by Observing Other Agents Performing a Task: A Sequential Pattern Mining Approach 279
Philippe Fournier-Viger, Roger Nkambou, Engelbert Mephu Nguifo, Usef Faghihi

A Multi-agent System Architecture for Personal Support during Demanding Tasks 285
Tibor Bosse, Rob Duell, Mark Hoogendoorn, Michel Klein, Rianne van Lambalgen, Andy van der Mee, Rogier Oorburg, Alexei Sharpanskykh, Jan Treur, Michael de Vos

Planning and Scheduling

The Selective Traveling Salesman Problem with Regular Working Time Windows 291
Hu Qin, Andrew Lim, Dongsheng Xu

Scheduling for Dedicated Machine Constraint Using the Shortest Paths in Graphs 297
Huy Nguyen Anh Pham, Arthur Shr, Peter P. Chen, Alan Liu

Comparing Path Length by Boundary Following Fast Matching Method and Bug Algorithms for Path Planning ... 303
Chia Hsun Chiang, Jing-Sin Liu, Yo-Shing Chou

A Capacitated Vehicle Routing Problem with Toll-by-Weight Rule 311
Chenghao Shen, Hu Qin, Andrew Lim

Mining Interesting Knowledge

A Novel Multidimensional Scaling Technique for Mapping Word-Of-Mouth Discussions 317
B. John Oommen, Ole-Christoffer Granmo, Zuoyuan Liang

Comprehensible Knowledge Discovery Using Particle Swarm Optimization with Monotonicity Constraints 323
Chih-Chuan Chen, Chao-Chin Hsu, Yi-Chung Cheng, Sheng-Tun Li, Ying-Fang Chan

Incorporation of Adapted Real World Tournament Selection into Artificial Immune Recognition System 329
Shahram Golzari, Shyamala Doraisamy, Md. Nasir Sulaiman, Nur Izura Udzir

Competition State Visualization for Sales Record Mining 335
Yoshihiro Hayashi, Ryosuke Saga, Hiroshi Tsuji

Author Index 341

An Intelligent System for Analyzing Gene Networks from Microarray Data

Austin H. Chen* and Ching-Heng Lin

Abstract. High-throughput techniques, such as microarray experiments, have given biologists the opportunity to measure the expression levels of a huge amount of genes at the same time. How to utilize these huge amounts of data, however, has become a major challenge in the post-genomic research era. One approach utilizes a Bayesian network, a graphical model that has been applied toward inferring genetic regulatory networks from microarray experiments. However, a user-friendly system that can display and analyze various gene networks from microarray experimental datasets is now needed. In this paper, we developed a novel system for constructing and analyzing gene networks. Firstly, we developed five Bayesian network algorithms to construct gene networks of the yeast cell cycle from four different microarray datasets. Secondly, we implemented a user-friendly gene network analyzing system. GNAnalyzer is capable of generating gene networks of the yeast cell cycle from experimental microarray data but also analyzing the performance of gene networks for every algorithm. Thirdly, our system utilizes both the powerful processing abilities of MatLab and the dynamic interface of LabVIEW in a single platform.

1 Background

Recent progress in the field of molecular biology enables us to obtain huge amounts of data. High-throughput techniques, such as microarray experiments, have given biologists the opportunity to measure the expression levels of a huge amount of genes at the same time. As a result, the rapidly increasing amount of known sequence data, or massive gene expression data, requires computational

Austin H. Chen
Department of Medical Informatics, Tzu Chi University, 701, Sec. 3, Jhongyang Rd., Hualien City, Hualien County 97004, Taiwan
e-mail: achen@mail.tcu.edu.tw

Ching-Heng Lin
Graduate Institute of Medical Informatics, Tzu Chi University, 701, Sec. 3, Jhongyang Rd., Hualien City, Hualien County 97004, Taiwan
e-mail: whitesky0453@yahoo.com.tw

* Corresponding author.

B.-C. Chien and T.-P. Hong (Eds.): Opportunities and Challenges, SCI 214, pp. 1–6.
springerlink.com

effort to extract information from them. Since the information about the location of genes in the cell is now available due to the completion of the Human Genome Project, it is necessary to reveal the causal connections between different genes in order to fully understand the cell's dynamics.

Several significant studies have attempted to establish a method to infer a gene regulatory network from large-scale gene expression data. Two types of gene expression data are obtained as either time series or steady-state data. Recently, genome-wide gene expression time series microarray data relevant to the yeast cell cycle has been collected [2, 8]. Several analyzing techniques were developed for inferring gene networks from time series data, such as information theory [6], genetic algorithms [7], or simulated annealing. On the other hand, the steady-state data can be obtained by altering specific gene activities, such as gene knockouting in a variety of experimental animals. Previous efforts at modeling gene networks from steady-state gene expression data have generally fallen into one of two classes, either employing Boolean networks, which are restricted to logical relationships between variables, or using systems of differential equations to model the continuous dynamics of coupled biological reactions.

Using Bayesian networks in the construction of gene networks from gene expression data was initialized by the work of Friedman et al. [4]. A Bayesian network is a graphical model that finds probabilistic relationships among variables (i.e. genes) of the system. Bayesian networks successfully amalgamate probability theory and graph theory to efficiently model multidimensional probability distributions by searching for independent relationships in the data [1,5]. Currently, however, most Bayesian network algorithms were written either by Matlab, C++, or Java. There are needs to implement these algorithms into a graphical environment to provide more user-friendly interfaces. In this paper, we developed a system that can integrate these algorithms and powerful dynamic interfaces (such as LabVIEW) in a single platform. This novel design provided a visualization approach to display the key results. Meanwhile, a user-friendly system that can display and analyze various gene networks from microarray experimental datasets is urgently needed. In this study, our goal is to develop a gene network analyzing system (GNAnalyzer) that can generate the gene networks of the yeast cell cycle from experimental microarray data as well as analyze the performance of gene networks using five different Bayesian network algorithms.

2 Methods

We used three kinds of datasets in this study, including the Alarm network, the Asia network, and the yeast cell cycle network. The first two datasets were commonly used in Bayesian networks since the known structure of the Alarm and Asia networks can be used to compare the performance of different Bayesian network algorithms. The third dataset is S. cerevisiae cell cycle gene expression data collected by Spellman et al. [8]. This dataset contains four medium time series: 18, 24, 17 and 14 time series points for alpha, cdc15, cdc28 and elu respectively. In the assessment of a gene network, we use each of the three medium time

Table 1 Partial conditional probability tables of genes in CDC15 dataset

	RNR3					CDC5		
CLN2		-1	0	1	ACE2		-1	0	1
	-1	0	0	0.556		-1	1	0	0
	0	0.5	0.444	0.444		0	0	0.909	0.375
	1	0.5	0.556	0		1	0	0.091	0.625
	CLB2					CLN2			
CLN1		-1	0	1	CLB2		-1	0	1
	-1	0	0.143	0.6		-1	0	0.25	0.625
	0	0.286	0.714	0.3		0	0.125	0.375	0.375
	1	0.714	0.143	0.1		1	0.875	0.375	0

series: alpha, cdc15, and cdc28. The gene expression data in Spellman's experiment is first normalized into the value of log2. We then categorize these values into three classes based on Friedman's threshold value of 0.5 [4]. These classes are represented by 3 discrete values: under-expressed (-1), normal expressed (0), and over-expressed (+1). The results were compared with a known YPL256C sub network [3]. As an example of how to calculate the partial conditional probability among genes, the data calculated between gene CLN2 and gene RNR3 in CDC15 conditions is computed and shown on Table 1. From Table 1, the conditional probability of P(RNR3|CLN2) can be expressed as P(-1|-1) = 0.0, P(0|-1) = 0.0, P(1|-1) = 0.556, P(-1|0) = 0.5, P(0|0) = 0.444, P(1|0) = 0.444, P(-1|1) = 0.5, P(0|1) = 0.556, P(1|1) = 0. The conditional probability is 0.0 when both RNR3 and CLN2 are under-expressed as well as 0.444 and 0.0 when both RNR3 and CLN2 are normal expressed and over-expressed. The Bayesian Gene Networks are then generated from these values using five algorithms.

3 Results and Discussion

The cell cycle networks were used in this study to compare the performance for five different Bayesian network algorithms. K2 is the most widely used algorithm in Bayesian network structure learning. It is well known as a general method for inferring inter-node relations in a given node group based on a complete database free of missing data. The result of the cell cycle is shown in Figure 1.

The MWST algorithm was developed by Chow and Liu. This algorithm searches for an optimal tree structure by using the computed mutual information as edge weights. The MWST associates a weight to each connection, where each weight represents the mutual information between the two variables. When the weight matrix is created, the MWST algorithm gives an optimal tree structure. The result of the cell cycle is shown in Figure 2.

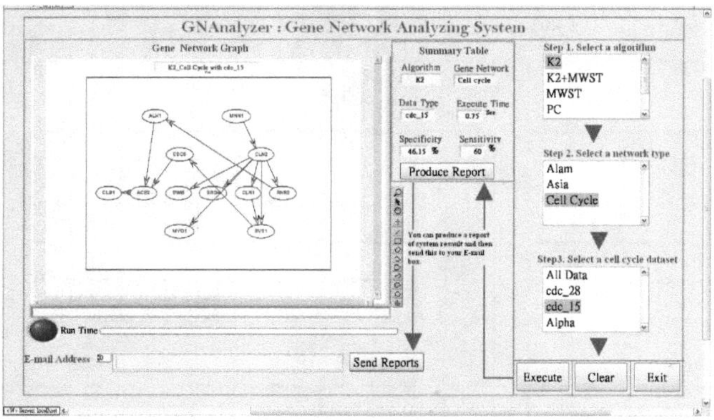

Fig. 1 A screen of the system interface after the execution is completed for the K2 algorithm and the Cell cycle network

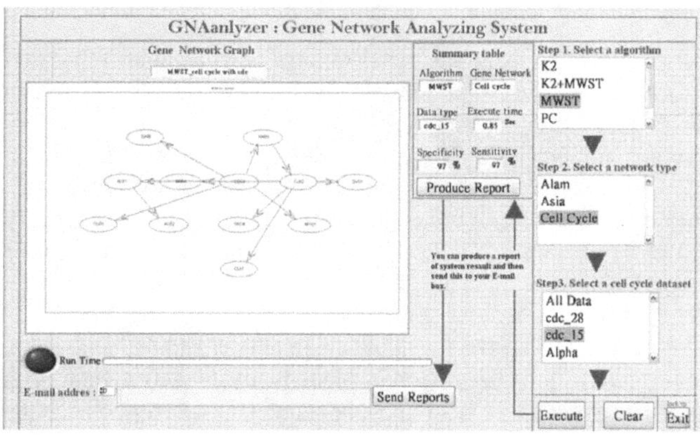

Fig. 2 A screen of the system interface after the execution is completed for the MWST algorithm and the Cell cycle network

4 Description of GNAnalyzer

In this study, we developed a gene network analyzing system (GNAnalyzer) that can generate the gene networks of the yeast cell cycle from experimental microarray data as well as analyze the performance of gene networks using five different Bayesian network algorithms. The main functions of the interface include: (1) Algorithm selection section: users select the desired algorithm. (2) Network selection section: users select a desired network. (3) Dataset selection section: only displayed if the cell cycle button is clicked. (4) Execute icon: users click to run the program. (5) Clear icon: users click to clear all selections and release memory space. (6) Gene network display section: displays the resulting gene network. (7) Status slide bar section: shows the current execute status. (8) Summary table: displays summaries of users' requests. (9) Report icon: users click to generate reports that include the network graph and summary table. (10) Send icon: the system will

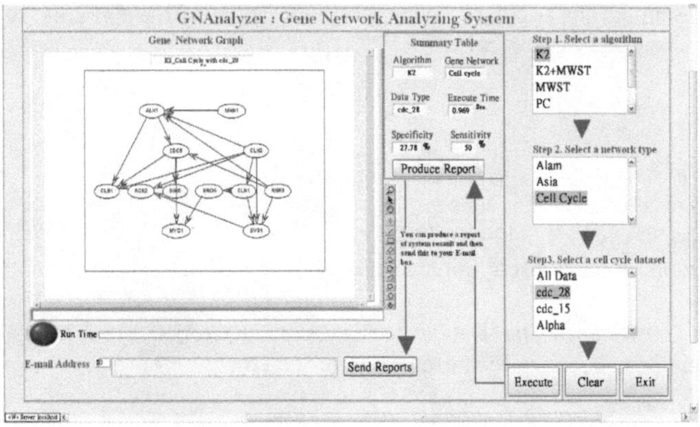

Fig. 3 A screen of the system interface when the execution is finished

send a report to the user through e-mail. (11) Exit icon: exits the interface. When the users select one of the five Bayesian Network algorithms, the available network will be automatically displayed. Furthermore, if the users select cell cycle, four dataset types will then be displayed. After selecting the data set and clicking the Execute button, the light will turn to green. The red light informs the user that the process is currently running. The slider bar on the right-hand side will show the execution status. Figure 3 shows the final computation time and the gene network for the selected conditions. The result column will display information for the users, including algorithm, network type, dataset type, computation time, sensitivity, and specificity. The Clear button is also provided in case the users wish to clear the information. When the Clear button is clicked, the memory space will be released. As one of the features in this system, we provide this function in order to avoid the memory overflow problems that usually found when the huge bioinformatics data such as biochip data is operated.

5 Conclusions

In this paper, we have developed a novel system to construct and analyze various gene networks from microarray datasets. Three aspects characterize the major contributions of this paper.(1) Five Bayesian network algorithm codes were developed and written to construct gene networks of the yeast cell cycle using the information from four different microarray datasets. (2) A gene network analyzing system, GNAnalyzer, consisting of several user-friendly interfaces was implemented. GNAnalyzer is capable of running Bayesian algorithms, constructing gene networks, and analyzing the performance of each network algorithm simultaneously. (3) The system utilizes both the powerful processing ability of MatLab and the dynamic interfaces of LabVIEW in a single platform. The system is designed to be extendible. Our next goal is to apply this technique to other real biomedical applications, such as human cancer classification and prognostic prediction.

Acknowledgements. The authors thank the National Science Council of the Republic of the China for their financial support of project NSC 97-2221-E-320-004.

References

1. Beal, et al.: A Bayesian approach to reconstructing genetic regulatory networks with hidden factors. Bioinformatics 21(3), 349–356 (2005)
2. Cho, R.J., et al.: A genome-wide transcriptional analysis of the mitotic cell cycle. Mol. Cell 2, 65–73 (1998)
3. Dejori, J.: Analyzing Gene-Expression Data with Bayesian Networks, MS The-sis, Elektro- und Biomedizinische Technik Technische Universität Graz (2002)
4. Friedman, N., Linial, M., Nachman, I., Pe'er, D.: Using Bayesian networks to Analyze Expression data. Journal of Computational Biology 7, 601–620 (2000)
5. Heckerman, D., Geiger, D., Chickering, D.: Learning Bayesian networks: The combination of knowledge and statistical data. Machine Learning 20(3), 197–243 (1995)
6. Liang, S., Fuhrman, S., Somogyi, R.: REVEAL: a general reverse engineering algorithm for inference of genetic network. In: Proc. Pacific Symp. Biocomputing 1998, pp. 18–29 (1998)
7. Morohashi, M., Kitano, H.: Identifying gene regulatory networks from time series expression data by in silico sampling and screening. In: Proc. 5th Euro. Conf. Artificial Life, pp. 477–486 (1999)
8. Spellman, P.T., et al.: Comprehensive Identification of cell cycleregulated genes of the yeast saccharomyces cerevisiae by microarray hybridization. Molecular Biology of the Cell 9, 3273–3297 (1998)

Predicting Residue-Wise Contact Orders in Proteins by Support Vector Regression with Parametric-Insensitive Model

Pei-Yi Hao and Lung-Biao Tsai

Abstract. A major challenge in structural bioinformatics is the prediction of protein structure and function from primary amino acid sequences. The residue-wise contact order (RWCO) describes the sequence separations between the residues of interest and its contacting residues in a protein sequence. RWCO provides comprehensive and indispensable important information to reconstructing the protein three-dimensional structure from a set of one-dimensional structural properties. Accurately predicting RWCO values could have many important applications in protein three-dimensional structure prediction and protein folding rate prediction, and give deep insights into protein sequence-structure relationships. In this paper, we developed a novel approach to predict residue-wise contact order values in proteins based on support vector regression with parametric insensitive model.

1 Introduction

A major challenge in structural bioinformatics is the prediction of protein structure and function from primary amino acid sequences. This problem is becoming more pressing now as the protein sequence-structure gap is widening rapidly as a result of the completion of large-scale genome sequencing projects [1,2]. As an intermediate but useful step, predicting a number of key properties of proteins including secondary structure, solvent accessibility, contact numbers and contact order is a possible and promising strategy, which simplifies the prediction task by projecting the three-dimensional structures onto one dimension, i.e. strings of residue-wise structural assignments [6].

Residue-wise contact order (RWCO) is a new kind of one dimensional protein structure representing the extent of long-range contacts, which is a sum of sequence separations between the given residue and all the other contacting residues [8,9,11]. RWCO provides comprehensive and indispensable important information

Pei-Yi Hao and Lung-Biao Tsai
Department of Information Management, National Kaohsiung University of Applied Sciences, Kaohsiung, Taiwan
e-mail: haupy@cc.kuas.edu.tw

B.-C. Chien and T.-P. Hong (Eds.): Opportunities and Challenges, SCI 214, pp. 7–12.
springerlink.com

to reconstructing the protein three-dimensional structure from a set of one-dimensional structural properties. Kinjo et al. first proposed a simple linear regression method to predict RWCO values and the local sequence information with multiple sequence alignments in the form of PSI-BLAST profiles was extracted using a sliding window scheme centered on the target residue [8]. Song et al. first adopt a support vector regression (SVR) algorithm to predict residue-wise contact order values in proteins, starting from primary amino acid sequences [11]. In this paper, a new SV regression algorithm, called *par-v*-SVR, is proposed by using a parametric insensitive loss function such that the corresponding insensitive zone of *par-v*-SVR can have arbitrary shape. This can be useful in situations where the noise is heteroscedastic, that is, where it depends on **x**.

2 SV Regression with Parametric Insensitive Model

Support Vector (SV) machines comprise a class of learning algorithms, motivated by results of statistical learning theory [12]. Originally developed for pattern recognition, they represent the decision boundary in terms of a typically small subset of all training examples, called the Support Vectors. In order for this property to carry over to the case of SV Regression, Vapnik devised the so-called ε-insensitive loss function [12]:

$$|y - f(\mathbf{x})|_{\varepsilon} = \max\{0, |y - f(\mathbf{x})| - \varepsilon\} \tag{2.1}$$

which does not penalize errors below some ε>0, chosen a priori. To motivate the new algorithm that shall be proposed, note that the parameter ε in original support vector regression (SVR) algorithm can be useful if the desired accuracy of the approximation can be specified beforehand. Besides, the ε-insensitive zone in the SVR is assumed to has a tube (or slab) shape. Namely, the radius of the insensitive zone is a user-predefined constant, and we do not care about the errors as long as they are inside the ε-insensitive zone. The selection of a parameter ε may seriously affect the modeling performance. In this paper, an new SV algorithm, called *par-v*-SVR, is derived to evaluate the interval regression model by using a new parametric-insensitive loss function, which automatically adjusts the interval to include all data [4]. The parametric-insensitive loss function is defined by

$$|y - f(\mathbf{x})|_{g} := \max\{0, |y - f(\mathbf{x})| - g(\mathbf{x})\} \tag{2.2}$$

where f and g are real-valued functions on the a domain R^n, $\mathbf{x} \in R^n$ and $y \in R$. The basic idea of SV regression is that a nonlinear regression function is achieved by simply mapping the input patterns $\mathbf{x}_i$ by Φ: $R^n \rightarrow$ F into a high-dimensional feature space F. Hence, the proposed *par-v*-SVR seeks to estimate the following two functions:

$$f(\mathbf{x}) = \langle \mathbf{w} \cdot \Phi(\mathbf{x}) \rangle + b \text{, where } \mathbf{w} \in F \text{, } \mathbf{x} \in R^n \text{, } b \in R \text{,}$$

$$g(\mathbf{x}) = \langle \mathbf{c} \cdot \Phi(\mathbf{x}) \rangle + d \text{, where } \mathbf{c} \in F \text{, } \mathbf{x} \in R^n \text{, } d \in R \text{.}$$

The problem of finding the **w**, **c**, b, and d that minimize the empirical risk $R^g_{emp}[f] = \frac{1}{N}\sum_{i=1}^{N} |y_i - f(\mathbf{x}_i)|_g$ is equivalent to the following optimization problem:

$$\underset{\mathbf{w},\mathbf{c},b,d,\xi_i,\xi_i^*}{\text{minimize}} \ \frac{1}{2}\|\mathbf{w}\|^2 + C\left(v \cdot \left(\frac{1}{2}\|\mathbf{c}\|^2 + d \right) + \frac{1}{N}\sum_{i=1}^{N} \left(\xi_i + \xi_i^* \right) \right)$$

subject to

$$\left(\langle \mathbf{w} \cdot \Phi(\mathbf{x}_i) \rangle + b \right) + \left(\langle \mathbf{c} \cdot \Phi(\mathbf{x}_i) \rangle + d \right) \geq y_i - \xi_i \tag{2.3}$$

$$\left(\langle \mathbf{w} \cdot \Phi(\mathbf{x}_i) \rangle + b \right) - \left(\langle \mathbf{c} \cdot \Phi(\mathbf{x}_i) \rangle + d \right) \leq y_i + \xi_i^* \quad \text{and} \quad \xi_i, \xi_i^* \geq 0 \quad \text{for } i=1,\ldots,N.$$

Using the Lagrangian theorem, the dual problem can be formulated as

$$\text{maximize} \begin{cases} \dfrac{-1}{2}\displaystyle\sum_{i=1}^{N}\sum_{j=1}^{N} (\alpha_i - \alpha_i^*)(\alpha_j - \alpha_j^*) \langle \Phi(\mathbf{x}_i) \cdot \Phi(\mathbf{x}_j) \rangle \\ -\dfrac{1}{2Cv}\displaystyle\sum_{i=1}^{N}\sum_{j=1}^{N} (\alpha_i + \alpha_i^*)(\alpha_j + \alpha_j^*) \langle \Phi(\mathbf{x}_i) \cdot \Phi(\mathbf{x}_j) \rangle + \sum_{i=1}^{N} (\alpha_i - \alpha_i^*) y_i \end{cases}$$

subject to (2.4)

$$\sum_{i=1}^{N} (\alpha_i - \alpha_i^*) = 0, \quad \sum_{i=1}^{N} (\alpha_i + \alpha_i^*) = C \cdot v, \quad \alpha_i, \alpha_i^* \in \left[0, \frac{C}{N} \right]. \quad \text{for } i=1,\ldots,N.$$

From the Karush-Kuhn-Tucker (KKT) conditions, parameters b and d can be computed as follows:

$$b = \frac{-1}{2}\left(\langle \mathbf{w} \cdot \Phi(\mathbf{x}_i) \rangle + \langle \mathbf{w} \cdot \Phi(\mathbf{x}_j) \rangle + \langle \mathbf{c} \cdot \Phi(\mathbf{x}_i) \rangle - \langle \mathbf{c} \cdot \Phi(\mathbf{x}_j) \rangle - y_i - y_j \right) \tag{2.5}$$

$$d = \frac{-1}{2}\left(\langle \mathbf{w} \cdot \Phi(\mathbf{x}_i) \rangle - \langle \mathbf{w} \cdot \Phi(\mathbf{x}_j) \rangle + \langle \mathbf{c} \cdot \Phi(\mathbf{x}_i) \rangle + \langle \mathbf{c} \cdot \Phi(\mathbf{x}_j) \rangle - y_i + y_j \right) \tag{2.6}$$

for some $\alpha_i, \alpha_j^* \in (0, C/N)$.

3 The Concept of Residue-Wise Contact Orders

Prediction of protein three-dimensional structure from primary sequence is the central problem in structural bioinformatics. One protein structural feature is of particular interest here, namely, residue-wise contact order (RWCO) which can be used to enhance protein fold recognition. The concept of residue-wise contact order (RWCO) was first introduced by Kinjo and Nishikawa [8,9]. The discrete RWCO values of the *i*-th residue in a protein sequence with M residues is defined by

$$RWCO_i = \frac{1}{M} \sum_{j:|j-i|>2}^{M} |i-j| \sigma(r_{i,j}) \quad \begin{cases} \sigma(r_{i,j}) = 1, & if\ r_{i,j} < r_d \\ \sigma(r_{i,j}) = 0, & if\ r_{i,j} \geq r_d \end{cases} \tag{3.1}$$

where $r_{i,j}$ is the distance between the C atoms of the *i*-th and *j*-th residues (C atoms for glycine) in the protein sequence. Two residues are considered to be in contact if their C atoms locate within a sphere of the threshold radius r_d. Note that the trivial contacts between the nearest and second-nearest residues are excluded. In order to smooth the discrete RWCO values, Kinjo et al. proposed a particular sigmoid function [8,9], which is given by

$$\sigma(r_{i,j}) = 1/\{1 + \exp[w(r_{i,j} - r_d)]\} \tag{3.2}$$

where w is a parameter that determines the sharpness of the sigmoid function. In the present study, for the sake of comparison, we set rd = 12 Å and w = 3, which was adopted by Kinjo et al. [8,9].

4 Experiments

In this experiment, we apply the proposed *par-ν*-SVR to predict residue-wise contact order values in proteins, starting from primary amino acid sequences. The Gaussian kernel $k(\mathbf{x},\mathbf{y}) = \exp\left(-\|\mathbf{x}-\mathbf{y}\|^2 / 2\sigma^2\right)$ is used here. The optimal choice of parameters C, ν and σ was tuned using a grid search mechanism. We used the same dataset previously prepared by Kinjo and Nishikawa [8,9], which included 680 protein sequences and was originally extracted from ASTRAL database version 1.65 [3]. There are a total of 120421 residues in this dataset. The protein chain names and their corresponding amino acid sequences, and the detailed RWCO information with a radius cutoff of 12Å can be found in [11]. To measure the performance of *par-ν*-SVR methods in this application, we calculated the Pearson's correlation coefficients (CC) between the predicted and observed RWCO values in a protein sequence as given by

$$CC = \sum_{i=1}^{N} (x_i - \bar{x})(r_i - \bar{r}) \Big/ \sqrt{\left[\sum_{i=1}^{N} (x_i - \bar{x})^2\right]\left[\sum_{i=1}^{N} (r_i - \bar{r})^2\right]} \tag{4.1}$$

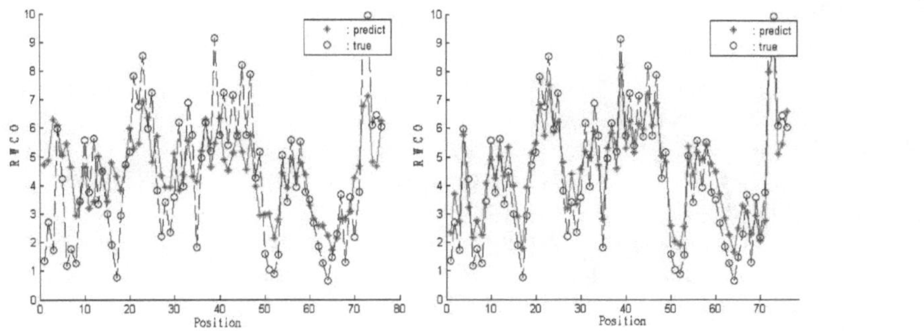

(a) (b)

Fig. 4.1 The predicted RWCO for protein d1n7oa2 obtained by (a) the original SVR and (b) the proposed *par-v*-SVR, respectively. RWCO values are used with a radius cutoff of 12 A. Observed and predicted RWCO are represented by solid and dashed lines, respectively

where x_i and r_i are the observed and predicted normalized RWCO values of the i-th residue, and $\overline{x}$ and $\overline{r}$ are their corresponding means. Here N is the total residue number in a protein. The root mean square error (RMSE) is also given by

$$RMSE = \sqrt{\frac{1}{N}\sum_{i=1}^{N}(r_i - x_i)^2} \tag{4.2}$$

The predicted performances were evaluated on the whole datasets by 5-fold cross-validation. In original SVR, the correlation coefficient (CC) between predicted and observed residue-wise contact order (RWCO) can reach 0.695, with normalized root mean square error (RMSE) less than 2.0545. In our approach, the use of parametric insensitive model can increase the correlation coefficient to 0.734 and decrease the root mean square error to 1.9719. To illustrate the performance in this study, fig 4.1 shows the predicted and observed RWCO values in protein d1n7oa2 obtained by the original SVR and the proposed *par-v*-SVR, respectively.

5 Conclusion

In the present study, we proposed a novel method to predict the RWCO profiles from amino acid sequences based on support vector regression with parametric insensitive model (*par-v*-SVR). Different from the linear regression approach, our method uses the non-linear radial basis kernel function (RBF) to approximate and determine the sequence-RWCO relationship. We compared our prediction accuracy with original SVR. The experimental results show that the proposed *par-v*-SVR is slightly better than the original SVR in predicting protein structural profile values and describing sequence-structure relationships.

References

1. Bairoch, A., Apweiler, R.: The SWISS-PROT protein sequence data bank and its supplement TrEMBL. Nucleic Acids Res. 28, 45–48 (2000)
2. Berman, H.M., et al.: The Protein Data Bank. Nucleic Acids Res. 28, 235–242 (2000)
3. Chandonia, J.M., et al.: The ASTRAL Compendium in 2004. Nucleic Acids Res. 32, D189–D192 (2004)
4. Hao, P.Y.: Shrinking the Tube: A New Support Vector Regression Algorithm with Parametric Insensitive Model. In: The 6th International Conference on Machine Learning and Cybernetics, Hong Kong, China (2007)
5. Jones, D.T.: Protein secondary structure prediction based on position-specific scoring matrices. J. Mol. Biol. 292, 195–202 (1999)
6. Kinjo, A.R., Nishikawa, K.: Recoverable one-dimensional encoding of three-dimensional protein structures. Bioinformatics 21, 2167–2170 (2005)
7. Kihara, D.: The effect of long-range interactions on the secondary structure formation of proteins. Protein Sci. 14, 1955–1963 (2005)
8. Kinjo, A.R., Nishikawa, K.: Predicting Residue-wise Contact Orders of Native Protein Structure from Amino Acid Sequence. arXiv.org. q-bio.BM/0501015 (2005)
9. Kinjo, A.R., Nishikawa, K.: Predicting secondary structures, contact numbers, and residue-wise contact orders of native protein structure from amino acid sequence using critical random networks. Biophysics 1, 67–74 (2005)
10. Plaxco, K.W., et al.: Contact order, transition state placement and the refolding rates of single domain proteins. J. Mol. Biol. 277, 985–994 (1998)
11. Song, J., Burrage, K.: Predicting residue-wise contact orders in proteins by support vector regression. BMC Bioinformatics 7, 425 (2006)
12. Vapnik, V.: The Nature of Statistical Learning Theory. Springer, New York (2000)
13. Yuan, Z., Huang, B.: Prediction of protein accessible surface areas by support vector regression. Proteins 57, 558–564 (2004)
14. Yuan, Z.: Better prediction of protein contact number using a support vector regression analysis of amino acid sequence. BMC Bioinformatics 6, 248 (2005)

SNP-Flankplus: SNP ID-Centric Retrieval of Flanking Sequences

Cheng-Hong Yang, Yu-Huei Cheng, Li-Yeh Chuang, and Hsueh-Wei Chang

Abstract. PCR is essential for many single nucleotide polymorphism (SNP) genotyping methods. However, the flanking sequences provided by dbSNP of NCBI are usually short and fixed length without further extension, thus making the design of appropriate PCR primers difficult. Here, we provide the system design and algorithm to describe a tool named "SNP-Flankplus" to provide a web environment for retrieval of SNP flanking sequences from both the dbSNP and the nucleotide databases of NCBI. SNP rsID# and ssID# are acceptable for retrieval of the SNP flanking sequences with adjustable lengths for at least sixteen organisms.

1 Introduction

Single nucleotide polymorphisms (SNPs) are the most common type of genetic variants. Polymerase chain reaction (PCR) is essential for fast mass duplication of DNA templates (Mullis and Faloona, 1987) for SNP genotyping. Some primer design tools, such as Primer design assistant (PDA) (Lin *et al.*, 2003) and Primer 3 (Rozen and Skaletsky, 2000), design the PCR primer set for the PCR-based genotyping. Usually, it is reliable for optimal primer design if an extendable flanking sequence for target site is available; however, the SNP flanking sequences

Cheng-Hong Yang and Yu-Huei Cheng
Department of Electronic Engineering, National Kaohsiung University of Applied Sciences, Taiwan
e-mail: chyang@cc.kuas.edu.tw, yuhuei.cheng@gmail.com

Li-Yeh Chuang
Department of Chemical Engineering, I-Shou University, Taiwan
e-mail: chuang@isu.edu.tw

Hsueh-Wei Chang
Faculty of Biomedical Science and Environmental Biology, Kaohsiung Medical University, Taiwan

Graduate Institute of Natural Products, College of Pharmacy, Kaohsiung Medical University, Taiwan

Center of Excellence for Environmental Medicine, Kaohsiung Medical University, Taiwan
e-mail: changhw@kmu.edu.tw

B.-C. Chien and T.-P. Hong (Eds.): Opportunities and Challenges, SCI 214, pp. 13–18.
springerlink.com

retrieved from NCBI dbSNP (Sherry *et al.*, 2003) are not extendable and sometimes short. Due to inherent size limitations of the PCR product, feasible primer sets are hard to design based on short template sequences.

For NCBI statistics, the numbers of RefSNP clusters (rs#), validated (rs#) and submissions (ss#) had reached 14,708,752, 6,573,786 and 55,949,029, respectively, for *Homo sapiens* as of dbSNP Build 129; these numbers are increasing steadily (dbSNP Summary, http://www.ncbi.nlm.nih.gov/SNP/snp_summary.cgi). However, currently available software programs often fail to update their internal databases simultaneously with the latest dbSNP data. Accordingly, both the availability for SNP records and the length of the corresponding full flanking sequences for these programs are out of date.

FESD (Kang *et al.*, 2005) provided a function of "SNPflank" to retrieve the flanking sequences for SNP IDs and provided customizable options for length, alternating case and reverse complement change for flanking sequences. However, it only accepted the rs# input for human SNPs and the web server is dysfunction. In order to provide template sequences for input SNP for genotyping experiments, such as TaqMan real-time PCR (De la Vega *et al.*, 2005), PCR-RFLP (Chang *et al*, 2006), and PCR-CTTP (Hamajima *et al.*, 2002), we propose the SNP-Flankplus to retrieve the flanking sequences of target SNPs for many organism genomes. Therefore, the SNP-Flankplus successfully retrieves appropriate flanking sequences of the target SNPs and implements with up-to-date data of the dbSNP database by on-line retrieval system.

2 Materials and Methods

The system design, algorithm and database of the program are described below.

2.1 System Design

The framework of this system is shown in Fig. 1. The system consists of four modules: (1) Input Module; (2) Query Module; (3) Locate Module; and (4) Output Module. They are explained below.

(1) Input Module
The Input Module accepts a query key from the users. Four input methods are available, such as Reference cluster ID (rs#) input, NCBI Assay ID (ss#) input, pasting input and file input.

(2) Query Module
The Query Module links to the database of NCBI dbSNP to query sequence information of SNPs, mainly for ss# mapping to rs#, available accession number, and SNP contig position.

(3) Locate Module
The core of the SNP-Flankplus system is the Locate Module. It uses the accession number and SNP contig position acquired by the Query Module to locate the position of the target SNP and to retrieve the desired flanking sequence of a specific length.

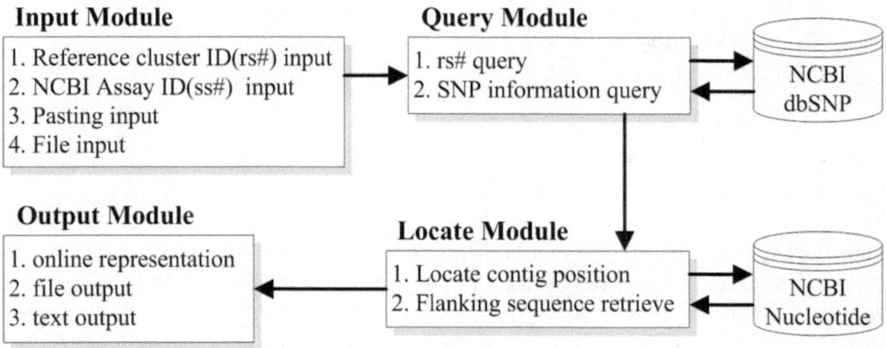

Fig. 1 The framework of SNP-Flankplus. Input Module accepts a query key from users and sends to Query Module. Subsequently, Query Module retrieves the sequence information for inputting SNP through the remote database of NCBI dbSNP and gives Locate Module an available accession number and SNP contig position. Finally, Locate Module provides the retrieval of desired flanking sequence of specific length and the results are provided by Output Module

(4) Output Module
The Output Module gets the flanking sequence and transforms it into a fasta format, and then outputs the data online, in either a file or text format.

2.2 *Algorithm*

This program adopts the sequences of accession numbers of the corresponding SNPs and the SNP contig position to obtain desired flanking sequence with specific length. In order to save memory space during reading the sequence of accession numbers, this system employs "block location way", which splits the sequence of the accession numbers into multiple blocks. A specific block is loaded into the memory to search the required sequence and is hit by the following algorithm:

```
if(SNP contig position % m * n == 0)
    block hit = SNP contig position / m * n;
else
    block hit = contig position / m * n + 1;
```

where m is the line length of the sequence of accession numbers in the fasta format and n is the block size having split. The symbols '%', '/', and '*' represent to get the remainder after division, the division operation, and the multiplication, respectively.

When the flanking length exceeds a block, some nearby blocks must be used, i.e. (block hit - d) or (block hit + d). d is the size of extending blocks and is calculated by the following algorithm:

```
if (flanking length / 2 > (SNP position in the block))
    d = (flanking length / 2) / m * n;
if ((SNP position in the block - 1) < (flanking length / 2) % (m * n))
    d++;
```

2.3 Database

The source databases are retrieved on-line from NCBI dbSNP and Nucleotide. All data are constantly updated. Sixteen organisms are included, such as *Anopheles gambiae*, *Apis mellifera*, *Bison bison*, *Bos indicus x bos taurus*, *Bos taurus*, *Caenorhabditis elegans*, *Canis familiaris*, *Danio rerio*, *Gallus gallus*, *Homo sapiens*, *Macaca mulatta*, *Monodelphis domestica*, *Mus musculus*, *Oryza sativa*, *Pan troglodytes* and *Rattus norvegicus*.

3 Results

SNP-Flankplus was developed for on-line retrieval of data from well-developed public-domain databases such as dbSNP and Nucleotide of NCBI. When a SNP ID or multiple SNP IDs (rs# or ss#) quer are entered, SNP-Flankplus is programmed to access NCBI dbSNP on-line to obtain the respective SNP information, and then accesses the NCBI Nucleotide database to retrieve the desired SNP flanking sequences based on their contig accessions. The input and output of the SNP-Flankplus software is described below:

Input
In SNP-Flankplus, four types of inputs are included, such as: (1) Single Reference cluster ID (rsID#); (2) Single NCBI Assay ID (ssID#); Multiple SNP ID rsID# and ssID# with (3) a copy-and-paste manner and (4) file uploading. Inputting with single or multiple SNP IDs (rs# or ss#) is acceptable to retrieve the SNP information. For ssID# input, the system is able to automatically find its corresponding rsID#, and then its SNP information is retrieved. The SNP information provides allele information, submitted SNPs and other data for this RefSNP Cluster. Users can choose the flanking length freely for the primer set design. Two flanking length options are available: (1) the system provides the default lengths of 300, 400, 500, 600, 700, 800, 900 and 1000 bps, and (2) the user-defined maximum length for the corresponding contig accession (Fig. 2A).

Output
The output for flanking sequence is presented in fasta format with on-line window and file and/or text. SNP ID information in the NCBI dbSNP is retrieved such as SNP rsID#, allele name, chromosome position of SNP, contig position of SNP, organism source, contig accession and sequence corresponding position, SNP type, sequence type, and case sensitivity. The length of maximum flanking is limited to the nature of the corresponding contig accession number. Three types of SNP flanking sequences are available, such as: (1) SNP types with general nucleotides, alleles, and IUPAC formats, (2) sequence types with original, reverse, complementary, antisense sequences, and (3) case sensitive types with upper case and lower case (Fig. 2B).

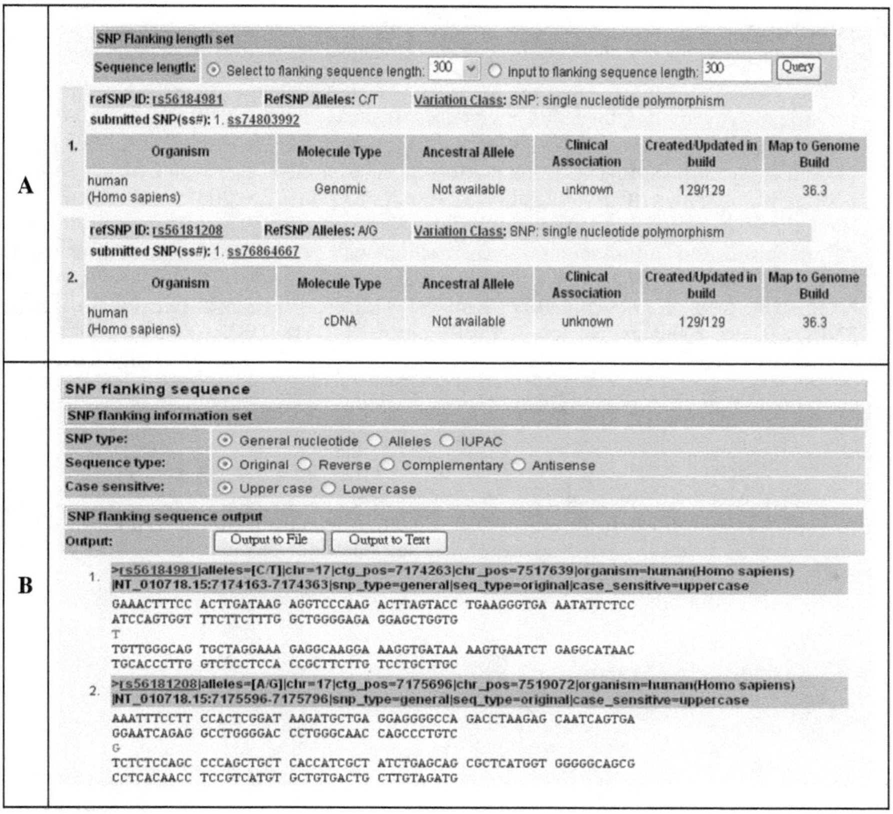

Fig. 2 (A) SNP information containing allele information, submitted SNPs and other data for this RefSNP Cluster. The information is based on dbSNP of NCBI. The options "System default length" and "User definited length" can be selected to obtain the desired flanking length for the design of feasible primer sets. (B) SNP type, sequence type and case sensitive type are employed to change the orientation and the surface of SNP flanking sequence. Output as a file or as plain text can be selected

4 Conclusion

SNP genotyping can determine the allele of a known polymorphism in target sequences. PCR-based SNP genotyping is commonly applied to save cost and simplify genotyping assay. A template sequence is required for primer design in SNP genotyping. SNP-Flankplus provides an available template sequence for primer set design in a PCR assay. A real-time update mechanism is employed, and two SNP ID types (rs# and ss#) for sixteen organisms can be entered to obtain the latest SNP information and sequence. A maximum flanking length can be retrieved based on the corresponding contig accession number.

Acknowledgements. This work is partly supported by the National Science Council in Taiwan under grant NSC97-2622-E-151-008-CC2, NSC96-2221-E-214-050-MY3, and the KMU-EM-97-1.1b.

References

1. Chang, H.W., Yang, C.H., Chang, P.L., Cheng, Y.H., Chuang, L.Y.: SNP-RFLPing: restriction enzyme mining for SNPs in genomes. BMC Genomics 7, 30 (2006)
2. De la Vega, F.M., Lazaruk, K.D., Rhodes, M.D., Wenz, M.H.: Assessment of two flexible and compatible SNP genotyping platforms: TaqMan SNP genotyping assays and the SNPlex genotyping system. Mutatation Research 573, 111–135 (2005)
3. Hamajima, N., Saito, T., Matsuo, K., Tajima, K.: Competitive amplification and unspecific amplification in polymerase chain reaction with confronting two-pair primers. J. Mol. Diagn. 4, 103–107 (2002)
4. Kang, H.J., Choi, K.O., Kim, B.D., Kim, S., Kim, Y.J.: FESD: a functional element SNPs database in human. Nucleic Acids Research 33, D518–D522 (2005)
5. Lin, C.Y., Chen, S.H., Lo, C.Z., Cho, C.S., Hsiung, C.A.: Primer Design Assistant (PDA): a web-based primer design tool. Nucleic Acids Research 31, 3751–3754 (2005)
6. Mullis, K., Faloona, F.: Specific synthesis of DNA in vitro via a polymerase catalyzed chain reaction. Methods Enzymol 155, 335–350 (1987)
7. Rozen, S., Skaletsky, H.: Primer3 on the WWW for general users and for biologist programmers. Methods Mol. Biol. 132, 365–386 (2000)
8. Sherry, S.T., Ward, M.H., Kholodov, M., Baker, J., Phan, L., Smigielski, E.M., Sirotkin, K.: dbSNP: the NCBI database of genetic variation. Nucleic Acids Research 29, 308–311 (2001)

Tumor Classification on Mammographies Based on BPNN and Sobel Filter

Enrique Calot, Hernan Merlino, Claudio Rancan, and Ramon Garcia-Martinez

Abstract. Breast cancer is a very common disease and the cause of death of many people. It has been proven that prevention decreases the death rate, but the costs of diagnosis and image processing are very high when applied to all the population with potential risk. This paper studies an existent computer aided diagnosis method using neural network and improves its detection success rate from 60% to 73%. This improvement is achieved due to the use of image and statistical operators over concentric regions around the tumor boundaries.

1 State of the Art

Breast cancer is the most common cancer among women worldwide representing a 31% of all tumors in the female population [1, 11]. Screening by mammography alone, with or without physical examination of the breasts, plus follow-up of individuals with positive or suspicious findings, will reduce mortality from breast cancer by up to one-third among women aged from 50 to 69 years. Unfortunately, mammography is an expensive test that requires great care and expertise both to perform and in the interpretation of results. It is therefore currently not a viable option for many countries [8, 9, 11]. Computer systems set on the classification of tumors found on mammographies may be of greater help during the diagnosis process in general hospitals [2, 5].

2 Addressed Problem in This Paper

Tumor classification issue has been addressed using neural networks in a previous work by [5] with a success rate of 60%. In the present work some complementary workarounds had been done to improve that results.

Enrique Calot, Hernan Merlino, and Ramon Garcia-Martinez
University of Buenos Aires, School of Engineering, Intelligent Systems Laboratory
Paseo Colon 850, (C1063ACV) Buenos Aires, Argentina
e-mail: {ecalot,rgarciamar}@fi.uba.ar

Hernan Merlino, Claudio Rancan, and Ramon Garcia-Martinez
University of La Plata, Computer Science School, PhD Program
Calles 50 y 120, (B1900) La Plata, Buenos Aires, Argentina
e-mail: hmerlino@gmail.com, crancan@itba.edu.ar

B.-C. Chien and T.-P. Hong (Eds.): Opportunities and Challenges, SCI 214, pp. 19–24.
springerlink.com

3 Proposed Workarounds

The first workaround was the use of *The Digital Database for Screening Mammography* (DDSM), a compilation made by the *University of South Florida* (USF) as a test-and-train database [6, 7].

Improvements available in this database

- Provides images with a higher resolution.
- Contains a bigger number of images.
- Has additional information at disposal. *i.e.* tumor boundaries.

The second workaround is the use of an image filter allowing gradient changes to perform efficient boundary detection.

The third workaround is the analysis of the anomaly as is, and not the whole breast. In order to do that the additional tumor boundary information provided by the database must be used. In the production stage, when new acquired images are to be classified, a method to automatically shape the tumors is to be used. The implementation of such algorithm is not in the scope of this paper.

3.1 Generating Boundary Regions

A region is defined as a section of the image surface with a specified distance to the boundary. Having known the shape of the anomaly, it is possible to generate a set of concentric regions and another one for the core of the tumor. This work has used 10 regions inside and outside the boundaries and another one in the middle as the tumor core. The usage of regions is very important because the boundary gradient is normal to the region width and providing a good set up for the Sobel operator explained below.

Having a list of pixels belonging to the boundary there are many possible algorithms to calculate adjacent regions. One possible way is applying the Bellman-Ford [3] or the even faster Dijkstra [4] algorithms considering the shape boundary bitmap as a graph where each pixel has a connection weighted 1 to the four immediately adjacent ones and the boundary pixels as start points with null distance between them. Since both weights are non-negative numbers, the Dijkstra algorithm should be preferred. This way to separate regions is very fast and may be stopped when the desired number of regions is reached. There is no need to process all the pixels. The most important disadvantages of this algorithm are that it uses the taxicab 1-norm metric as distance measure and the boundary curve must be a closed shape.

Another way to achieve region separation is applying a Gaussian filter to the shape bitmap. With the resultant image it is possible to define brightness intervals with respective regions associated to them. The Gaussian radius should be considerably as big as to generate distant regions. This algorithm is using a Euclidean

2-norm measure of distance but requires more time to calculate the convolution between the big-enough Gaussian matrix and the whole input image.

3.2 Sobel Operator

Sobel operator is an image filter capable of transforming an image into two others representing the increment of brightness in each pixel compared to the surrounding ones. Both generated images may be represented either in Cartesian coordinates *(x,y)* or in polar ones (module, argument). As a result, the Sobel module image will show brighter pixels on abrupt changes and the Sobel argument image will show the direction of this change [10].

Malignant tumors (cancer) spread themselves to other placements (metastases) and to adjacent tissues (local invasion) causing potential danger. In the other hand, benign tumors do not spread to other locations, having more defined boundaries. The fact that cancer invades adjacent tissues makes their boundaries not so abrupt in images and therefore much more sensitive to a Sobel module filter.

The proposed workaround calculates mean bright, variance and size for a specified region and image. In figure 1, cases *a*, *b*, *c* and *d* have the same bright mean. Furthermore, cases *b* and *c* have the same variance, bigger than *a*; this shows that the variance may be important. Even though, the case *b* and *c* have the same variance, the Sobel module mean is bigger in case *c*, because there are more abrupt changes. Exaggerating an example of a tumor, *b* should correspond to a benign tumor, with a defined boundary and *c* or *d* to a malignant one, with larger borders in a same-sized surface or branches spreading to adjacent tissues.

Some final generated images are shown as figures 2 to 4, where the original, Sobel argument and Sobel module are shown respectively. Note that the original image is not directly the one obtained from the database, it has been pre-processed to drop out the radiographic labels.

After images are generated, they are separated in different regions according to the shape of the tumor and then the mean, variance and size of each resultant region is calculated. The input of the back propagation neural network should be the real numbers obtained in this process.

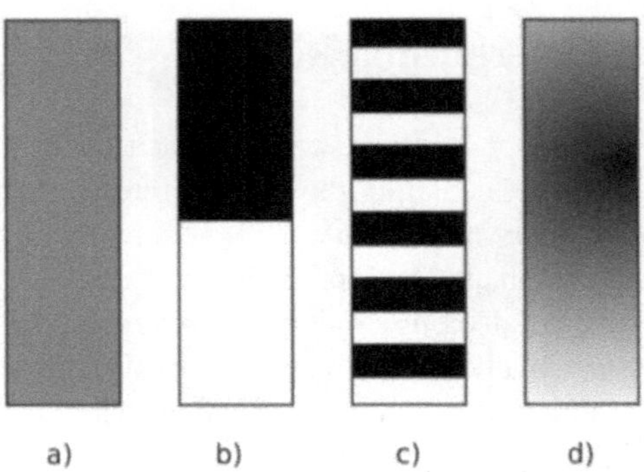

Fig. 1 Bright distribution over a region

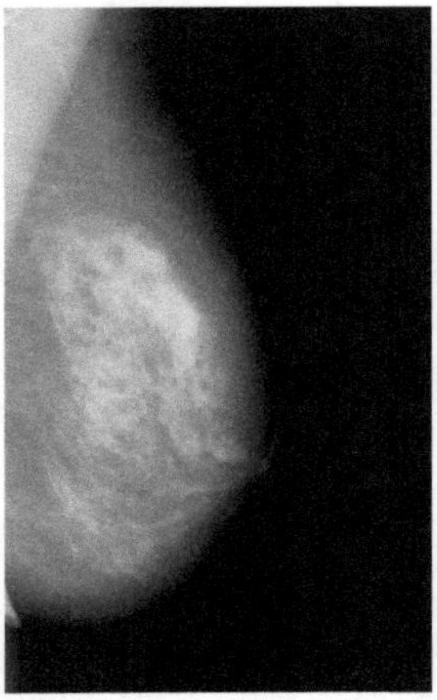

Fig. 2 Original image

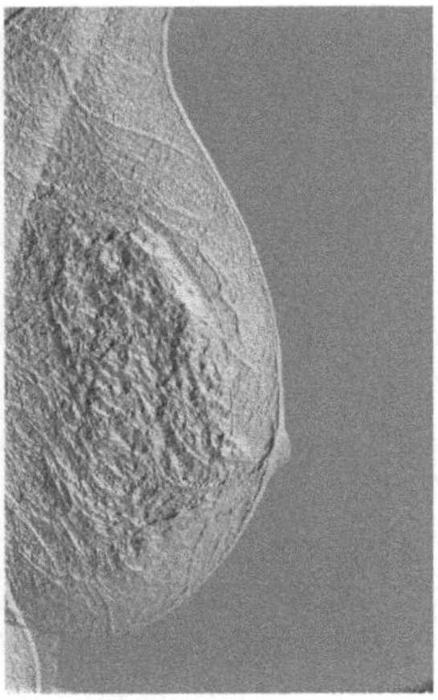

Fig. 3 Sobel argument

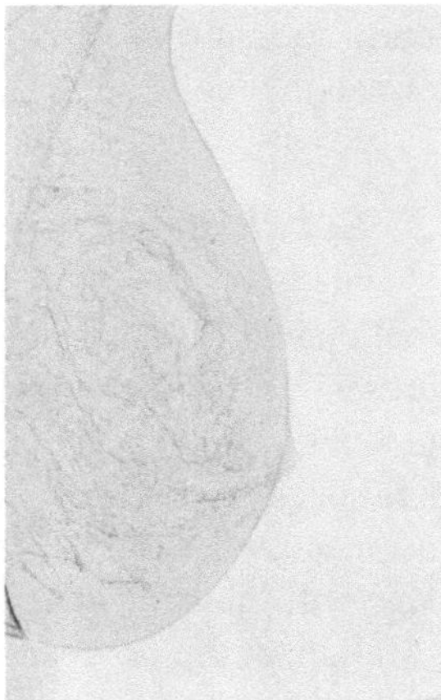

Fig. 4 Sobel module

4 Experiment Description

Experiments have been performed with images from the DDSM and their preprocessing was the one described in the previous work by [5].

Lowest and highest bright values have been detected and readjusted using linear conversion to generate a grayscale image of 16 bits.

After this process, each image with shaped anomalies was applied the Sobel filter to generate both polar components.

Different back propagation neural network input setups had been tested including or excluding the mean, variance and size of each region in each of the three images (original, Sobel module and argument).

For each configuration, one third of the data had been kept back for the verification process. The remaining two thirds were used to train the network. After training the verification was performed running the network with the kept-back third of the data to compare the results with the previously known ones obtaining a success ratio.

5 Experimental Results

After suppressing the error of each study considering it as a *True/False* answer and then calculating the success rate the achieved value was 73%, independently of its error interval. It was found a correlation between the wrong-detected images and bad-shaped tumors on the input database.

It was also observed that an argument Sobel filter and a variance were not necessary in any of the images and it has an overloading element for the network and, in some cases, adding noise and deteriorating the results.

The best results achieved had used 10 layers for each side of the boundary.

For the tested data set, the setup with a higher success rate was a neural network with three layers, having 12 neurons in the middle hidden one.

The average timing to process 271 images, including decompression and Sobel operator was of 42 minutes and the average training time was inside an interval between 3 a 5 minutes. The computer used was an average-fast computer sold in 2008.

6 Conclusions

Stated workarounds achieved better success rates than the previous work, that is a 73% over a 60%.

The shape and boundary type of the tumors is important and the Sobel filter (only the module part) proved to be useful to detect them. The variance applied to all the images and the Sobel argument image had shown no improvement after the classification process.

Developing an algorithm to improve bound accuracy should increase dramatically the success rates.

References

1. AMA, Consenso Nacional Inter-Sociedades sobre Cáncer de Mama: Pautas para el Diagnóstico y Manejo de las Lesiones Mamarias Subclínicas. Asociación Médica Argentina (2006)
2. Antonie, M., Zaïene, O., Coman, A.: Application of data mining techniques for medical image classification. In: Proceedings of the Second International Workshop on Multimedia Data Mining, San Francisco (2001)
3. Bellman, R.: On a Routing Problem. Quarterly of Applied Mathematics 16(1), 87–90 (1958)
4. Dijkstra, E.W.: A note on two problems in connexion with graphs. Numerische Mathematik 1, S269–S271 (1959)
5. Ferrero, G., Britos, P., García-Martínez, R.: Detection of Breast Lesions in Medical Digital Imaging Using Neural Networks. In: Debenham, J. (ed.) IFIP International Federation for Information Processing. Professional Practice in Artificial Intelligence, vol. 218, pp. 1–10. Springer, Boston (2006)
6. Heath, M., Bowyer, K., Kopans, D., Kegelmeyer, W.P., Moore, R., Chang, K., MunishKumaran, S.: Current status of the Digital Database for Screening Mammography. In: Digital Mammography. Proceedings of the Fourth International Workshop on Digital Mammography, pp. 457–460. Kluwer Academic Publishers, Dordrecht (1998)
7. Heath, M., Bowyer, K., Kopans, D., Moore, R., Kegelmeyer, W.P.: The Digital Database for Screening Mammography. In: Yaffe, M.J. (ed.) Proceedings of the Fifth International Workshop on Digital Mammography, pp. 212–218. Medical Physics Publishing (2001) ISBN 1-930524-00-5
8. Selman, S.: Data Mining of Digital Mammograms Will Aid in War against Cancer (2000), http://www.gatech.edu (accessed March 27, 2008)

9. Smith, R.A., Caleffi, M., Albert, U.S., Chen, T.H., Duffy, S.W., Franceschi, D., Nystrom, L.: Breast cancer in limited-resource countries: early detection and access to care. Breast J. 12(suppl. 1), S16–S26 (2006)
10. Sobel, I., Feldman, G.: A 3x3 Isotropic Gradient Operator for Image Processing. Presented at the Stanford Artificial Project (1968)
11. WHO. Screening for Breast Cancer (2008), `http://www.who.int/cancer/detection/breastcancer/en/index.html` (accessed October 26, 2008)

An Adaptive Biometric System Based on Palm Texture Feature and LVQ Neural Network

Chen-Sen Ouyang, Ming-Yi Ju, and Han-Lin Yang

Abstract. We propose an adaptive biometric system based on the palm texture feature and LVQ2 neural network. The user's palm image is acquired by a scanner and preprocessed to be a labeled palm contour in the binary image format. Then, the positions of 12 feature points are identified speedily and roughly on the contour and refined to be more precise with a proposed correction mechanism. By referring the positions of feature points, six subimages of five fingers and the palm are obtained and transformed into six feature vectors with a modified texture descriptor of LFP (local fuzzy pattern). We employ the LVQ2 to learn the prototypes of feature vectors of each user. Therefore, an unknown user's palm feature vector is compared with prototypes to identify or verify his identity.

1 Introduction

Biometric systems have attracted much of attention in recent years since there have been many applications in many areas, such as security control, entrance control, identity identification and verification, etc. Biometric means using measurable physiological and/or behavioral features to identify or verify a user's identity. Physiological features (e.g. fingerprint, iris, face, etc.) and behavioral features (e.g. voiceprint, handwriting, signature, etc.) have been employed in many biometric systems and

Chen-Sen Ouyang and Han-Lin Yang
Department of Information Engineering, I-Shou University, Kaohsiung County 840, Taiwan R.O.C
e-mail: ouyangcs@isu.edu.tw

Ming-Yi Ju
Department of Information Engineering, National University of Tainan, Tainan 700, Taiwan R.O.C
e-mail: myju@mail.nutn.edu.tw

B.-C. Chien and T.-P. Hong (Eds.): Opportunities and Challenges, SCI 214, pp. 25–30.
springerlink.com

possess high recognition rate. However, most of them encounter the problems of high cost of equipments, high computation complexity, inconvenience for feature acquirement, and personal privacy.

Recently, many researchers have paid their attention to view the palm as the source of feature extraction. The advantages of palm features are high public acceptance, low cost of equipments, and high accuracy level. Therefore, we focus on the approaches based on palm features in this paper. Ribaric et al. [5] proposed a biometric identification system by extracting the eigenpalm and eigenfinger features from subimages of palm and fingers. However, the feature point extraction and K-L transformation used in this approach are not robust enough, especially in the condition that the palm in the image is inclined or illumination change. Wu et al. [7] proposed an approach for palm line extraction and matching with chaining coding. However, the coding method is too sensitive to the extracted palm lines. A hierarchical identification of palmprint using the line-based Hough transform is proposed by Li and Leung [3]. Wu and Qiu [6] proposed a hierarchical palmprint identification method using the hand geometry and gray-scale distribution features. Both of these two approaches are too sensitive to the rotation and position of palm in the image.

From the discussion above, we can conclude that three important problems should be solved. The first one is the palm position and rotation in the image should be automatically detected. The second one is the feature extraction should be robust enough in different conditions. The third one is a leaning mechanism is needed for the feature learning of different users, so as to reduce the stored features and the complexity and computation power in the matching phase.

2 Our Approach

In our system, there are four steps in the user registration phase, i.e., palm image preprocessing, palm feature point extraction, palm feature extraction, and LVQ2 learning. We describe the four steps in detail as follows.

The main purpose of palm image preprocessing is to transform the scanned grey-scale palm image into a binary image of palm contour and label the points on the palm contour in order. Firstly, the scanned image are resized into an image with 352×485 size by the bilinear interpolation [1], as shown in figure 1(a). Secondly,

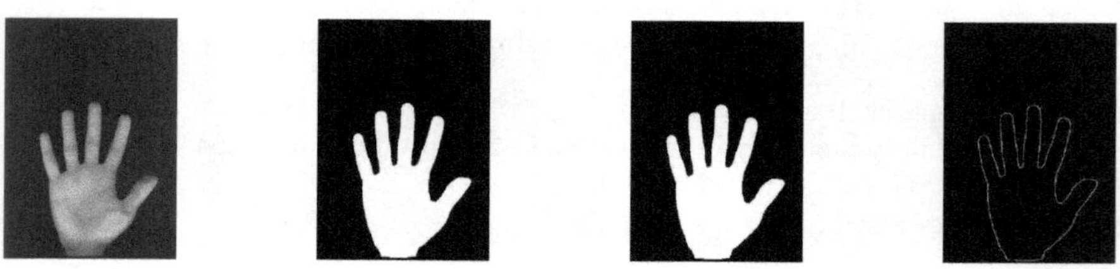

(a) (b) (c) (d)

Fig. 1 Palm image preprocessing: (a) Acquired image; (b) Binary image after thesholding; (c) Image after median filtering; (d) Contour image after Laplacian filtering

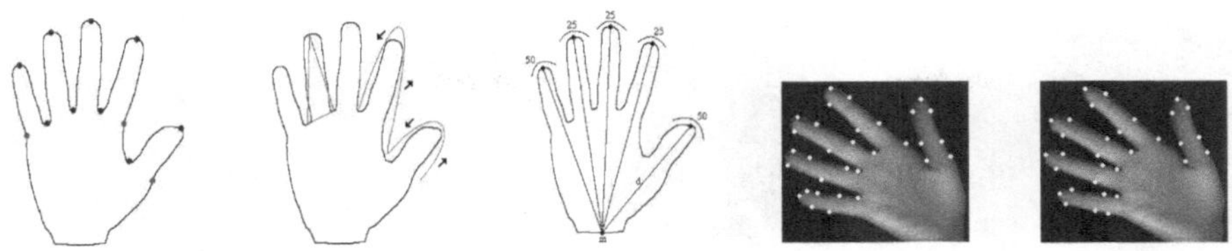

(a) (b) (c) (d) (e)

Fig. 2 Feature Point Extraction: (a) Ideal positions of feature points; (b) The search order for initial feature points; (c) The adopt neighborhoods of initial fingertip feature points; (d) Initial positions of feature points of an inclined palm. (e) Corrected result

we set a threshold of grey level to transform the resized grey-scale image into a binary image, as shown in figure 1(b). The threshold we set is the average of grey values in the whole palm image. Note that there are some wrong classified pixels caused by noise, especially in the background area. Therefore, we employe a 3×3 median filter to cancel these noise points in the third sub-step. Figure 1(c) shows a resultant binary image. Apparently, the noise points in the background area are canceled. Fourthly, the Laplacian filtering for edge detection is employed to find out the contour of the palm, as shown in figure 1(d). Finally, the points on the contour are labeled in order.

The main purpose of palm feature point extraction is to identify the 12 feature points located on the palm contour. The ideal result is shown in the figure 2(a). Note that 9 points are located on the fingertips and the valleys between the fingers, while the other 3 points are located at positions which are symmetric to the points in the valleys of the correspond fingers (thumb finger, index finger, and little finger). The 5 points located on five fingers and 4 points located on the valleys between the fingers are identified by the local minima and maxima, respectively, on the contour according to the y values of the sequence of contour point positions. Figure 2(b) shows the order for searching. To identify the other 3 points, we employ the property that the distance between each of these points and its corresponding fingertip is the same as the distance between the fingertip and the valley-point on the other side of the corresponding finger. Therefore, we have the initial positions of 12 feature points roughly. However, the initial positions may be not so precise, especially when the palm in the acquired image is inclined as shown in figure 2(d). To correct the initial positions to be more precise, we employ a simple property that the distance between the middle contour point at the wrist and each fingertip is the longest by comparing with the neighboring contour points of the fingertip. Therefore, we calculate the distances between the middle contour point at the wrist and all neighboring contour points of each initial fingertip feature point. Then, the corresponding fingertip feature point is corrected to the point with the largest distance. Through our experimental experience, the number of neighboring points are set as 50 for the thumb fingertip and little fingertip, and 25 for the other three fingertips, as shown in figure 2(c). Figure 2(e) shows the correction result of figure 2(d). Apparently, the initial positions of feature points are corrected to be more precise and reasonable.

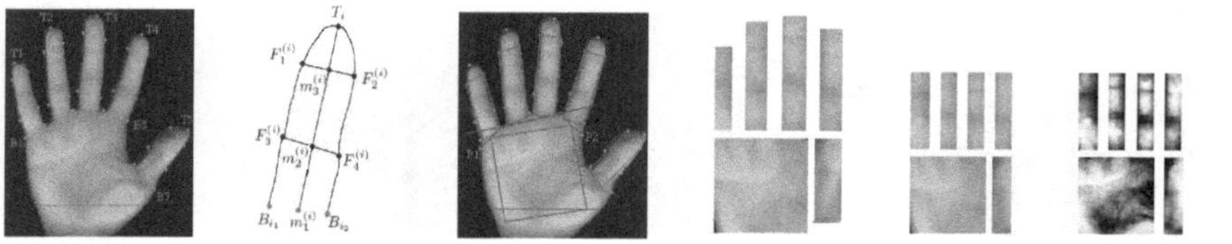

(a) (b) (c) (d) (e) (f)

Fig. 3 Subimage Extraction: (a) Positions of 12 feature points; (b) Reference points for finger subimage extraction; (c) Reference points for palm subimage extraction; (d) Six obtained subimages; (e) Normalized subimages; (f) Resultant subimages after lighting normalization

In the third step, we focus on the feature extraction from the palm image. Let the 5 fingertip feature points and the other 7 feature points be denoted as $T_1, T_2, \ldots, T_5$ and $B_1, B_2, \ldots, B_7$, respectively. The figure 3(a) shows the 12 feature points. For each finger with the fingertip T_i and the (B_{i_1}, B_{i_2}), we firstly obtain the middle point $m_1^{(i)}$ of the line segment $B_{i_1} - B_{i_2}$. The distance between the middle point $m_1^{(i)}$ and the fingertip T_i is defined as the length d_i of the corresponding finger. As shown in figure 3(b), two additional point pairs $(F_1^{(i)}, F_2^{(i)})$ and $(F_3^{(i)}, F_4^{(i)})$ are determined at the one-third and two-thirds, respectively, of d_i, and the middle points $m_3^{(i)}$ and $m_2^{(i)}$ of $F_1^{(i)} - F_2^{(i)}$ and $F_3^{(i)} - F_4^{(i)}$ are also determined accordingly. The line connecting the middle points $m_2^{(i)}$ and $m_3^{(i)}$ is viewed as the line of symmetry for the sub-image. The length of the subimage is set as five-sixths of the length d_i. The width of the subimage is determined by the distance between the point $F_1^{(i)}$ and the corresponding contour point in the other side of the line of symmetry. Besides, we have to obtain the subimage of the palm which is the inner surface of the hand between the wrist and the fingers. The subimage of the palm is defined as a square region with two of its corners placed on the middle points of the two line segments $P_1 - B_2$ and $B_4 - P_2$. Note that the points P_1 and P_2 are determined by rotating the line segments $B_1 - B_2$ and $B_4 - B_5$ with 30° and 40°, respectively. The obtained six subimages of the five fingers and palm should be resized by bilinear interpolation method for size normalization. The thumb and little finger subimages are resized to 16×64, while the other finger subimages are resized to 14×64. Besides, the palm subimage is resized to 64×64. Then, we perform a procedure of histogram equalization [1] on each subimage for lighting normalization.

To extract the feature vectors for the six subimages, we use a modified version of LFP texture descriptor [4]. In the original version of LFP, a LFP histogram for each pixel should be calculated by aggregating the LFPs of all pixels in the corresponding circular neighborhood. To save the computation power, we aggregate the LFPs of all pixel in each subimage directly. Therefore, each subimage is transformed into a 2^P-dimensional feature vector which describes the texture distribution of the corresponding subimage. Note that P is a parameter related to the LFP.

Finally, we use a LVQ2 [2] to obtain representative prototypes of different users. The learned prototypes of different users are stored in a registration database.

3 Experimental Results

To demonstrate the advantages of our proposed system, we present some experimental results in this section. There are 20 users and 15 palm images are acquired from each user by putting his palm on the scanner for scanning in several possible conditions, like normal position, different rotations, illumination change, different positions of fingers, etc. For each user, we randomly choose 5 images for training and the other 10 for testing.

In the first part, we present two experimental results to demonstrate the effectiveness of our correction mechanism. In the first experiment, two palm images are acquired in normal and inclined conditions, respectively. Figure 4 shows the correction results. Apparently, most of finger feature points are identified at incorrect initial positions. After correction, we obtain more precise positions of these feature points. The second experiment tests the rotation tolerance of our correction mechanism. Therefore, we consider different rotation angles from 0° to 50° of the palm. We can see that the correction mechanism still works very well even when the rotation reaches to 40°. However, it is fail when the rotation reaches to 50°. As mentioned earlier, we can improve the problem by choosing more neighboring points for each initial fingertip feature point. However, more computation power is needed.

In the second part, we compar the recognition rates, time of feature extraction, and time of matching on the samples of the test dataset with these two methods, as shown in Table 1. Our method presents a better recognition rate. Besides, the time taken by our method is more in the feature extraction, however, less in the search.

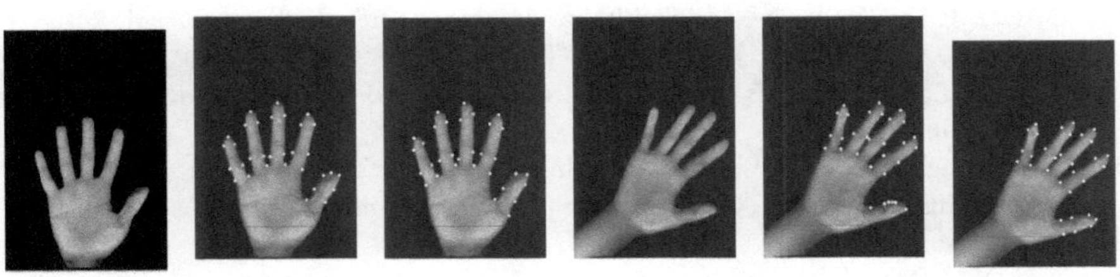

(a) (b) (c) (d) (e) (f)

Fig. 4 Correction results: (a) palm image 1; (b) Initial result 1; (c) Corrected result 1; (d) palm image 2; (e) Initial result 2; (f) Corrected result 2

Table 1 Comparison on recognition rate, time of feature extraction, and time of matching

Recognition rate		Time of feature extraction		Time of matching	
K-L	LFP	K-L	LFP	K-L	LFP
88.50%	91.00%	1.003 (sec.)	1.187 (sec.)	0.070 (sec.)	0.027 (sec.)

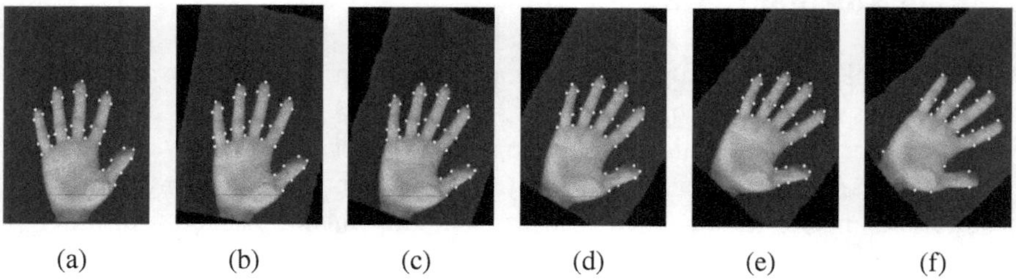

Fig. 5 Results of rotation test with (a) 0°; (b) 10°; (c) 20°; (d) 30°; (e) 40°; (f) 50°

4 Conclusion and Future Works

We have presented an adaptive biometric system based on the palm texture feature and LVQ2 neural network. A correction mechanism is integrated into the feature point extraction to improve the tolerance of palm rotation. A palm feature extraction with a modified version of our proposed LFP texture descriptor is employed to increase the robustness. Besides, we employ a LVQ2 neural network to learn the prototypes of each registered user. Therefore, our proposed system has better performance. In our future works, we consider to extend the feature point extraction to be tolerable for any rotation of the palm. Besides, we will extend our database by collecting more user's palm images.

References

1. Gonzalez, R.C., Woods, R.E.: Digital Image Processing, 3rd edn. Prentice Hall, Reading (2007)
2. Hagan, M.T., Demuth, H.B., Beale, M.H.: Neural Network Design. PWS Publishing, Boston (1996)
3. Li, F., Leung, M.K.H.: Hierarchical identification of palmprint using line-based hough transform. In: Proc. of 18th Conference on Pattern Recognition, Hong Kong, China, pp. 149–152 (August 2006)
4. Liu, C.-Y.: A Robust Texture-Based Background Subtraction for Moving Object Detection in Video Sequences. Thesis, I-Shou University, Kaohsiung, Taiwan (2006)
5. Ribaric, S., Fratic, I.: A biometric identification system based on eigenpalm and eigenfinger features. IEEE Transactions on Pattern Analysis and Machine Intelligence 27(11), 1698–1709 (2005)
6. Wu, J., Qiu, Z.D.: A hierarchical palmprint identification method using hand geometry and grayscale distribution features. In: Proc. of 18th Conference on Pattern Recognition, Hong Kong, China, pp. 409–412 (August 2006)
7. Wu, X., Zhang, D., Wang, K.: Palm line extraction and matching for personal authentication. IEEE Transactions on Systems, Man, and Cybernetics-Part A: Systems and Humans 3(1), 978–987 (2006)

A Binarization Approach for Wafer ID Based on Asterisk-Shape Filter

Hsu Wei-Chih, Yu Tsan-Ying, and Chen Kuan-Liang

Abstract. The binarization of wafer ID image is one of the key techniques of wafer ID recognition system and its results influence the accuracy of the segmentation of characters and their identification directly. The process of binarization of wafer ID is similar to that of the car license plate characters. However, due to some unique characteristics, such as the unsuccessive strokes of wafer ID, it is more difficult to make of binarization of wafer ID than the car license plate characters. In this paper, a wafer ID recognition scheme based on asterisk-shape filter is proposed to cope with the serious influence of uneven luminance. The testing results show that our proposed approach is efficient even in situations of overexposure and underexposure the wafer ID with high performance.

1 Introduction

It is important for Assembly / Test process of semiconductor manufacturing to link to a wafer database by the wafer identification (wafer ID), which is inscribed by a laser scribe system. In order to trace wafer IDs easily, the wafer ID should be recognized automatically. The binarization stage plays an important role in wafer ID recognition.

The binarization techniques can be categorized as global thresholding and local thresholding. Global thresholding [1-3] selects a single threshold value from the histogram of the entire image. Local thresholding [4, 5] uses localized gray-level information to choose multiple threshold values; each is optimized for a small region in the image. Global threshoding is simpler and easier to implement but its result relies on good (uniform) illumination. Local thresholding methods can deal

---

Hsu Wei-Chih and Chen Kuan-Liang
Department of Computer and Communication, National Kaohsiung
First University of Science and Technology, Kaohsiung, Taiwan (R.O.C)

Yu Tsan-Ying
Institute of Engineering Science and Technology, National Kaohsiung
First University of Science and Technology, Kaohsiung, Taiwan, (R.O.C)

Department of Electrical Engineering, Kao Yuan University, Lu Chu, Taiwan, (R.O.C)
e-mail: weichih@ccms.nkfust.edu.tw, u9315902@ccms.nkfust.edu.tw, yotnyg@mail.nsysu.edu.tw

B.-C. Chien and T.-P. Hong (Eds.): Opportunities and Challenges, SCI 214, pp. 31–37.
springerlink.com

with non-uniform illumination but they are slow. More recent studies on this subject can be found in B. Gatos, I. Pratikakis, and S. J. Perantonis [6] and L. M. Sheikh, I. Hassan [7].

The binarization of wafer ID recognition is very similar to that of car license plate recognition (LPR). However, because of some wafer ID images' special features, the binarization scheme must be modified in order to enhance recognition performance. Some special features are described as follows.

1) Usually, both the brightness and contrast of the wafer ID images are not as sharp as those of the license plate images. This will mislead character recognition.
2) There are other disturbances which are caused by uneven luminance, and caused by wafer processes.
3) The wafer scribes are often unsuccessive. Each character is composed of unsuccessive horizontal scribes caused by laser scribe.

The remainder of this paper is organized as follows. Section 2 presents the algorithm in detail; Section 3 shows the experimental results. Finally, conclusion and the future work will be addressed.

2 The Algotithm of Binarization Approach

The binarization stage plays an important role in wafer ID recognition. The binarization stage affects the accuracy of character recognition. If a poor binarization algorithm is used, accurate character regions cannot be obtained and the characters cannot be extracted from the wafer ID correctly. While a highly accurate binary image is produced, the character segmentation process will go smoothly and the characters can be obtained from the wafer ID efficiently. The flowchart of our binarization approach is shown in Fig. 1 and in the remainder of this paper the detail of the binarization algorithm will be explored.

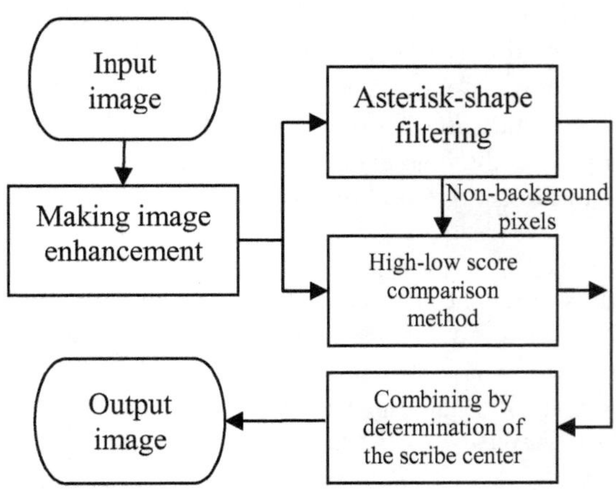

Fig. 1 Flowchart of the image binarization algorithm

2.1 Image Enhancement

Insufficient exposure, overexposure or uneven luminance will lead to low contrast between the wafer ID string and background. Therefore, we have to make image enhancement to improve the contrast of wafer ID image

We define a gray level *br* as background reference in which the number of the pixels with gray level less or equal to *br* is 40% of total pixels. We also define a level *sr* as stroke reference in which the number of the pixels with gray level greater or equal to *sr* is 90% of total pixels. If the difference of the gray level between stroke reference(*sr*) and background reference(*br*) is less than δ_1, it means that the wafer ID image is low contrast. In other words, we have to enhance image by enlarging the difference to δ_1. The enhancement equation is described in the following:

$$r = \frac{sr + br}{2} \tag{1}$$

$$f(x) = r + \frac{\delta_1}{sr - br}(x - r) \tag{2}$$

Note that the difference is not allowed to be more than δ_1, otherwise the noises will increase after enhancement. The threshold δ_1=25 is determined by experiments.

2.2 Image Binarization

Once we have separated the wafer ID string from background, we employ two stages to carry out binarization. At first stage, two methods are applied. The first one called the asterisk-shape filtering will remove luminance disturbance. The second method called the high-low score comparison method primarily aims to eliminate the noisenear wafer scribe edges. At the second stage, we compare and combine the two output images resulted from the asterisk-shape filter and the high-low score comparison method, respectively. At the final stage, the unsuccessive vertical edges are compensated

The asterisk-shape filtering. Because of background and luminance disturbance effect, the closer to wafer ID scribe center, the higher the gray level will be. Furthermore, the direction of wafer ID scribe is a very important feature. The direction of standard wafer ID scribes usually can be classified into vertical direction, horizontal direction, slope -1 line, and slope +1 line. For the measured pixel, judging from the gray level of the pixels distributing in its corresponding four directions, we can determine whether the measured pixel is foreground pixel or not. Since the scribe width is between 4 and 7 pixels, we set the window size as 9×9. Let the measured pixel as the window center. We divide the 81 samples of 9×9 window into upper, lower, left and right regions. There exist three flags whose values are true or false determined by the specified equations for each region.

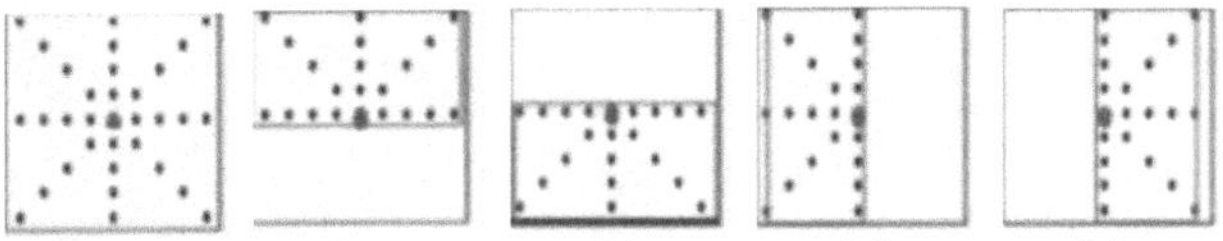

Fig. 2 (a) The four distribution directions of the asterisk-shape filter window (b) The upper region (c) The lower region (d) The left region (e) The right region

For the measured pixel, each flag is to indicate which direction's pixels are continuously black. The upper region is related to flags of $flag_{right}$, $flag_{upper\_righ}$, $flag_{upper}$, $flag_{upper\_left}$ and $flag_{left}$ as depicted in Fig. 2(b), the lower region $flag_{left}$, $flag_{lower\_left}$, $flag_{lower\_right}$ and $flag_{right}$ as depicted in Fig. 2(c); the left region $flag_{upper}$, $flag_{upper\_left}$, $flag_{left}$, $flag_{lower\_left}$ and $flag_{lower}$ as depicted in Fig. 2(d); the right region $flag_{upper}$, $flag_{upper\_right}$, $flag_{right}$, $flag_{lower\_right}$ and $flag_{lower}$ as depicted in Fig. 2(e). If any one of these regions whose related flags are all "true", then the measured pixel will be set to 0, else the measured pixel will remain the original value. As for

$$flag_{right} = \begin{cases} true & if \quad \sum_{k=1}^{4} f(x+k,y) - f(x,y) < \delta_2 \quad or \quad \sum_{k=1}^{4} f(x+k,y) = 0 \\ false & otherwise \end{cases} \tag{3}$$

$$flag_{upper\_right} = \begin{cases} true & if \quad \sum_{k=1}^{4} f(x+k,y+k) - f(x,y) < \delta_2 \quad or \quad \sum_{k=1}^{4} f(x+k,y+k) = 0 \\ false & otherwise \end{cases} \tag{4}$$

where f is the orginal image, $f(x,y)$ is the measured pixels and δ_2 is a threshold, determined as Eq.(3). $flag_{upper}$, $flag_{upper\_left}$, $flag_{left}$, $flag_{lower\_left}$, $flag_{lower}$, $flag_{lower\_right}$, $flag_{right}$, and $flag_{upper\_right}$ are available from the same reason. Sometimes the variance of the gray level near wafer scribe edges is too large, noise can not be removed. The high-low score comparison method is to solve this problem.

$$\delta_2 = \begin{cases} \delta_4 & if \quad sr - br \geq \delta_1 \ \ and \ \ sr - br \leq \delta_3 \\ \delta_5 & otherwise \end{cases} \tag{5}$$

where sr and br are the stroke and background reference obtained in previous section , and the threshold δ_3, δ_4 and δ_5 are 50, 20 and 15 determined by experiments.

The high-low score comparison method. The high-low score comparison method aims to deal with those noises near to wafer scribe edges but not eliminated out by the asterisk-shape filter. As before, the window size is set to 9×9, and the measured pixel is set at the center.

Assume $G(x,y)$ is the image after the asterisk-shape filtering. Let the $HighScore_{avg}$ be the average of the gray level of the pixels which are higher than that of $G(x,y)$ in the window. And let the $LowScore_{avg}$ be the average of the gray level of the pixels in the window which are lower than that of $G(x,y)$ in the window.

In order to separate out the pixels of the higher gray level from the background, Eq.(7). and Eq.(8). are evaluated and checked. Utilizing this criteria, the measured pixels near wafer scribe edges will be classified into wafer ID or background. This method is implemented as follows:

$$H(x,y)=0 \quad \textit{if} \quad G(x,y)=0 \tag{6}$$

$$HighScore_{avg} = Average\left(\sum_{i=-4}^{4}\sum_{j=-4}^{4} f(x+i,y+j)\right)$$

$$\textit{if} \quad G(x.y) < \sum_{i=-4}^{4}\sum_{j=-4}^{4} f(x+i,y+j)\ \textit{and}\ \ G(x,y) \neq 0 \tag{7}$$

$$LowScore_{avg} = Average\left(\sum_{i=-4}^{4}\sum_{j=-4}^{4} f(x+i,y+j)\right)$$

$$\textit{if} \quad G(x,y) > \sum_{i=-4}^{4}\sum_{j=-4}^{4} f(x+i,y+j)\ \textit{and}\ \ G(x,y) \neq 0 \tag{8}$$

where f is the original binary image, G the image after the asterisk-shape filtering, H the output image, δ_6 the threshold, (x,y) the measured pixel index.

If $G(x,y) > LowScore_{avg} + \delta_6$ and $G(x,y) > (LowScore_{avg} + HighScore_{avg})/2$, then the measured pixels are set to 255, otherwise they are set to 0. The threshold δ_6 is set as 15 by experiments.

Combining by determination of the scribe center. Since the two images G and H resulting from the asterisk-shape filter and the high-low score comparison method respectively are not the same for all pixels, they are combined in some way. These processing steps are described as follows. While we zoom in to observe the cross section of a wafer ID character image, the pixels near the scribe center are brighter than those far away from the scribe center. Based on this fact and the phenomenon, the directions of standard wafer scribes can be classified into four types: vertical, horizontal, slope +1 line, slope -1 line. By the asterisk-shape filtering method, reserved pixels have indicated the direction for wafer scribes of G.

The direction for wafer scribes at the measured pixel is obtained in step 2. The cross direction of wafer scribes at the measured pixel can be also acquired. The original gray image is checked in this step. Base on the fact that the width of a scribe is always between 4 and 7 pixels; along with cross direction of wafer scribes at the measured pixel and within 5 pixels distance from the measured pixel, the pixels with the maximum gray level value are regarded as the center of the wafer scribes. If the distance between the measured pixel and the center of the wafer scribe is less than or equal 2 pixels, then the gray level value of the measured pixel is set to 255, otherwise 0.

Fig. 3 Some test images

3 Experimental Results

We present in this section several tests images. The representative experimental results are shown in Fig. 3. Experiments involving two types of original wafer ID image: wafer ID image with overexposure, wafer ID image with underexposure. In Fig. 3 type a is the image with overexposure, type b is the image with underexposure. In each type, a1 and b1 are original wafer ID images, a2 and b2 are the images after enhancement and asterisk-shape filtering, a3, and b3 are the images after high-low score comparing and a4 and b4 are the final binarized images after combining by determination of the scribe center. From Fig. 3 we can directly see that our approach allows to differentiating the characters from the background efficiently.

With the proposed binarization method, 92% of 125 sample vehicle wafer ID images are correctly recognized. There are 1316 characters in 121 wafer ID images which step in recognition stage successfully. Total 1308 ones are successfully segmented out. The ratio of character recognition is high to 99.39%. The results show that the approach is promising and the binarization rate is close to 100%.

4 Conclusion

In this paper, a binarization approach based on asterisk-shape filter for wafer ID images is proposed. The proposed approach can effectively enhance the performance of wafer ID recognition. It enhances the wafer ID image at first and original binarization based on asterisk-shape filter and high-low score comparison method are applied in the following. Finally, final binarization result is obtained by combining the determination of the scribe center. Experimental results presented in the paper show that the algorithm provides good results even in situations of uneven luminance, overexposure and underexposure. As a result, we believe that the proposed technique is an attractive alternative to currently available methods for binarizing wafer ID images. We may apply this technique to other applications, such as a car license-plate recognition and the container ID numbers recognition.

References

1. Kittler, J., Illingworth, J.: On Threshold Selection Using Clustering Criteria. IEEE Transactions on Systems, Man, and Cybernetics 15, 652–655 (1985)
2. Brink, A.D.: Thresholding of digital images using two-dimensional entropies. Pattern Recognition 25, 803–808 (1992)
3. Yan, H.: Unified formulation of a class of image thresholding techniques. Pattern Recognition 29, 2025–2032 (1996)
4. Sauvola, J., Pietikainen, M.: Adaptive document image binarization. Pattern Recognition 33, 225–236 (2000)
5. Trier, O.D., Jain, A.K.: Goal-Directed Evaluation of Binarization Methods. IEEE Transactions On Pattern Analysis And Machine Intelligence, 1191–1201 (1995)
6. Gatos, B., Pratikakis, I., Perantonis, S.J.: Adaptive degraded document image binarization. Pattern Recognition 39, 317–327 (2006)
7. Sheikh, L.M., Hassan, I., Sheikh, N.Z., Bashir, R.A., Khan, S.A., Khan, S.S.: An adaptive multi-thresholding technique for binarization of color images. In: Proceedings of the 9th WSEAS International Conference on Computers table of contents (2005)

Feature Particles Tracking for the Moving Object

Tao Gao, Zheng-guang Liu, and Jun Zhang

Abstract. For particle filtering tracking method, particle choosing was random to some degree according to the dynamics equation, which may cause inaccurate tracking results. To compensate, an improved particle filtering tracking method was presented. The motion region was detected by redundant discrete wavelet transforms method (RDWT), and then the key points were obtained by scale invariant feature transform. The matching key points in the follow-up frames obtained by SIFT method were used as the initial particles to improve the tracking performance. Experimental results show that more particles centralize in the region of motion area by the presented method than traditional particle filtering, and tracking results are more accurate and robust of occlusion.

1 Introduction

Detecting and tracking moving objects from video sequences is one of the important tasks of video surveillance system. Recently, many approaches have been proposed in this field. Ref.[1-2] use a template matching method to track the target. The blobs correspond to the moving target in the video sequences. But the method is difficult in handling scale change of the target, and threshold is subjectively determined with less robustness. Ref.[3] uses a snake model based tracking method which can reduce computational complexity and improve tracking accuracy, but it is sensitive to initialization and difficult for actual application. Ref.[4-5] present a mean-shift method for motion tracking. Mean-shift method manifests high efficiency for target tracking with low complexity. But as a hill climbing algorithm, it may fall into a local minimum and lose the motion target when occlusion occurs. Ref.[6] uses Particle filtering to track moving targets; it is a successful numerical approximation technique for Bayesian sequential estimation with non-linear, non-Gaussian models. The basic Bayesian filtering is a

Tao Gao, Zheng-guang Liu, and Jun Zhang
School of Electrical Engineering and Automation, Tianjin University,
Tianjin, 300072, China
e-mail: gaotao09@yahoo.cn

B.-C. Chien and T.-P. Hong (Eds.): Opportunities and Challenges, SCI 214, pp. 39–44.
springerlink.com

recursive process in which each iteration consists of a prediction step and a filtering step. In this paper, the positions of particles are determined by key points obtained by scale invariant feature transform (SIFT) to improve the tracking efficiency, and we organize the paper as follows. A brief introduction of RDWT motion detection is given in Section 2. Scale invariant feature transform and the improved particle filtering tracking method is described in Section 3. Experimental results are reported in Section 4. Conclusions are summarized in Section 5.

2 Motion Region Detection

In this paper, motion region are detected by redundant discrete wavelet transforms (RDWT) [7], which conquer the drawback of time-domain methods. The RDWT is an approximation to the continuous wavelet transform that removes the down-sampling operation from the traditional critically sampled DWT to produce an over-complete representation. The shift-variance characteristic of the DWT arises from its use of down-sampling; while the RDWT is shift invariant since the spatial sampling rate is fixed across scale. As a result, the size of each sub-band in an RDWT is the exactly the same as that of the input signal. Because the coefficients of the sub-bands of the redundant wavelet transform are highly correlated, this paper uses redundant wavelet transforms to obtain the motion area. First, if the two adjacent frames are f_1 and f_2, equation (1) is used to obtain the $MAS(x, y)$:

$$MAS\,(x,y)=\sum_{j=J_0}^{J_1}\left(\begin{array}{l}\left|LL_1^{(j)}(x,y)-LL_2^{(j)}(x,y)\right|+\left|LH_1^{(j)}(x,y)-LH_2^{(j)}(x,y)\right|\\+\left|HL_1^{(j)}(x,y)-HL_2^{(j)}(x,y)\right|+\left|HH_1^{(j)}(x,y)-HH_2^{(j)}(x,y)\right|\end{array}\right) \tag{1}$$

The J_0 and J_1 are the starting and ending scales. We can obtain the motion area according to T which is the threshold which can be obtained automatically by OTSU method [8], and then the mathematical morphology is used to remove noise points. The binary motion mask obtained from the redundant wavelet transforms can be considered as the original mask of the moving object. As the inner district is usually flat and the characteristic is not obvious, this paper uses an assimilation method [9] to fill the mask. Figure 1 shows the motion region detection result.

Fig. 1 Motion region detection

3 Feature Particles Tracking

3.1 Key Points Obtained

After getting the motion region, we use scale invariant feature transform (SIFT) to obtain key points in the object region. As SIFT transforms image data into scale-invariant coordinates relative to local features, an important aspect of this approach is that it generates large numbers of features that densely cover the image over the full range of scales and locations [10-11]. For image matching and recognition, SIFT features are first extracted from a set of reference images and stored in a database. A new image is matched by individually comparing each feature from the new image to this previous database and finding candidate matching features based on Euclidean distance of their feature vectors.

The SIFT detector extracts from an image a collection of frames or key points. These are oriented disks attached to blob-alike structures of the image. As the image translates, rotates and scales, the frames track these blobs and thus the deformation. By canonization, i.e. by mapping the frames to a reference (a canonical disk), the effect of such deformation on the feature appearance is removed. The SIFT descriptor is a coarse description of the edge found in the frame. Due to canonization, descriptors are invariant to translations, rotations and scalings and are designed to be robust to residual small distortions. Figure 2 shows the key points SIFT matching result of a moving object.

Fig. 2 SIFT matching result

3.2 Active Particle Filtering Combined with SIFT

Particle filtering [12-15] essentially combines the particles at a particular position into a single particle, giving that particle a weight to reflect the number of particles that are combined to form it. This eliminates the need to perform redundant computations without skewing the probability distribution. Particle filtering accomplishes this by sampling the system to create N particles, then comparing the samples with each other to generate an importance weight. After normalizing the weights, it resamples N particles from the system using these weights. This process greatly reduces the number of particles that must be sampled, making the system less computationally intensive. For active tracking, we initializes the state space for the first frame automatically by using the matching key points obtained by SIFT. A second-order auto-regressive dynamics is chosen on the parameters by

SIFT matching to represent the state space (x, y).The following steps depict the overall structure of our tracking system combined with SIFT and particle filtering (SI_P).

(1) Sample initiation. The motion region is obtained by RDWT, and the color probability distribution is $q = \{q(u)\}_n$, $q(u) = f\sum_{i=1}^{I} k(\frac{\|x_i\|}{a})\delta(h(x_i) - u)$. Where I is the number of pixel in the particle region, δ is Kronecker residual sub-function, a is the size of particle region. k is the contour function of Epanechnikov kernel, f is normalization factor. Combined with SIFT key points and current motion region, initial state sample set is $S_0 = \{sift(X_0^{(i)}), 1/N\}_{i=1}^{N}$.
(2) By substitutable selection method [16], N samples are extracted form S_{t-1} according to w_{t-1}^{i}. Compute the normalized accumulation weights: $c_{t-1}^{0} = 0$, $c_{t-1}^{i} = c_{t-1}^{i-1} + w_{t-1}^{i}$, $c_{t-1}^{'i} = c_{t-1}^{i} / c_{t-1}^{N}$, to generate the uniform random number r at $[0,1]$. Search the minimum k for $c_{t-1}^{'k} \geq r$, and let $S_{t-1}^{'i} = S_{t-1}^{k}$.
(3) The dynamics equation is $S_t = AS_{t-1} + R \cdot rank$, where A is state transition matrix, and $R \cdot rank$ is a random Gaussian matrix: $A = \begin{bmatrix} 1 & \sigma \\ 0 & 1 \end{bmatrix}$,

$$R = \alpha \begin{bmatrix} \sigma^3/3 & \sigma^2/2 \\ \sigma^2/2 & \sigma \end{bmatrix}$$, $\sigma = 3$, $\alpha = 0.35$. $S_{sift,t}$ can be obtained by endowing the SIFT matching key points at time t to S_t.
(4) Compute the Bhattacharyya $\rho[p_t, q]$ between candidate sample and target mask.
(5) Recompute the sample weights $W_t^i = \frac{1}{\sqrt{2\pi\sigma}} \exp(-(1-\rho[p_t,q])/2\sigma^2) \cdot W_{t-1}^{i}$ of $S_{sift,t}$.
(6) The central location of the moving object at time t can be obtained by the average of weighted samples: $E(S_t) = \sum_{i=1}^{N} W_t^i S_{sift,t}^i / \sum_{i=1}^{N} W_t^i$.

4 Experimental Results

We compare the tracking results between SI_P and traditional particles filtering, Mean-shift tracking method showed in figure 3. The video is sampled at a

(a) SI_P

(b) Traditional particle filtering tracking

(c) Mean-shift

Fig. 3 Tracking results comparison (Frames 49, 102, 171, 188, 196, 201, 239, and 247)

resolution of 320×240 and a rate of 25 frames per second. The algorithms are tested on a 1400 MHz Celeron CPU, and software environment is VC++ 6.0. Initial particle number is 300. From the results we can see that when the scale of the tracked person changes drastically, particles still locate in the region of person by SI_P; while for traditional particle filtering, particles obviously deviate from the person which causes inaccurate results. The blue cross sign shows the particle, and red curve shows the motion track. We can also find that the robustness of occlusion by SI_P, while Mean-shift loses the target when occlusion occurs.

5 Conclusions

In this paper, a novel moving object tracking method based on particle filtering and SIFT key points is presented. First, the motion region is detected by redundant discrete wavelet transforms, and scale invariant feature transform is used to extract the key points of object; then according to SIFT matching, initial positions of particles can be obtained. By actively choosing the particles, tracking performance can be significantly improved and without the influence of occlusion.

References

1. Zhang, L., Han, J., He, W., Tang, R.S.: Matching Method Based on Self-adjusting Template Using in Tracking System. Journal of Chongqing University 6, 74–76 (2005)
2. Magee, D.R.: Tracking Multiple Vehicles Using Foreground, Background and Motion Models. Image and Vision Computing 22, 143–155 (2004)
3. Liu, H., Jiang, G., Li, W.: A Multiple Objects Tracking Algorithm Based on Snake Model. Computer Engineering and Applications 42(7), 76–79 (2006)
4. Comaniciu, D., Ramesh, V.: Mean Shift and Optimal Prediction for Efficient Object Tracking. In: Proc. of the IEEE International Conference on Image Processing, vol. 3, pp. 70–73 (2000)
5. Comaniciu, D., Ramesh, V., Meer, P.: Kernel-based Object Tracking. IEEE Transactions on Pattern Analysis and Machine Intelligence 25(5), 564–577 (2003)
6. Hue, C., Le Cadre, J., Perez, P.: Tracking Multiple Objects with Particle Filtering. IEEE Transactions on Aerospace and Electronic Systems 38, 313–318 (2003)
7. Gao, T., Liu, Z.-g., Zhang, J.: BDWT based Moving Object Recognition and Mexico Wavelet Kernel Mean Shift Tracking. Journal of System Simulation 20(19), 5236–5239 (2008)
8. Otsu, N.: A Threshold Selection Method from Gray-Level Histogram. IEEE Trans. SMC 9(1), 62–66 (1979)
9. Gao, T., Liu, Z.-g.: Moving Video Object Segmentation based on Redundant Wavelet Transform. In: Proc.of the IEEE International Conference on Information and Automation, pp. 156–160 (2008)
10. Lowe, D.G.: Distinctive Image Features from Scale-Invariant Keypoints. International Journal of Computer Vision 60(2), 91–110 (2004)
11. Lowe, D.G.: Object Recognition from Local Scale-Invariant Features. In: Proc. of the International Conference on Computer Vision, pp. 1150–1157 (1999)
12. Arulampalam, S., Maskell, S., Gordon, N., Clapp, T.: A Tutorial on Particle Filters for On-Line Nonlinear/ Nongaussian Bayesian Tracking. IEEE Trans. Signal Process. 50(2), 174–188 (2002)
13. Pitt, M., Shephard, N.: Auxiliary Particle Filters. J. Amer. Statist. Assoc. 94(446), 590–599 (1999)
14. Doucet, A., Vo, B.-N., Andrieu, C., Davy, M.: Particle Filter for Multi-Target Tracking and Sensor Management. In: Proc. of the Fifth International Conference on Information Fusion, pp. 474–481 (2002)
15. Sidenbladh, H.: Multi-target Particle Filtering for the Probability Hypothesis Density. In: Proc. of the Sixth International Conference on Information Fusion, pp. 1110–1117 (2003)
16. Reckleitis, I.: A Particle Filter Tutorial for Mobile Robot Localization. In: Proc. of the International Conference on Robotics and Automation, vol. 42, pp. 1–36 (2003)

Break-Segment Detection and Recognition in Broadcasting Video/Audio Based on C/S Architecture

Xiangdong Wang, Xinhui Li, Yuelian Qian, Ying Yang, and Shouxun Lin

Abstract. A novel scheme for break-segment detection and recognition in broadcasting video/audio based on client/server architecture is proposed, where break-segments are referred to as segments inserted between or within main programs such as commercials, upcoming program announcements, head leader and closing credits of programs, etc. At the server-end, a break-segment database is initially generated and thereafter updated by results of repetition detection. The clients get the latest break-segment database from the server through network and recognize the break-segments in the database. For both repetition detection and recognition, only the audio signal is processed using the feature EEUPC and corresponding similarity measurement and matching algorithm, achieving much higher efficiency than current methods. Experimental results demonstrate high accuracy and efficiency of the proposed framework.

1 Introduction

This paper focuses on detection and recognition of break-segments in broadcasting video/audio, where break-segments are referred to as segments inserted between or within main programs, like commercials, head leader and closing credits of

Xiangdong Wang, Xinhui Li, Yuelian Qian, and Shouxun Lin
Institute of Computing Technology, Chinese Academy of Sciences,
Beijing 100190, China

Xinhui Li
Graduate University of Chinese Academy of Sciences, Beijing 100085, China

Ying Yang
China Agricultural University, Beijing 100083, China
e-mail: {xdwang,lixinhui,ylqian,yyang,sxlin}@ict.ac.cn

B.-C. Chien and T.-P. Hong (Eds.): Opportunities and Challenges, SCI 214, pp. 45–51.
springerlink.com

programs, etc. Automatic detection and recognition of break-segments are useful for a variety of applications. For example, for TV views or digital recorder users, automatic filtering of break-segments brings better experience, and recognition of head leaders and closing credits leads to convenient location of specific programs.

While there is yet little work focused on break-segment detection and recognition, there has been much work on relative topics like commercial detection. There are mainly three kinds of methods for commercial detection: Classification-based methods classify video shots into commercials or ordinary programs using heuristic rules or statistical classifiers [1, 2, 3]; methods based on repetition detection detect commercials as visual and acoustical repetition segments [4, 5, 6]; and recognition-based methods maintain a database of known commercials and match them in the video/audio [7]. The classification-based methods are relatively unreliable, since there is no distinct difference between ordinary programs and commercials. The main problem of repetition detection is that large amount of data should be acquired and used, making it unsuitable for applications for end-users. And the main challenge for recognition-based methods is the availability of the commercial database. Moreover, most methods cannot achieve real-time performance due to low efficiency of visual features used.

In this paper, we propose a novel scheme of break-segment detection and recognition in broadcasting video/audio, as shown in Fig.1. The system is divided into the server-end and the client-end, which are connected by networks. At the server-end, a break-segment database is initially generated and thereafter updated using results of repetition detection. The clients get the latest break-segment database through network and recognize the break-segments in the database. This framework solves both the problems of methods based on repetition detection and recognition: By using the repetition detection in the server-end, processing of large data is avoided for end-users, and the break-segment database is updated continuously to cover almost all break-segments. For both repetition detection and recognition, only the audio signal is processed using the feature EEUPC and corresponding similarity and matching algorithm, which achieves higher efficiency than visual features used by current methods.

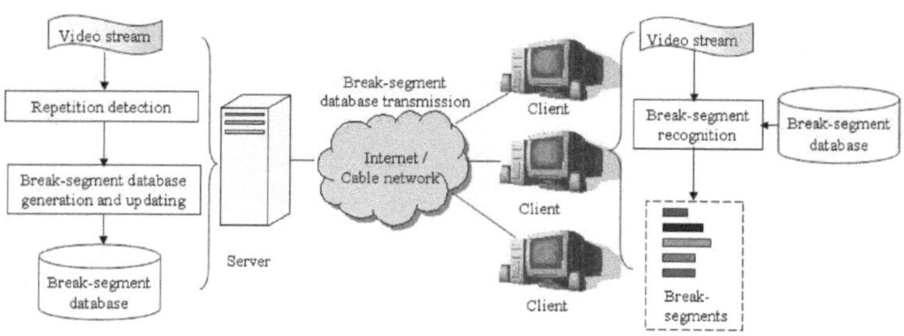

Fig. 1 Framework for shot segmentation and classification

2 Detection and Recognition of Break-Segment Using EEUPC

As mentioned in Section 1, most current approaches of commercial detection use visual features and are quite time-consuming. Our method only uses the audio signals by using a novel audio feature EEUPC proposed in our earlier work [7], which is very simple and brings both high efficiency and accuracy.

EEUPC (Energy Envelope Unit Position and Confidence) is an audio feature based on energy envelope, the main idea of which is to segment the energy envelope into units containing one major peak each, and use the positions of units and confidence of segmentation as feature of the audio. Therefore, the EEUPC feature of an audio segment looks like $U = (u_1, c_1), (u_2, c_2), ..., (u_n, c_n)$, where u_i is the frame number of the ith segmentation position and c_i is corresponding confidence value. The extraction of EEUPC is much simpler than features such as MFCC [7]. Based on EEUPC, a similarity measurement of two audio segments is proposed [7], which considers the EEUPC of one audio segment as reference and EEUPC of the other as detection results, and then calculate detection precision, recall and use the F-value (which can be seen as an average of precision and recall) as similarity.

2.1 Break-Segment Detection Based on Audio Repetition Detection

Given an audio stream represented by EEUPC as U=$\{(u_1, p_1)$,(u_2, p_2),...,$(u_n, p_n)\}$ candidates of repeated segments can be detected as segments with high similarity. In fact, it can be inferred from the definition of similarity based on EEUPC that corresponding units of matching segments have almost the same length. Therefore, matching can be conducted only from units that with similar lengths. To quickly find units with similar lengths, the list of all units is sorted according to length, and units with similar length become neighbors. If by this method the unit between u_i and u_{i+1} and the unit between u_j and u_{j+1} are neighbors satisfying $|(u_{i+1} - u_i) - (u_{j+1} - u_j)| < T_{Len}$, where T_{Len} is a pre-determined threshold (5 in our work), then segmentation positions after u_{i+1} and are added to the match one by one and the similarity based on EEUPC is calculated for each time. The process is stopped whenever the similarity drops below a threshold Tsim (0.6 in our work). The procedure of the algorithm is shown in Fig. 2.

Since there may be false alarms and the boundaries may be not accurate in the result, the candidate matches need to be verified and refined. In our work, Euclidean distances between MFCC vectors of any two frames of the two matched segments are calculated and the percentage of low distance values is thresholded for

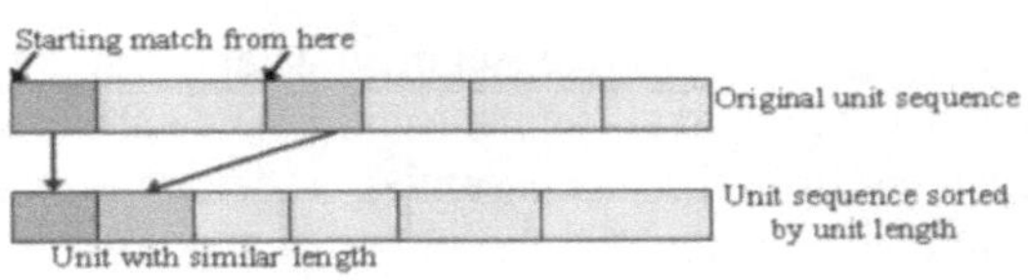

Fig. 2 Procedure of the algorithm for repetition detection based on EEUPC

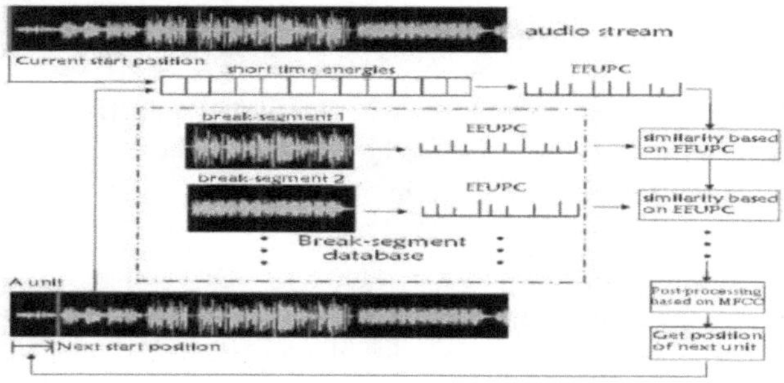

Fig. 3 Procedure of fast break-segment recognition based on EEUPC

verification. Despite low efficiency of MFCC extraction and distance calculation, this can be done quickly due to small number of candidate matches.

2.2 *Fast Break-Segment Recognition Using EEUPC*

In the client-end, for real-world application, fast break-segment recognition is especially needed, since the break-segment database may contains thousands of break-segments which are all required to be searched in the purposed audio stream. In our work, this problem is solved by using EEUPC for fast audio matching [7].

The whole procedure of break-segment recognition at client-end is shown in Fig. 3. EEUPCs of all known break-segments are stored in the break-segment database, and the EEUPCs are compared one by one to real-time audio stream. For each break-segment at each time of comparison, a current match segment is selected. Then, similarity function is calculated between the two segments, and they are decided to be a match if the value of similarity function exceeds a threshold.

3 Break-Segment Database and Related Operations

In the approach proposed in this paper, the break-segment database is a file containing features and information of all known break-segments. The information recorded in the database includes break-segment name, category, etc.

The procedure of generation of the break-segment database is shown in Fig. 4. After repetition detection, manual verification and refinement of candidate break-segments obtained are conducted. Meanwhile, manual annotation is also done to give each break-segment a name and a category, as mentioned above. Finally, features of EEUPC are extracted for each break-segment and written into the database file in conjunction with information of break-segments.

The updating of break-segment database at the server-end includes insertion of newly detected break-segments into the database and deletion of "outdated" break-segments from the database, the procedure of which is shown in Fig. 5. For the monitored video/audio, searching of the current break-segment database is also

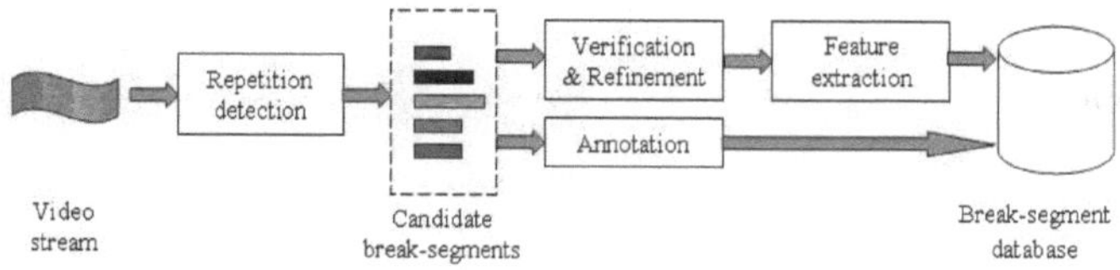

Fig. 4 Procedure of generation of the break-segment database

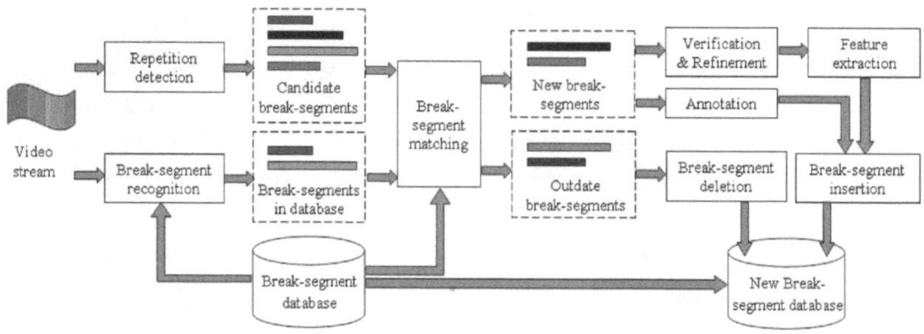

Fig. 5 Procedure of updating of the break-segment database

conducted in addition to repetition detection, and the results of repetition detection and recognition are compared and matched. Only break-segment candidates that do not intersect with recognized ones are inserted to the database. Also, the break-segments that have not appeared within a week are deleted in the database.

4 Experimental Results and Analysis

A series of experiments were conducted simulating applications at both server-end and client-end. The test data are TV broadcast videos of 20 hours, recorded from the channel Hunan-TV of China. The video is stored in MPEG-1 format, while the processed signal is 16KHZ, 16-bit, mono wav audio. All experiments were run on a PC with a Pentium IV 2.4GHz CPU and 512 RAM.

Results of repetition detection at the server-end are shown in Table 1. A recall rate of 97.8% and precision rate of 98.8% are achieved. Due to the fact that repeated segments are not all break-segments, the precision rate for break-segment detection is about 88%, but the recall rate is quite high (98.8%) and false-alarms can be further verified manually, as explained in Section 3. Most importantly, the processing time

Table 1 Results of repetition detection

	Detection of repeated segments	Detection of break-segments
Recall	97.7915%	98.7816%
Precision	98.7916%	88.2462%
Time	25 minutes 23 seconds	

Table 2 Recogntion results by old and new break-segment databases

	Old database	New database
Recall	50.7757%	97.7095%
Precision	99.2518%	99.1292%

including feature extraction is only about 25 minutes for 20- hour data, which is much faster than current methods.

To evaluate performance of break-segment recognition, we used a break-segment database containing 456 break-segments. The break-segment recognition system was tested using 5-hour video drawn from the 20-hour test set. A recall rate of 97.7% and precision rate of 98.2% are achieved, and the processing time is less than 4 minutes (203.61 seconds), which is only 0.01 times real-time.

To simulate the procedure of break-segment database updating, we used the break-segment database described above as an "old" database, and used the result of repetition detection on the 20-hour test data to update it. The system gave 221 candidate new segments, and by manually deleting 31 non-break-segments, 190 new break-segments were inserted to the database. Then, the 20-hour test data was recognized using the updated database and the result was compared to that by the old database. As shown in Table 2, the high recall rate demonstrates that the updating has inserted almost all new break-segments.

5 Conclusions

In this paper, a scheme for break-segment detection and recognition based on C/S architecture is proposed. A break-segment database is initially generated and thereafter updated by repetition detection at the server-end, and the client uses the latest break-segment database got from the server to recognize break-segments. For both repetition detection and recognition, the audio feature EEUPC and corresponding similarity measurement and matching algorithm are used to achieve higher efficiency than current methods. This framework incorporates the advantages of repetition detection and recognition-based methods and makes real-world application feasible, which are demonstrated by experimental results.

References

1. Li, Y., Kuo, C.-C.J.: Detecting Commercial Breaks in Real TV Program based on Audiovisual Information. In: SPIE Proc. on IMMS, vol. 421 (2000)
2. Duan, L., Wang, J., et al.: Segmentation, Categorization, and Identification of Commercials from TV Streams Using Multimodal Analysis. In: Proc. ACM MM 2006, Santa Barbara (2006)
3. Hua, X., et al.: Robust Learning-Based TV Commercial Detection. In: Proc. ICME (2005)

4. Duygulu, P., Chen, M., Hauptmann, A.: Comparison and Combination of Two Novel Commercial Detection Methods. In: Proc. ICME 2004, pp. 1267–1270 (2004)
5. Gauch, J.M., Shivadas, A.: Identification of New Commercials Using Repeated Video Sequence Detection. In: Proc. ICIP, pp. 1252–1255 (2005)
6. Covell, M., Baluja, S., et al.: Advertisement Detection and Replacement using Acoustic and Visual Repetition. In: Proc. International Workshop on Multimedia Signal Processing (2006)
7. Zhao, D., Wang, X., et al.: Fast Commercial Detection based on Audio Retrieval. In: Proc. ICME 2008, Hannover (2008)

A Time Series Case-Based Predicting Model for Reservation Forecasting

Tsung-Hsien Tsai and Sheryl E. Kimes

Abstract. This study addresses how to construct sales forecasting models by using restaurant reservation data. The issues of how to retrieve booking patterns, search for influential parameters, and divide samples for training, validating, and testing are discussed. Regression and Pick Up models, which are common practice, are also built as benchmarks. We used data from a mid-sized restaurant to show that the proposed Time Series Case-Based Predicting model can significantly outperform the benchmarks in all testing cases.

Keywords: Time Series Case-Based Predicting, Advanced Booking Model, Sample Selection, Revenue Management, Data Mining.

1 Introduction

Demand forecasting is important because it provides input information for the efficient and profitable operation of business enterprises. In revenue management applications, forecasting is essential for resource allocation and overbooking. The benefit of improving forecasting accuracy in revenue management has also proven to be significant [1].

Data used for revenue management forecasting has two dimensions to it: when the reservation was booked and when the service took place. The booking information gives additional detail that can be used to update the forecast. Without this information, the forecast would be based solely on the historical information on the daily number of customers served.

Three forecasting approaches have been identified [2]. Historical booking models use historical arrival data to predict the future. For instance, all historical final sales data are used to project future sales [3]; Exponential Smoothing and Autoregressive Integrated Moving Average (ARIMA) fall into this category. Advanced booking models use information on when customers placed their reservation to develop forecasts; Pick Up and Regression models are frequently used for this purpose. Combination methods use a weighted average of the historical and

Tsung-Hsien Tsai and Sheryl E. Kimes
School of Hotel Administration, Cornell University,
14850 Ithaca, United States of America

B.-C. Chien and T.-P. Hong (Eds.): Opportunities and Challenges, SCI 214, pp. 53–58.
springerlink.com

advanced booking models [4]. Much of the sales forecasting research uses historical booking models but few papers have studied advanced booking models. The aim of this study was to construct an advanced booking model with the potential to improve predictive accuracy.

In this study, we proposed a novel advanced booking model based on the concept of pattern retrieval. A four-stage procedure was provided to project forecasts; furthermore, the issue of sample selection was discussed. We tested models on reservations data from a 100-seat restaurant.

2 Booking Data

A restaurant manager may open seats for reservations via different channels several weeks (or months) before a specific service date. Customers who do not make reservations and show up in the restaurant are categorized as walk-ins. When customers make their reservations, the restaurant's computer system records the reservation date and the number of seats reserved. If we accumulate this volume of reservations over the whole booking period plus walk-ins, then we can obtain the volume of final sales.

We used 20 months of detailed reservations data from a 100-seat restaurant to develop our model. The restaurant has a busy lunch period and tracks their reservations and walk-ins with a reservation system called OpenTable.com. OpenTable.com is used by over 8000 restaurants worldwide and seats over 2 million customers each month. Restaurants can use OpenTable to track both their reservations (regardless of whether they are made online or over the phone) and walk-in business. The data includes information on when the reservation was made, the date and time of service and the number of people in each party. This provided us with the necessary information to develop booking curves.

The graph of the complete booking data for each service date shows the number at which reservations are received (Fig. 1.). DBS (-1) represents the number of walk-ins plus all reserved customers on a service date; DBS (0) is the number of all reserved customers on a service date; DBS (k>0) represents the number of accumulated reservations k days before a service date. Historical booking models use only DBS (-1) data for model construction while advanced booking models use all booking data which availability depends on DBS (k).

We used our booking data to develop booking curves for the restaurant by service date. Average booking curves were developed by day of week (Fig. 1.) and showed that on average most reservations were made within 3 days of service. Without the influence of cancellations, the booking curves are all monotonic increase, and the phenomenon implies that the reliability of information improves over time. While the averages are helpful, the daily booking curves showed quite a bit of week-to-week variation (Fig. 2.). Fig. 1. also showed weekday and weekend effects although weekday demand was fairly similar. The variation of final sales also rendered valuable information (Fig. 3.). The data was non-stationary, and seasonality may be the driver causing this phenomenon which sales were generally higher in May than those in other months. In addition, there was no significant trend and cycle was not also expected during the research period.

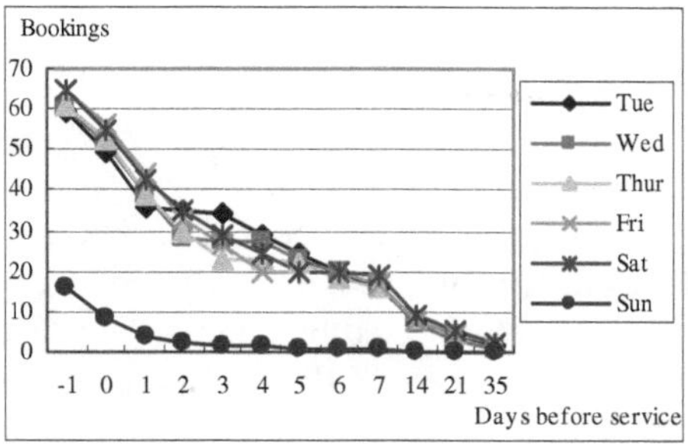

Fig. 1 Average Booking Curves

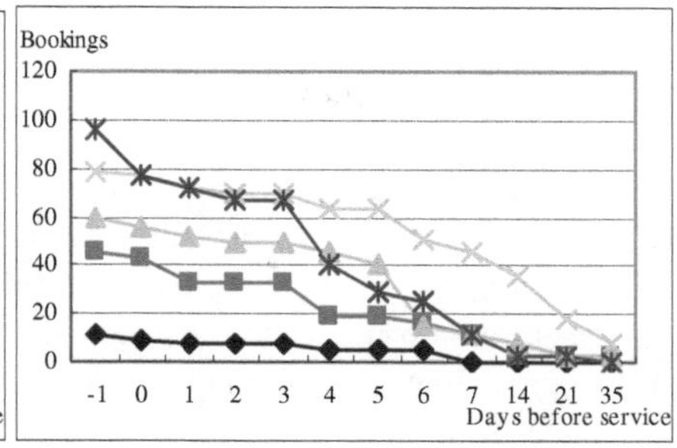

Fig. 2 Weekly variation of Tuesdays

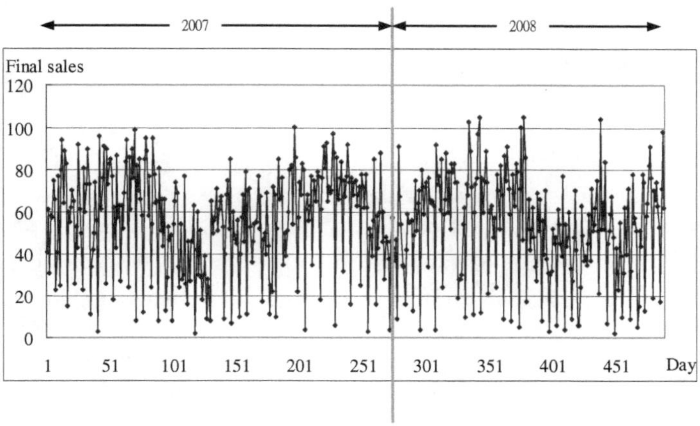

Fig. 3 The volume of final sales over the research period

3 Time Series Case-Based Predicting Model

In the following, $x_{j,k}$ is the number of accumulated reservations for service date *j* at booking point *k*. Forecasts are developed for the final sales of service date *j*, $x_{j,-1}$. The data collection point (DCP), *m*, is the point where the computer system collects data and updates forecasts. It is common to have updates once for a reservation several weeks before the service date and more frequent recalculations as the service date approaches. In this study, the DCPs used were -1, 0, 1, 2, 3, 4, 5, 6, 7, 14, 21, and 35 in terms of the data we collected.

CBP is composed of four stages, the first being similarity evaluation, which calculates similarity between booking curves in the database and the booking patterns of a targeted service date. The calculations are updated each time new reservation information becomes available at each DCP. CBP also incorporates a temporal characteristic that considers information at each DCP to be of exponential importance with a parameter α to show the reliability of information over DCPs. Equation (1) computes the distance between the service date *j* and a booking curve *i* at DCP (k) (k indicates the current DCP, and *t* is the first DCP).

$$D_k(j,i) = \sum_{m=k}^{t} (x_{j,m} - x_{i,m})^2 (\frac{1}{m})^{\alpha} \,. \quad (1)$$

The second phase is to select the most similar booking curves in terms of the calculation in the first stage. CBP ranks the similarity of all booking curves and selects ten most similar samples based on the patterns of the targeted service date.

The third step is to integrate the final sales of the selected booking curves. Instead of calculating a simple average for these final sales, CBP incorporates the influence of similarity and also adds an adaptive term to enhance the importance of the current reservation information (Equation (2)). β and γ are parameters to show the effects of similarity and adaptability, respectively.

$$\hat{x}_{j,-1}^{k} = \sum_{s=1}^{10} \left(\frac{\frac{1}{D_k(j,s)}}{\sum_{s=1}^{10} \frac{1}{D_k(j,s)}} \right)^{\beta} \left(\frac{x_{j,k}}{x_{s,k}} \right)^{(\frac{1}{k})^{\gamma}} x_{s,-1} \,. \quad (2)$$

The last step is to search for three parameters in Equations (1) and (2). It is difficult to apply a conventional gradient-based algorithm. As a result, we applied the Hooke-Jeeves algorithm [5], which is a direct search method, to find α, β, and γ. The weakness of direct search algorithms is the possibility to stick into local minima. In this study, the multi-start strategy, which tries different initial seeds, was applied to select the most possible global minima.

Another focus of this study was to investigate what samples in the database should be used for searching a suitable combination of parameters. Before going forward, we divided the collected booking curves into three categories: training, validating, and testing samples. In Regression and Pick up models [3], the current practice is to use all data except testing samples for estimating parameters. The parameters are then used to forecast and the predictive performance is evaluated by using testing samples. In CBP, the calibrating procedure is different because it tries to decide parameters by matching patterns between training and validating samples. One possibility is to set training samples as the base and minimize mean square errors (MSE) of the validating ones. The obtained parameters are then verified by using testing samples. The problem is how to decide training and validating samples so that a valid combination of parameters can be obtained.

We tested three mechanisms of sample division for the CBP parameter search. The first method is the sequential method (SEQ) and uses the month of data immediately before the testing sample as the validating set. The concept of this approach is to use the most recent booking trends. Randomly (RAN) selecting a certain number of samples (about one month) from all but testing data is another way to select validating samples. The idea of this approach is to learn patterns from different time periods. The last approach is to take the same month of the previous year (LAST) as the validating month. For example, May in 2008 is taken as the validating samples while predicting sales in May 2009. The logic behind this approach is to learn the patterns with the same time factor from the past.

4 Empirical Study

We tested the performance of CBP and also the above three division mechanisms. In order to have an overall consideration of time effects, we first saved one-year daily data for both training and validating purposes (2/07~1/08) and tested the performance of next month (2/08). Once the performance was computed, the testing samples were included into the training and validating database (2/07~2/08), and the model was re-optimized to test the performance of the next month (3/08). The procedure was repeated dynamically until all testing samples were exhausted (9/08). The purpose is to verify CBP's performance dynamically.

We next studied the method to select validating samples, and LAST obtained the best predictive accuracy of the three approaches. Fig. 4. shows the average improvement of MSE by using LAST in comparison with using SEQ and RAN, respectively (Equation (3)). LAST outperformed SEQ in 6 out of 8 cases and beat RAN in 5 out of 8 cases. More importantly, LAST usually was significantly better than the other two alternatives. This was because LAST seizes seasonal effects properly; SEQ and RAN systematically under- or overestimate in some situations.

Another important observation was to see whether the proposed CBP model and LAST procedure could outperform conventional Regression and Pick Up models. Fig. 5. displays the average improvement of MSE (formula is analogous to Equation (3))by using the proposed CBP in comparison with using Regression and Pick Up models. It is apparent that the proposed CBP can significantly outperform two conventional benchmarks. This result shows the value of CBP and its potential for predicting sales by using reservation data.

$$\frac{MSE^{LAST} - MSE^{Base}}{MSE^{Base}} \times 100\% \cdot \quad (3)$$

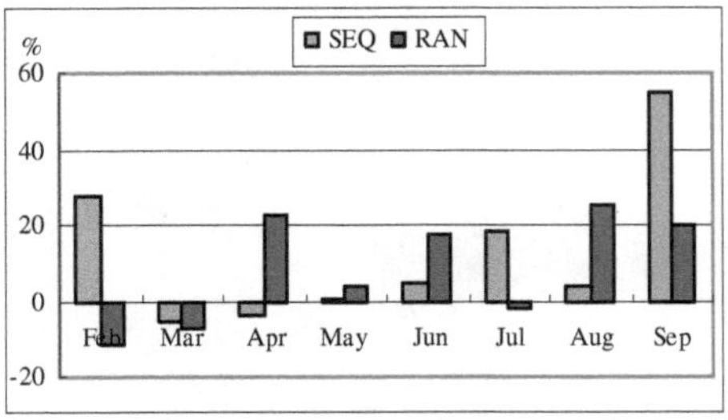

Fig. 4 Average improvement of LAST

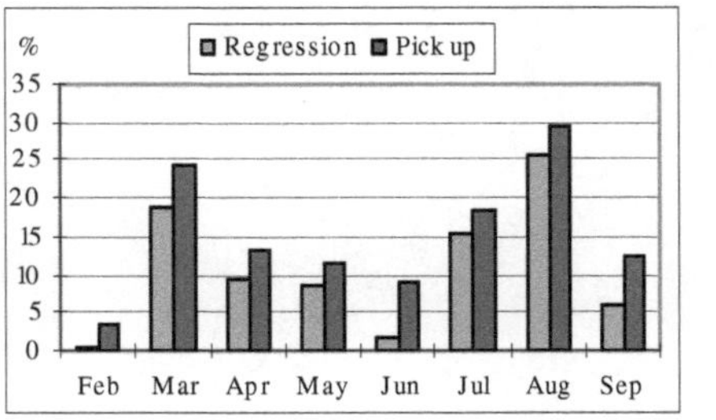

Fig. 5 Average improvement of CBP

5 Conclusions

Forecasting accurately is important for making correct decisions in daily operations. Regression and Pick Up are two common models used for arrival or sales forecasting. In the empirical study, we demonstrated that the proposed CBP model with careful selection of samples leads to a better combination of parameters and results in better predictive accuracy in comparison with the benchmarks.

In this study, the variation of final sales seems to be non-stationary, as shown in Fig. 3. As a result, the conclusions obtained in this study may be only valid for problems with similar data characteristics. It would be interesting to study what would happen if final sales have a linear trend or other periodic patterns, and how we should redesign the model so that CBP can still maintain its edge.

References

1. Lee, A.O.: Airline Reservations Forecasting: Probabilistic and Statistical Models of the Booking Process. Massachusetts Institute of Technology Press, Boston (1990)
2. Wickham, R.R.: Evaluation of Forecasting Techniques for Short-term Demand of Air Transportation. Massachusetts Institute of Technology Press, Boston (1995)
3. Weatherford, L.R., Kimes, S.E.: A Comparison of Forecasting Methods for Hotel Revenue Management. International Journal of Forecasting 19, 401–415 (2003)
4. Rajopadhye, M., Ghalia, M.B., Wang, P.P., Baker, T., Eister, C.V.: Forecasting Uncertain Hotel Room Demand. Information Sciences 132, 1–11 (2001)
5. Himmelblau, D.M.: Applied Nonlinear Programming. McGraw-Hill, U.S.A. (1972)

A Filtering Approach for Mining Frequent Itemsets

Jen-Peng Huang* and Huang-Cheng Kuo

Abstract. Many efficient association rule mining algorithms have been proposed in the literature. In this paper, we propose an algorithm FRM (Mining Frequent Itemsets by Frequent-Related Mechanism). Most of the studies adopt an Apriori-like candidate set generation-and-test approach. However, candidate generation is still costly when there exist a large number of long patterns. FRM scans database only four times and it does not adopt the Apriori-like approach in mining process. It uses the frequent-related mechanism to generate the itemsets which are the most possible to be frequent and it eliminates a great number of infrequent itemsets. So FRM is very suitable to mine the databases whose record length is very long.

1 Introduction

Association rule mining finds interesting association or correlation relationships among a large set of data items. With massive amounts of data continuously being collected and stored in databases, many industries are becoming interested in mining association rules from their databases. For example, the discovery of interesting association relationships among huge amounts of business transaction records can help catalog design, cross-marketing and other business decision making processes. Market basket analysis is one of the powerful methods which aim at finding regularities in the shopping behaviors of customers of supermarkets, mail-order companies, and on-line shops. This process analyzes customer buying habits by finding associations between the different items that customers place in their "shopping baskets." The discovery of such associations can help retailers develop marketing strategies by gaining insight into which items are frequently purchased together by customers. Such information can lead to increased sales by helping retailers to do selective marketing and plan their shelf space.

Jen-Peng Huang
Department of Information Management
Southern Taiwan University

Huang-Cheng Kuo
Department of Computer Science and Information Engineering
National Chiayi University

* Corresponding author.

B.-C. Chien and T.-P. Hong (Eds.): Opportunities and Challenges, SCI 214, pp. 59–63.
springerlink.com

Frequent-pattern mining plays an essential role in mining association rules [1, 2, 3, 5, 6, 7, 9], correlations[5], bag database[4], and many other important data mining tasks.

Definition 1. We call the number of items in an itemset its size, and call an itemset of size k a k-itemset.

Definition 2. A k-subset is a sub-itemset of size k.

2 FRM: Mining Frequent Itemsets by Frequent-Related Mechanism

We propose a new algorithm FRM (Mining Frequent Itemsets by Frequent-Related Mechanism) which uses the frequent-related mechanism to generate those itemsets which are very possible to be frequent. Besides, FRM uses a hash based technique, Hash MAP, which is similar to Hash Table to store the sub-itemsets in order to increase the access efficiency.

Definition 3. Given a *k*-itemset *X* and a set of *k*-itemset $X_s = \{S_1, S_2, \ldots, S_m\}$ where S_i, S_j are the sub-itemsets of *X*, $S_i \neq S_j$, $1 \leqq i < j \leqq m \leqq 2^k\text{-}1$ then X_s is called a *decompositional set* of *X*.

Lemma 1. If *X* is a *k*-itemset then there are $2^k\text{-}1$ distinct subsets of *X*.

Definition 4. Given *k*-itemset *X* and $X_C = \{S_1, S_2, \ldots, S_m\}$ where X_C is a decompositional set of X , $S_i \neq S_j$, $1 \leqq i < j \leqq m$, if $m=2^k\text{-}1$ then X_C is called the *complete decompositional set* of *X*.

The *complete decompositional sets* of all transaction records are very huge, so it is almost impossible to use them as the candidate itemsets for mining frequent itemsets is nearly impossible. Instead of generating *complete decompositional set*, FRM uses the frequent-related mechanism to reduce the number of candidate itemsets.

Definition 5. Given two items *x* and *y*, if the combined itemset *xy* is a frequent 2-itemsets then *x* and *y* are *frequent-related.* Otherwise, *x* and *y* are *infrequent-related.*

Definition 6. Given an itemset *X* and an item *y*, if every item of *X* is *frequent-related* to y then *X* and *y* are *frequent-related.* Otherwise, *X* and *y* are *infrequent-related.*

By the Apriori property we know that any superset of an infrequent itemset is not frequent, so we can use the frequent 1-itmesets to trim database. The pseudo-code of FRM is shown in Fig. 1.

Lemma 2. If an itemset X and an item y are not frequent-related, then the combined itemsets Xy is not frequent.

```
Algorithm FRM      // FRM algorithm
Input: DB_data  // DB_data: Database
       min_sup  // min_sup: minimum support threshold
       CMap     // a hash map to store the candidate itemsets
       LMap     // a hash map to store the frequent itemsets
Output: the frequent itemset

CMap = Null;  LMap= NULL; //initialize CMap and LMap
scan DBdata to get the frequent 1-itemsets, and save in Ln[1];
Use the frequent 1-itemsets to shorten the DB_data and save in NewDB;
scan DBdata to get the frequent 2-itemsets, and save in Ln[2];
NewDB=TrimRecord(NewDB, Ln[2]); //Use Apriori property to trim the DB
While (there is any record r in NewDB){
    For (i=1; i<=Length(r);i++) {
        For (every itemset X in  CMap]) do  {
          if (r[i] is frequent-related to X) then
              Append itemset (x r[i]) to CMap; //Append the new combined itemset to CMap
          }
        }
       Append r[i] to CMap; //Append the item r[i] itself to CMap
    }
}
For (every itemset X in CMap) do {
    if Support(X) > min_sup then  //check the support of itemset X
       Append itemset X to LMap;  //Append the frequent itemset to LMap
}
return(LMap); //return the frequent itemsets
```

Fig. 1 The pseudo-code of FRM

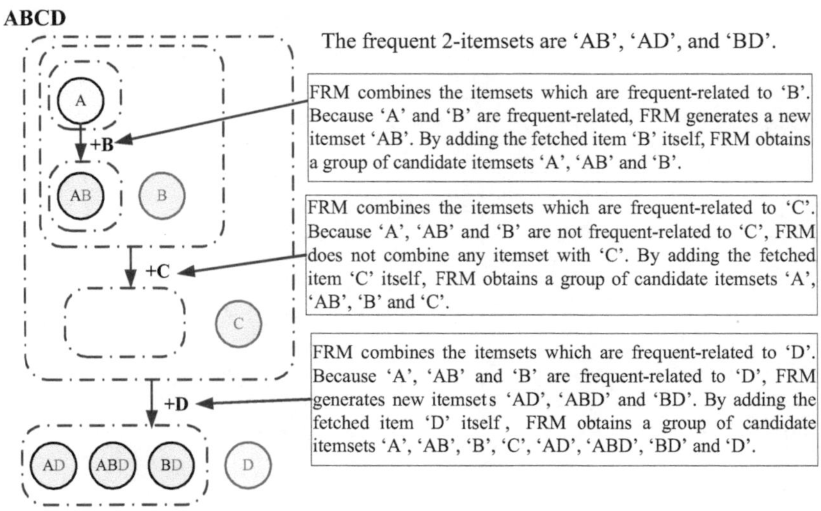

Fig. 2 The processes of generating itemsets by frequent-related mechanism

Instead of generating *complete decompositional sets* of all transaction records, the frequent-related mechanism of FRM uses Lemma 2 to reduce the number of candidate itemsets. For each transaction record of database FRM goes through the items of the record sequentially and uses the frequent-related mechanism to generate a group of candidate itemsets which are very possible to be frequent.

Here we use an example which is shown in Fig. 2 to explain the processes of generating itemsets with frequent-related mechanism.

3 Experimental Results

This section compares the experimental performance of FRM, FP-growth[3] and OP[8]. The three algorithms are implemented with Java language running under J2SDK1.4.2 environment. All experiments are performed on Intel Pentium IV 2.8GHz PC machine with 512MB memory. The operating system is Windows 2000 Server. Synthetic datasets are generated using publicly available synthetic data generation program of IBM Quest data mining project at http://www.almaden.ibm.com/cs/quest/, which has been used in most association rules mining studies. Besides, we use the BMS-POS dataset, and the Retail market basket dataset which are publicly available at http://fimi.cs.helsinki.fi/ as the other test datasets. The experimental results are shown in Fig. 3.

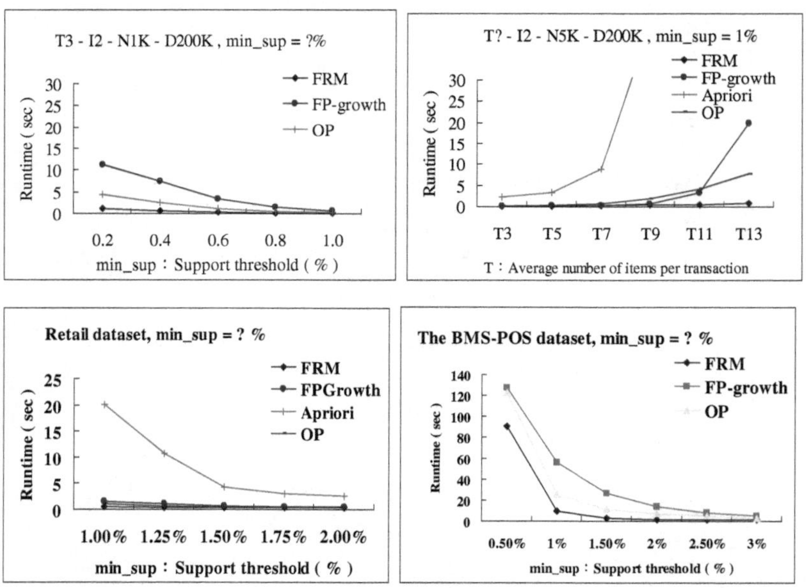

Fig. 3 Experimental result

4 Conclusion

FRM mainly uses the frequent-related mechanism to increase performance of discovering all of the frequent itemsets. In the mining processes, it can avoid generating a great number of candidate itemsets via the frequent-related mechanism, and then increases the efficiency and the utility rate of memory.

The advantages of FRM are as follows:

(1) FRM scans the database only four times to finish the mining task. Hence, FRM can avoid wasting a good deal of unnecessary I/O time, and then increase the efficiency.

(2) In mining frequent itemsets, FRM does not adopt the Apriori-like candidate set generation-and-test approach. It can use the frequent-related mechanism to filter out a huge number of candidate itemsets so it can save a great deal of time of generation of candidate itemsets, and then increase the efficiency.

(3) The framework of FRM is easy to be implemented.

Acknowledgments. This work was partially supported by grant NSC 92-2213-E-218 -020 and NSC 93-2213-E-218-012 from the National Science Council, Taiwan.

References

1. Agrawal, R., Imieliski, T., Swami, A.: Mining association rules between sets of items in large databases. In: Proceedings of the 1993 ACM SIGMOD International Conference on Management of Data, pp. 207–216. ACM Press, Washington (1993)
2. Brin, S., Motwani, R., Ullman, J.D., Tsur, S.: Dynamic itemset counting and implication rules for market basket data. In: Proceedings of the 1997 ACM SIGMOD International Conference on Management of data, pp. 255–264. ACM Press, Tucson (1997)
3. Han, J., Pei, J., Yin, Y., Mao, R.: Mining frequent patterns without candidate generation: A frequent-pattern tree approach. Data Mining Knowledge Discovery 8, 53–87 (2004)
4. Hsu, P.-Y., Chen, Y.-L., Ling, C.-C.: Algorithms for mining association rules in bag databases. Information Sciences 166, 31–47 (2004)
5. Huang, J.-P., Chen, S.-J., Kuo, H.-C.: An efficient incremental mining algorithm-QSD. Intelligent Data Analysis 11(3), 265–278 (2007)
6. Huang, J.-P., Lan, G.-C., Kuo, H.-C., Hong, T.-P.: A decomposition approach for mining frequent itemsets. In: Third International Conference on Intelligent Information Hiding and Multimedia Signal Processing (IIHMSP 2007), vol. 2, pp. 605–608 (2007)
7. Lin, D.I., Kedem, Z.: Pincer-Search: A new algorithm for discovering the maximum frequent set. In: Proceedings of the 6th International Conference on Extending Database Technology: Advances in Database Technology, pp. 105–119 (1998)
8. Liu, J., Pan, Y., Wang, K., Han, J.: Mining frequent itemsets by opportunistic projection. In: Proceedings of the eighth ACM SIGKDD International Conference on Knowledge Discovery and Data Mining, pp. 229–238. ACM Press, Edmonton (2002)
9. Park, J.S., Chen, M.-S., Yu, P.S.: Using a hash-based method with transaction trimming for mining association rules. IEEE Transactions on Knowledge and Data Engineering 9, 813–825 (1997)

Mining Data Streams with Skewed Distribution by Static Classifier Ensemble

Yi Wang, Yang Zhang, and Yong Wang

Abstract. In many data stream applications, the category distribution is imbalanced. However, current research community on data stream mining focus on mining balanced data streams, without enough attention being paid to the study of mining skewed data streams. In this paper, we proposed an clustering-sampling based ensemble algorithm with weighted majority voting for learning skewed data streams. We made experiments on synthetic data set simulating skewed data streams. The experiment results show that clustering-sampling outperforms under-sampling, and that compared with single window, the proposed ensemble based algorithm has better classification performance.

1 Introduction

Recently, there are many successful algorithms for coping with data streams [1, 2], and most of these studies focus on mining balanced data streams. However, for many real-life data stream applications, the probability that we observe positive examples is much less than the probability that we could observe negative ones. For example, the online credit card fraud rate of US is 2% in 2006 [3]. While the cost of misclassifying a credit card fraud will impose thousands of dollars loss to the bank. Hence, it's necessary for us to study the skewed data streams.

Yi Wang and Yang Zhang
College of Information Engineering, Northwest A&F University, P.R. China
e-mail: {wangyi2000, zhangyang}@nwsuaf.edu.cn

Yong Wang
School of Computer, Northwest Polytechnical University, P.R. China
e-mail: wangyong@nwpu.edu.cn

B.-C. Chien and T.-P. Hong (Eds.): Opportunities and Challenges, SCI 214, pp. 65–71.
springerlink.com

In this paper, we propose clustering-sampling to deal with imbalanced data set. Compared with under-sampling, the clustering-sampling could reserve more useful information, which may be discarded by under-sampling. Furthermore, we propose an ensemble based algorithm with weighted majority voting to cope with skewed data streams.

This paper is organized as following. Section 2 reviews the related work. Section 3 shows our proposed algorithms. The experiment setting and results are shown in section 4, followed by conclusion and future work in section 5.

2 Related Work

Currently, several strategies have been proposed by the research community for handling imbalanced data sets: (1) To resize training sets. Drummond *et al.* [5] concluded that under-sampling outperforms over-sampling after detailed experiment with these two algorithms. (2) To emphasize on cost sensitive learning [4]. (3) The ensemble approach. Guo *et al.* [6] used boosting and data generation method to improve performance of the skewed data set mining.

In recent years, many algorithms specifically tailored towards mining from data streams have been proposed. Using a sampling strategy based on Hoeffding bounds, the VFDT algorithm efficiently induces a decision tree in constant time [7]. Meanwhile, it is generally believed that ensemble classifier could have better classification accuracy than a single classifier [1, 2].

To the best of our knowledge, only Jing *et al.* [3] proposed an ensemble framework to classify skewed data streams. By preserving all the positive examples in the past batches, this framework improves the classification of the positive class.

3 Classifying Skewed Data Streams

3.1 Sampling Skewed Data Streams

The basic sampling methods include under-sampling and over-sampling. Although the study result shows that both under-sampling and over-sampling can somehow improve the classification for the positive class, they have several drawbacks [5].

In our study, we propose to employ K-Means clustering algorithm for selecting negative examples for representing negative class.

Algorithm 1 gives our clustering-sampling algorithm for sampling skewed data streams. In this algorithm, we set the number of clusters k to the size of positive examples in batch i of the streams. The negative examples are clustered into k clusters, and the centroid of each cluster is used as negative example for representing negative class.

Algorithm 1. Clustering-sampling for sampling skewed data streams.

Input:

batch $B_i = POS_i \cup NEG_i$ for batch i with imbalanced category distribution;
//POS_i: the set of positive examples in B_i;
//NEG_i: the set of negative examples in B_i;

Output:

data set S_i with balanced category distribution for batch B_i

$k = |POS_i|$;
$CLUSTER = KMeans(NEG_i, k)$;
$NEG = \{centroid(c) | c \in CLUSTER\}$;
$S_i = POS_i \cup NEG$;
return S_i;

Algorithm 2. Learning algorithm for classifying skewed data streams.

Input:

batch $B_i = POS_i \cup NEG_i$ for batch i with imbalanced category distribution;
the ensemble of classifiers trained on previous batches of data, E_{i-1};
the maximum size of classifier ensemble, z;

Output:

the classifier ensemble, E_i;

$S_i = Sampling(B_i)$;
$c_i = Learn(S_i)$;
if $|E_{i-1}| < z$ **then**
  $c_i.weight = 1$;
  $E_i = E_{i-1} \cup \{c_i\}$;
else
  $j = \operatorname{argmin}_{c_j \in E_{i-1}}(AUC(c_j))$;
  if $AUC(c_i) > AUC(c_j)$ **then**
    $E_i = E_{i-1} \cup \{c_i\} - \{c_j\}$;
    for each $c_h \in E_i$ **do**
      $c_h.weight = AUC(c_h)$;
    end for
    for each $c_h \in E_i$ **do**
      $c_h.weight = NormalizeWeight(E_i, c_h)$;
    end for
  end if
end if

3.2 *Ensemble Based Learning Algorithm*

Suppose the incoming data streams is partitioned into sequential batches, B_1, B_2, $\ldots, B_i, \ldots$, with B_i being the most up-to-date batch. We train a base classifier, c_i, by an arbitrary successful binary learner from the data set S_i, which is sampled from B_i by some certain sampling algorithm. The AUC index, which represents area under ROC curve [8], is employed as the metric to evaluate the base classifier.

In algorithm 2, the function $Sampling(B_i)$ is an arbitrary sampling algorithm; $Learn(S_i)$ is an arbitrary binary learning algorithm; $AUC(c_i)$ returns the AUC value of base classifier c_i; and $NormalizeWeight(E_i, c_h)$ returns the normalized weight of base classifier c_h in ensemble E_i.

In the testing phase, the weighted majority voting strategy is employed to combine the classification result that is outputted by the base classifiers in E_i.

4 Experiments

In our experiment, we construct two different models: (1) The classifier is constructed on the most up-to-date batch B_i (*Single Window*). (2) The ensemble classifier is constructed following algorithm 2 (*Ensemble Model*).

In the testing phase, the batch B_{i+1} is used as the test data, and three sampling methods are experimented, under-sampling (*UnderS*), clustering-sampling (*ClusterS*), and no-sampling (no sampling is performed, *NoS*). Meanwhile, three algorithms, C4.5, Naive Bayes, and linear SVM, are employed as base classifier.

4.1 Data Sets

The experiments with synthetic data used a changing concept based on a rotating hyperplane [2, 3, 7]. A hyperplane in d-dimensional space is the set of points x that satisfy

$$\sum_{i=1}^{d} a_i x_i = a_0 \tag{1}$$

Here, x_i is the i-th coordinate of x. Examples with $\sum_{i=1}^{d} a_i x_i > a_0$ are labeled as positive examples, and examples with $\sum_{i=1}^{d} a_i x_i < a_0$ negative examples. Hyperplanes are useful for simulating time-changing concepts because the orientation and the position of the hyperplane can be changed in a smooth manner by changing the magnitude of the weights [7]. We choose the value of a_0 so that the hyperplane cuts the multidimensional space into two parts of different volume, with skewness ratio r.

In our study, we simulate concept drifting by a series of parameters. Let's write n for the number of examples in each batch, $n \in N$; k for the number of dimensions whose weights are involved in concept drifting, $k \in N$; t for the magnitude of the changing of weights $a_i, \ldots, a_k$, $t \in R$; and s_i, $1 \le i \le k$, for the direction of change for each weight a_i, $s_i \in \{-1, 1\}$.

4.2 Evaluation Measures

In our study, we use the mean squared error to evaluate the quality of probability estimation [3]. Let's write T for the set of testing examples, MSE is defined as:

Table 1 MSE for Ensemble Model

$r(\%)$	C4.5			Naive Bayes			SVM		
	NoS	UnderS	ClusterS	NoS	UnderS	ClusterS	NoS	UnderS	ClusterS
1	0.8701	0.0880	**0.0845**	0.8541	0.1031	**0**	1	**0.0081**	0.0673
5	0.6947	0.1264	**0.0448**	0.5257	0.0470	**9.97E-10**	1	**0.0089**	0.0092
10	0.5423	0.1439	**0.0657**	0.4063	0.0514	**2.20E-05**	1	0.0069	**0.0049**
15	0.4442	0.1439	**0.0844**	0.3334	0.0597	**9.23E-04**	0.9793	0.0069	**0.0046**
20	0.3888	0.1601	**0.1046**	0.2839	0.0701	**0.0056**	0.8196	0.0092	**0.0067**

Table 2 MSE for Single Window

$r(\%)$	C4.5			Naive Bayes			SVM		
	NoS	UnderS	ClusterS	NoS	UnderS	ClusterS	NoS	UnderS	ClusterS
1	0.9382	0.3243	**0.2752**	0.8770	0.2686	**0**	1	**0.1135**	0.2955
5	0.8052	0.3163	**0.1446**	0.5589	0.0933	**1.06E-14**	1	**0.0249**	0.0393
10	0.6957	0.3196	**0.1815**	0.4242	0.0712	**6.03E-05**	1	0.0210	**0.0112**
15	0.6159	0.3172	**0.2124**	0.3462	0.0743	**9.02E-04**	0.9946	0.0207	**0.0116**
20	0.5667	0.3314	**0.2477**	0.2918	0.0791	**0.0058**	0.9029	0.0235	**0.0144**

$$MSE = \frac{1}{|T|} \sum_{t_i \in T} (f(t_i) - p(+|t_i))^2 \qquad (2)$$

Here, $f(t_i)$ is the output of the classifier, which is the estimated posterior probability of testing example t_i; $p(+|t_i)$ is the true posterior probability of t_i.

4.3 Experiment Results

There are two group of data streams with 100 batches are generated. In group A, we set $d = 50$, $k = 30$, $t = 0.1$, $n = 1000$ and $z = 10$, with skewness ratio r being 1%, 5%, 10%, 15%, and 20%. In group B, we set $d = 50$, $k = 30$, $t = 0.1$, $r = 10\%$ and $z = 10$, with batch size n being 500, 1000, and 2000, respectively.

Sampling Methods. We make experiments on the group A to compare the performance of our clustering-sampling method with other sampling methods.

From table 1 and table 2, it is obvious that clustering-sampling outperforms under-sampling and no-sampling. This is because clustering-sampling samples examples by clustering, which helps to reserve useful information.

Ensemble model vs. Single window. We make experiments on the group B to compare the classification performance of different models.

It could be observed from table 3 and table 4 that the ensemble model outperforms single window for most of the cases. With the increasing of n, the performance of classifiers are also improving. The reason is that for a certain skew distribution, with increasing of n, there are more and more positive examples for learning the target concept.

Table 3 MSE for Under-sampling

n	C4.5		Naive Bayes		SVM	
	Ensemble	Single	Ensemble	Single	Ensemble	Single
500	**0.1360**	0.3310	**0.0599**	0.1136	**0.0221**	0.0769
1000	**0.1439**	0.3196	**0.0514**	0.0712	**0.0069**	0.0210
2000	**0.1406**	0.2862	**0.0502**	0.0540	**0.0043**	0.0044

Table 4 MSE for Clustering-sampling

n	C4.5		Naive Bayes		SVM	
	Ensemble	Single	Ensemble	Single	Ensemble	Single
500	**0.0698**	0.1916	**1.75E-05**	2.18E-05	**0.0194**	0.1056
1000	**0.0657**	0.1815	**2.20E-05**	6.03E-05	**0.0049**	0.0112
2000	**0.0632**	0.1555	4.45E-05	**4.15E-05**	0.0021	**0.0002**

5 Conclusion and Future Work

In this paper, we propose the clustering-sampling based ensemble algorithm to tackle the problem of mining data streams with skewed distribution. The experiment results demonstrate that our ensemble algorithm outperforms the single window remarkably.

The proposed algorithm could only cope with binary classification tasks. As many real-life application, such as network intrusion detection, is characterized as multi-class classification tasks. In the future, we plan to study multi-classification of skewed data streams.

Acknowledgements. This research is supported by the National Natural Science Foundation of China (60873196) and Young Cadreman Supporting Program of Northwest Agriculture and Forest University (01140301).

References

1. Street, W.N., Kim, Y.S.: A streaming ensemble algorithm for large-scale classification. In: Proceedings of the 7th ACM SIGKDD International Conference on Knowledge Discovery and Data Mining, pp. 377–382. ACM, New York (2001)
2. Wang, H., Fan, W., Yu, P.S., Han, J.: Mining concept-drifting data streams using ensemble classifiers. In: 9th ACM SIGKDD Int. Conf. on Knowledge Discovery and Data Mining, pp. 226–235. ACM, New York (2003)
3. Gao, J., Fan, W., Han, J., Yu, P.S.: A general framework for mining concept-drifting data streams with skewed distributions. In: Proc. 2007 SIAM Int. Conf. Data Mining (SDM 2007), Minneapolis, MN (2007)
4. Elkan, C.: The foundations of cost-sensitive learning. In: Proceedings of the 17th international joint conference on artificial intelligence (IJCAI 2001), pp. 973–978 (2001)

5. Drummond, C., Holte,R.: C4.5, class imbalance, and costsensitivity: why undersampling beats over-sampling. In: Proceedings of the ICML 2003 Workshop: Learning with Imbalanced DataSets II (2003)
6. Hongyu, G., Viktor Herna, L.: Learning from Imbalanced Data Sets with Boosting and Data Generation: The DataBoost-IM Approach. Newsletter of the ACM Special Interest Group on Knowledge Discovery and Data Mining 6(1) (2004)
7. Domingos, P., Hulten, G.: Mining High Speed Data Streams. In: 6th ACM SIGKDD Int. Conf. on Knowledge Discovery and Data Mining, pp. 71–80. ACM, New York (2000)
8. Fawcett, T.: An introduction to ROC analysis. Pattern Recognition Letters 27(8), 861–874 (2006)

Accumulative Influence Weight Collaborative Filtering Recommendation Approach

Nan Li and Chunping Li

Abstract. Memory-based collaborative filtering algorithms are widely used in practice. But most existing approaches suffer from a conflict between prediction quality and scalability. In this paper, we try to resolve this conflict by simulating the "word-of-mouth" recommendation in a new way. We introduce a new metric named influence weight to filter neighbors and weight their opinions. The influence weights, which quantify the credibility of each neighbor to the active user, form accumulatively in the process of the active user gradually provides new ratings. Therefore, when recommendations are requested, the recommender systems only need to select the neighbors according to these ready influence weights and synthesize their opinions. Consequently, the scalability will be significantly improved without loss of prediction quality. We design a novel algorithm to implement this method. Empirical results confirm that our algorithm achieves significant progress in both aspects of accuracy and scalability simultaneously.

1 Introduction

Memory-based collaborative filtering approaches are widely used in practice to help people cope with the problem of information overload. But most of them suffer from problems such as poor prediction quality or poor scalability.

Recently many methods [5] [3] have been proposed to improve the prediction quality by alleviating data sparsity. The evaluations of these approaches showed that it is really effective. But some extra effort required in these approaches, such as smoothing the missing data, makes the scalability of them worse.

Nan Li
School of Software, Tsinghua University, Beijing 100084, China
e-mail: ln8209@gmail.com

Chunping Li
School of Software, Tsinghua University, Beijing 100084, China
e-mail: cli@tsinghua.edu.cn

B.-C. Chien and T.-P. Hong (Eds.): Opportunities and Challenges, SCI 214, pp. 73–78.
springerlink.com

Most approaches aiming at good scalability achieve their goals by taking advantage of some model-based strategies [5] or conducting some precomputation offline [4] in order to reduce the searching space of users (items) or similarity computation online. But these methods either would limit the diversity of users (items) or could not keep pace with the change in the user-item matrix. Thus they would definitely bring negative influence to prediction quality.

Unfortunately, there seems to be a conflict between prediction quality and scalability in the general framework of most existing memory-based approaches.

In this paper, we try to resolve this conflict by simulating the "word-of-mouth" recommendation in a new way. We introduce a new metric named influence weight to filter neighbors and weight their opinions. The influence weights, which quantify the credibility of each neighbor to the active user, form accumulatively in the process of the active user gradually provides new ratings. Therefore, when recommendations are requested, the recommender systems only need to select the neighbors according to these ready influence weights and synthesize their opinions. Consequently, the scalability will be significantly improved without loss of prediction quality. We design a novel algorithm to implement this method. Empirical results confirm that our algorithm achieves significant progress in both aspects of accuracy and scalability simultaneously.

2 Related Work

Memory-based collaborative filtering approaches, which include user-based and item-based, search the most similar neighbors of the active user in the entire user-item database whenever they make predictions. User-based approaches [2] [5] predict for the active user based on the opinions of similar users, and item-based approaches [1] [4] make prediction based on the information of similar items.

Recently many methods have been proposed to improve the prediction quality by alleviating data sparsity problem. [5] proposes a novel framework for collaborative filtering which combines the strengths of memory-based approaches and model-based approaches in order to enable recommendation by groups of closely related individuals. [3] proposes an effective missing data prediction algorithm which predict the missing data by exploiting information of both user neighbors and item neighbors whose similarities are all higher than some thresholds.

To improve scalability, [4] proposes item-based collaborative filtering recommender algorithm which reduces online computation by utilizing the relatively static relationships between items. [5] exploits clustering techniques to reduce the searching space of potential neighbors and improve the scalability consequently.

3 "Word-of-Mouth" Recommendation

When a person, Tom, thinks about whether going to watch a new movie, he would ask friends with similar movie taste for advices. And then Tom will synthesize these advices according to the credibility of each friend. The credibility of each friend formed according to the validity of his past advices. After watching this movie, Tom will obtain his own opinion. Then he would adjust the credibility of his friends based on his opinion. The credibility of a friend will increase if his advice accord with Tom's opinion. Otherwise his credibility will decrease according to the deviation between his advice and Tom's opinion. It is natural that the lager the deviation is, the more credibility the giver loses.

We could quantify this credibility and use it as a metric to select neighbors and weight their opinions. Since the more similar with the active user a neighbor is, the larger his credibility would be, the concept of credibility actually reflect the similarity between each pair of users from another angle. Thus it may play the role as good as or even better than the similarity metrics such as Pearson Correlation Coefficient (PCC) or Vector Similarity (VS). More importantly, the credibility has two advantages over those broadly used similarity metrics.

On the one hand, to generate a recommendation, those algorithms using similarity metrics have to compute similarity scores tens of thousands of times, but when next recommendation is requested by the same user, they have to do the same heavy work again even if only a small portion of the user-item matrix has changed. In contrast, the credibility form accumulatively in the process of the active user continually provides new ratings. And the adjustment to the credibility would get done once and for ever whenever a new rating is provided. This means the overall work of a system adopting the credibility is approximately linear with the number of the ratings no matter how many recommendations would be requested. It is obviously that this would improve the efficiency significantly.

On the other hand, the limitation of responding latency requires those heavy computations to be completed in real time, which is exactly the bottleneck of performance and scalability in most existing memory-based approaches. Moreover, the recommender systems would be jammed by these real-time tasks in the rush hours but be free in most other time, which result in that the systems have to waste quite a lot of resource. In contrast, adjusting the credibility is not a real-time task. It could be conducted on background when the active user is providing new ratings. Or the systems could schedule these tasks freely just before next recommendation is requested by this user. When a recommendation is requested, all the necessary data are ready and the systems only need to conduct some simple retrieval and computation online. In this way, the systems could not only achieve very short responding latency to provide better user experience but also balance their load effectively to take use of the resource efficiently.

4 Accumulative Influence Weight Algorithm

We design a novel memory-based algorithm named Accumulative Influence Weight (AIW) to implement the method discussed above. We first formally define a metric named influence weight to quantify the credibility described above. AIW maintain a table named IW table to record the influence weights between each pair of existing users. When a new user registers, AIW add him into the IW table and set all the original influence weights relating to him 0. Then, whenever this user provides a new rating, AIW would adjust the influence weights between him and all the other existing users who have rated the same item with a zero-sum mechanism as follows:

Whenever a user u provides a new rating $r_{u,i}$ on item i,
1. select the users who have rated on i and divide them into two subsets:
S_1 contains the users whose rating on i are unequal to $r_{u,i}$
S_2 contains the users whose rating on i are equal to $r_{u,i}$
2. for each user a in S_1:
decrease the influence weight of a to u according to the deviation between $r_{a,i}$ and $r_{u,i}$,adjust the influence weight of u to a accordingly , and accumulate the deviations
3. for each user b in S_2
increase the influence weight of b to u with an average share of the accumulation of deviations and adjust the influence weight of u to b accordingly

In this paper, AIW directly convert the numerical value of the rating into the value of the increase or decrease on influence weight. Formally, the influence weights relating to u will be adjusted as follows:

$$IW_{a,u} = \begin{cases} IW_{a,u} - |r_{a,i} - r_{u,i}| & a \in S_1 \\ IW_{a,u} + \frac{\sum\limits_{b \in S_1} |r_{b,i} - r_{u,i}|}{|S_2|} & a \in S_2 \end{cases} \tag{1}$$

where $IW_{a,u}$ denotes the influence weight of a to u.

When predicting for the active user u on item i, AIW would firstly retrieve the IW table and select neighbors with the Top-N largest positive influence weights to u. If some selected credible neighbors have not rated on i, AIW would predict those missing data. The procedure of this smoothing is a little different: AIW would choose all the neighbors with a positive influence weight in order to alleviate the impact of the sparsity to the prediction quality of the missing data.

Both the prediction for missing data and the active user would use the weighted sum equation as follows:

$$P_{u,i} = \frac{\sum\limits_{a \in S(u)} (IW_{a,u} * r_{a,i})}{\sum\limits_{a \in S(u)} IW_{a,u}} \tag{2}$$

where $S_{(u)}$ contains the selected neighbors according to their influence weights, $P_{u,i}$ denotes the prediction for u on item i and the $IW_{a,u}$ denotes the influence weight of a to u.

After predicting all the items which the active user has not rated, AIW will select the ones with the biggest prediction values to generate a recommendation.

Table 1 MAE comparison with benchmarks under different sparsity

Num. of Training Users	100			200			300		
Ratings Given for Active users	10	20	30	10	20	30	10	20	30
AIW	0.823	0.801	0.787	0.808	0.783	0.764	0.795	0.774	0.753
EMDP	0.906	0.834	0.794	0.883	0.818	0.778	0.865	0.802	0.765
UBPCC	0.87	0.827	0.797	0.855	0.815	0.789	0.841	0.807	0.782
IBVS	0.887	0.838	0.805	0.863	0.816	0.792	0.849	0.805	0.783

5 Experiment and Evaluation

We use Movielens (http://www.grouplens.org/) dataset in our experiments. AIW here normalize the ratings involved by subtracting the mean value of all the ratings the same user has provided before adjusting influence weights and predicting for the active user. The normalization would convert the original discrete ratings to continuous values; consequently the "equal rating" in the basic algorithm is changed into "the abstract deviation of two ratings is less than 0.5".

1. Prediction Quality

We follow the evaluation procedures described in [3] to test our algorithm and compare it with other state-of-the-art approaches: Effective Missing Data Prediction (EMDP) [3], standard user-based PCC(UBPCC) and item-based VS(IBVS). We vary the number of items in the profiles of the active users from 10, 20 to 30 and then predict for all the rest items they had rated in order to test the prediction quality under different data sparsity. In consideration of both accuracy and performance, we set the neighbor size of AIW to 20.

Table 1 show that AIW outperforms those benchmarks in various configurations.

2. Performance and Scalability

We first randomly select a number of users and all their ratings to form a sequence according to the timestamps of these ratings. Then we divide this sequence into a series of user sessions based on an assumption that every single session last no

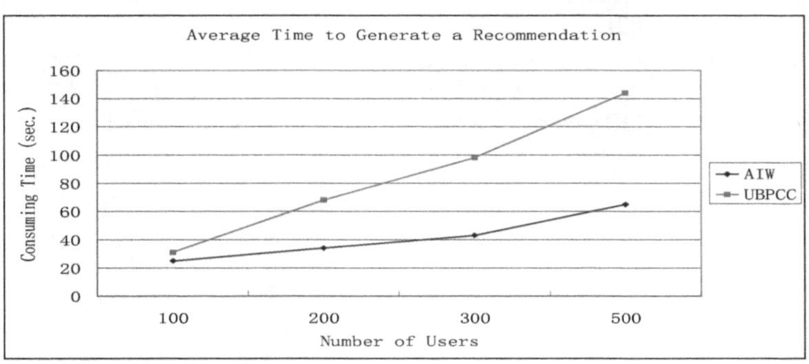

Fig. 1 Average recommendation time comparison

longer than 1 hour. And we assume that in each session the user would request only one recommendation, which means predicting for all the existing items this user has not rated. At last we apply the algorithms to deal with this interaction sequence and record the total consuming time. In this way, we could get an average consuming time for an algorithm to generate a recommendation. Then we gradually increase the number of the selected users to evaluate the scalability with different scales.

Fig.1 shows that the consuming time of AIW to generate a recommendation is less than UBPCC in all scales. In addition, the consuming time of AIW increases much more gently as the scale expands. This confirms that AIW is more scalable.

6 Conclusion

In this paper, we simulate the "word-of-mouth" recommendation in a novel way. The online computation of similarity, which is the bottleneck of scalability in most existing approaches, is completed gradually in the form of accumulation of credibility. The algorithm designed accordingly achieves significant improvement in both aspects of accuracy and scalability simultaneously. It confirms this method is effective to resolve the conflict between improving prediction quality and scalability.

References

1. Deshpande, M., Karypis, G.: Item-based top-N recommendation algorithms. ACM Transactions on Information Systems (TOIS) 22(1), 143–177 (2004)
2. Herlocker, J.L., Konstan, J.A., Riedl, J.: An algorithmic framework for performing collaborative filtering. In: Proceedings of the 22nd annual international ACM SIGIR conference on Research and development in information retrieval, pp. 230–237 (1999)
3. Ma, H., King, I., Lyu, M.R.: Effective missing data prediction for collaborative filtering. In: SIGIR 2007: Proceedings of the 30th annual international ACM SIGIR conference on Research and development in information retrieval, pp. 39–46. ACM, New York (2007)
4. Sarwar, B., Karypis, G., Konstan, J., Reidl, J.: Item-based collaborative filtering recommendation algorithms. In: Proceedings of the 10th international conference on World Wide Web, pp. 285–295 (2001)
5. Xue, G.R., Lin, C., Yang, Q., Xi, W.S., Zeng, H.J., Yu, Y., Chen, Z.: Scalable collaborative filtering using cluster-based smoothing. In: Proceedings of the 28th annual international ACM SIGIR conference on Research and development in information retrieval, pp. 114–121 (2005)

RAD: A Radar-Alike Data-Clustering Algorithm for Large Databases

Cheng-Fa Tsai and Jiun-Huang Ju

Abstract. The popularity of data analysis for business has created a heavy load on demand. Consequently, data mining and data clustering have become significant topics for revealing the implied information recently. This investigation hence presents a novel data clustering algorithm that can be applied to a large database efficiently. The proposed algorithm, called RAD because it is a Radar-alike data-clustering algorithm for large databases, attains this aim because its computation time rises linearly as the data size increases. Experimental results indicate that RAD performs clustering quickly and with fairly good clustering quality and outperforms K-means, DBSCAN and IDBSCAN.

Keywords: data mining, data clustering, K-means, IDBSCAN.

1 Introduction

Advances in information technology over the past have inspired many network services and increased user expectations. Data classification, data clustering and association rule techniques in data mining have matured wide-ranging research in the field. However, most works do not related to large databases. Hence, how enterprises retrieve implied information when the amount of raw data increases rapidly is now a significant research topic.

This investigation develops RAD, a radar-alike data clustering algorithm based on slope, and compares it with K-means and IDBSCAN. Experimental results reveal that RAD is with better clustering quality than k-means, and much faster faster than

Cheng-Fa Tsai
Department of Management Information Systems, National Pingtung University of Science and Technology, 91201 Pingtung, Taiwan
e-mail: cftsai@mail.npust.edu.tw

Jiun-Huang Ju
Department of Management Information Systems, National Pingtung University of Science and Technology, 91201 Pingtung, Taiwan
e-mail: m9656004@mail.npust.edu.tw

B.-C. Chien and T.-P. Hong (Eds.): Opportunities and Challenges, SCI 214, pp. 79–84.
springerlink.com

DBSCAN and IDBSCAN. The rest of this paper is organized as follows. Section 2 describes K-means, DBSCAN and IDBSCAN. Section 3 then presents the proposed algorithm, called RAD. Section 4 explains the experiments and analysis results. Conclusions are finally drawn in Section 5, along with recommendations for future research.

2 Related Works

Current data clustering schemes can be categorized into partitioning, hierarchy, density-based, grid-based and hybrid as follows.

K-means [1], developed by MacQueen in 1967, is the most widely adopted partitioning algorithm; it is fast, simple to implement. The clustering process of K-means is to continuously reassign the cluster cores until the cores are steady. The main drawback of K-means is that it is easily trapped in local optima and unable to recognize arbitrary shapes. Moreover, the proposed RAD algorithm also belongs to this category.

Hierarchical clustering algorithms can be classified as (1) agglomerative or (2) divisive. In agglomerative algorithms, each data point is treated an individual, and is merged with other data points. Divisive start from the set of all data points, and splits this set into some coessential groups. Hierarchical algorithms include BIRCH [2], CURE [3] and ROCK [4].

DBSCAN [5] (Martin Ester *et al.*, 1996) and IDBSCAN [6] are density-based clustering algorithms. DBSCAN groups the data points which is with a specified area size (radius ε) and the number of points exceed a specified threshold ($MinPts$) into the same cluster. DBSCAN performs quite well because it can handle an unusual shape and filter noises out, but is time-consuming because of scanning every data point. Therefore, IDBSCAN improves on DBSCAN, by adopting a sampling measure (MBO, Marked Boundary Objects), and thus has a much shorter computation time than DBSCAN.

Grid-based clustering algorithms divide data points into some regularly shaped cells, and are efficient if the base grid cells are well designed. STING [7] and STING+ [8] are the examples of grid-based clustering algorithms.

Some algorithms, such as KIDBSCAN [9], ANGEL [10] and G-TREACLE [11], belong to more than one category.

3 The Proposed RAD Clustering Algorithm

As discussed earlier, K-means is fast but has an unstable clustering result because it randomly selects the initial cluster cores. Moreover, DBSCAN and IDBSCAN have excellent results, but are slower than most other algorithms, making them unsuitable for large-scale databases. Accordingly, the proposed algorithm, RAD, attempts to perform clustering as efficiently and appropriately as possible.

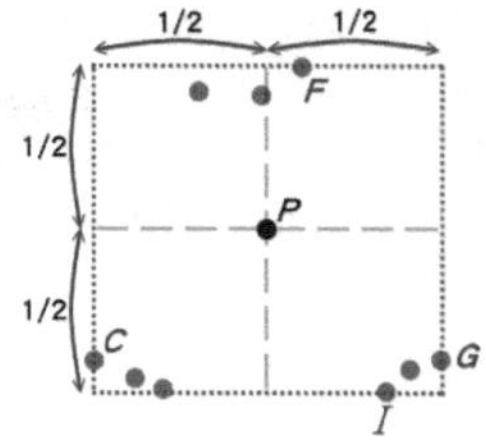

Fig. 1 The calculated virtual center

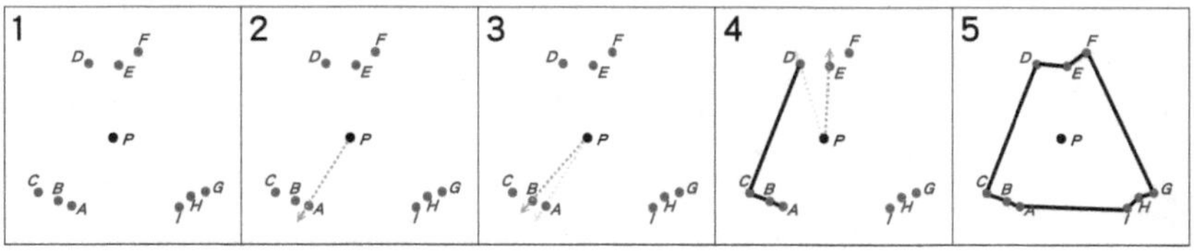

Fig. 2 The scanning concept of RAD

This section describes the principle of the proposed clustering algorithm, RAD. The algorithm first identifies the virtual center by "space center checking" through all objects. the object-scanning sequence is then determined by "radar-scanning". "triangle sorting and cutting" is then applied to search the biggest triangle areas containing two conjoint objects and the virtual center. The process eventually yields all clusters. The process is performed in four stages as follows:

(1) **Space center checking:** It is shown in Fig. 1. The space center as **virtual center** P is calculated with four boundary objects: the minimal object in terms of the x-axis, denoted as object C; the maximal object in terms of the x-axis, represented as object G; the minimal object in terms of the y-axis indicated as object I, and the maximal object in terms of the y-axis given as object F.

(2) **Radar-scanning:** This stage illustrates the scanning concept of RAD in Fig. 2. Consider a feature space with several objects (represented as A to I in Fig. 2) and a calculated space center (virtual center P in Fig. 2). Then, RAD selects an object randomly (e.g. object A) to perform clockwise scan and subsequently meets object B. The scanning process continues to scan the next object (object C), and the previously scanned two objects A and B are linked to establish their object scanning order. The scanning process is repeated until all of objects are linked, and the whole **sequence** is constructed.

(3) **Triangle sorting and cutting:** Each pair of objects is linked with the virtual center to form triangles once the sequence of the feature space is built. Fig. 3 presents triangle FGP, in which objects F and G are connected with virtual center P. Hence, n objects in a feature space produce n triangles. All triangles are then sorted according to area, and the links of the top K triangles (where K denotes the required number of clusters) are cut.

(4) **Division stage:** For a two-dimensional feature space, RAD first defines a virtual center of a minimal rectangle area formed by all objects, and sequentially scans all of objects like a radar. These objects are then linked in a scanning sequence. If $K = 3$, then RAD obtains the largest three triangles consisting of the pair of objects

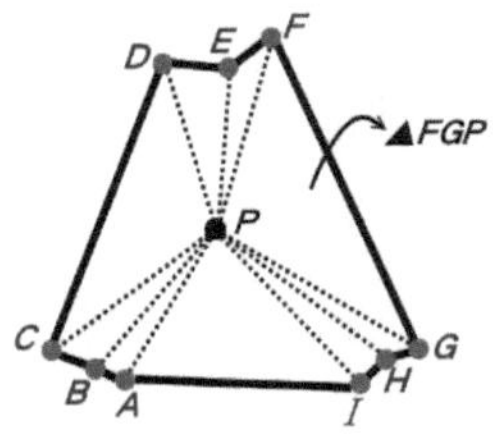

Fig. 3 The triangles consist of each two conjoint objects and the virtual center

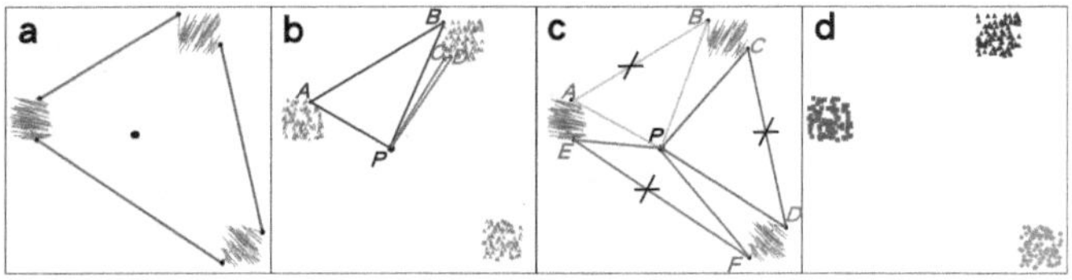

Fig. 4 The clustering demonstration of RAD

and the virtual center. The link of these triangles is then cut. Finally, *K* clusters are obtained as the clustering result.

The four diagrams in Fig. 4 show the complete clustering flow. (a) Consider a two-dimensional feature space whose sequence has been established. (b) The algorithm obtains a triangle *ABP* by linking the conjoint objects *AB* and the virtual center *P*, as well as another triangle *CDP* by connecting the conjoint objects *CD* and the virtual center *P*. Obviously, triangle *ABP* has a larger area than triangle *CDP*. (c) After comparing all triangles formed by pair of two conjoint objects and the virtual center *P*, the 3 largest triangle sets we derived as *ABP*, *CDP* and *EFP*. Finally, the link (*AB*, *CD* and *EF*) that is part of the largest triangle is cut. (d) The objects connected by each remaining link form a separate object cluster.

4 Experiment and Analysis

To verify the efficiency and accuracy of RAD, a series of 30 independent runs was performed for two data sets. The two tests involved five similar data sets with different sizes (10000, 50000, 100000, 500000 and 1000000 objects). The algorithm was implemented in the Java programming language using a computer with 1GB of RAM and an Intel 3.2 GHz CPU. The long computational time of density-based algorithms to it being abandoned in some tests, given as N/A (Not Available). The experimental results are listed below.

Data set 1 was composed of four isolated rectangles, as depicted in Fig. 5(a). Table 1 shows the experimental results. All algorithms clustered the data correctly, except that K-means made the wrong clustering result in some runs.

For data set 2, Fig. 6(a) illustrates the original feature space. These results demonstrate K-means performed far worse than the other algorithms. Table 2 summarizes the experimental results revealing that increasing the number of object had little effect on the speed of RAD, but caused the other algorithms, especially DBSCAN, to slow down rapidly.

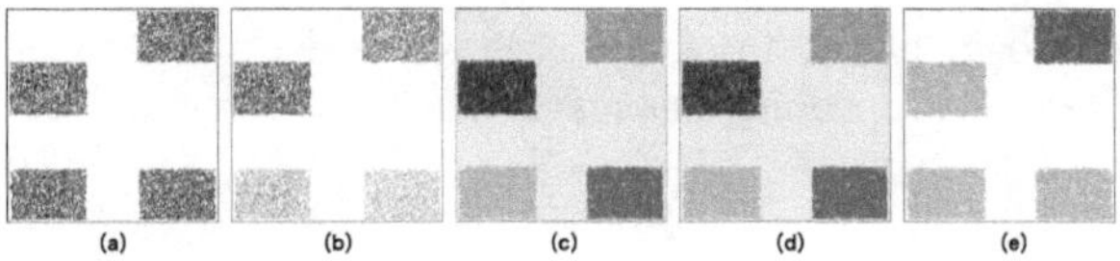

Fig. 5 Clustering result using data set 1 with different algorithms: (a) The original dataset (b) K-means, (c) DBSCAN, (d) IDBSCAN and (e) RAD

Table 1 The time-cost(in second) and ER(Error Rate, in percentage) comparison with different size of data set 1. The arguments of DBSCAN and IDBSCAN: ε=2 and *MinPts*=30

Algorithm	K-means		DBSCAN		IDBSCAN		**RAD**	
Data Size	TIME	ER	TIME	ER	TIME	ER	**TIME**	**ER**
10,000	0.01	3.23	20.39	0	1.77	0	**0.07**	**0**
50,000	0.09	2.58	578.45	0	57.48	0	**0.42**	**0**
100,000	0.18	1.63	N/A	N/A	514.88	0	**0.92**	**0**
500,000	1.15	3.26	N/A	N/A	N/A	N/A	**5.28**	**0**
1,000,000	3.33	5.77	N/A	N/A	N/A	N/A	**11.61**	**0**

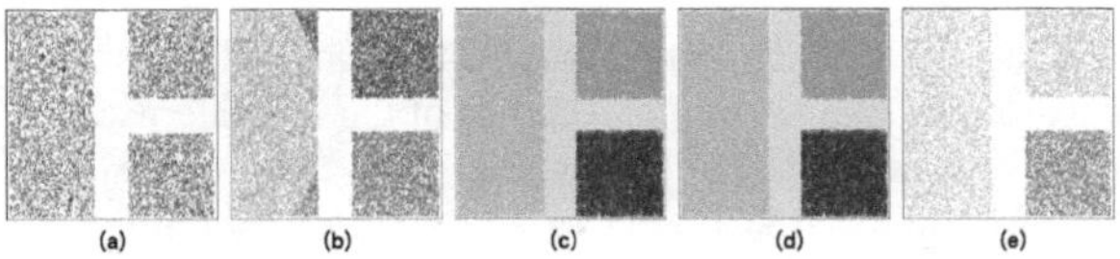

Fig. 6 Clustering result using data set 2 with different algorithm: (a) The original dataset (b) K-means, (c) DBSCAN, (d) IDBSCAN and (e) RAD

Table 2 The time-cost(in second) and ER(Error Rate, in percentage) comparison with different size of data set 2. The arguments of DBSCAN and IDBSCAN: ε=1.5 and *MinPts*=30

Algorithm	K-means		DBSCAN		IDBSCAN		**RAD**	
Data Size	TIME	ER	TIME	ER	TIME	ER	**TIME**	**ER**
10,000	0.03	6.09	18.35	0	2.57	0	**0.07**	**0**
50,000	0.29	6.93	538.05	0	118.77	0	**0.43**	**0**
100,000	0.43	6.71	N/A	N/A	646.78	0	**0.95**	**0**
500,000	2.99	6.55	N/A	N/A	N/A	N/A	**5.45**	**0**
1,000,000	7.1	6.74	N/A	N/A	N/A	N/A	**11.56**	**0**

5 Conclusions

This work proposes an innovative and efficient data clustering algorithm for large databases. The proposed algorithm, named RAD, performs data clustering by computing the slope of each link. The main advantage of RAD is that it has an excellent speed with large databases, because performs clustering in only one iteration. Therefore produces the same clustering result as DBSCAN and IDBSCAN, but much more rapidly. Furthermore, RAD produces much better clustering results than K-means.

Acknowledgements. The author would like to thank the National Science Council of Republic of China, Taiwan for financially supporting this research under contract no. NSC 96-2221-E-020-027.

References

1. McQueen, J.B.: Some Methods of Classification and Analysis of Multivariate Observations. In: Proceedings of the 5th Berkeley Symposium on Mathematical Statistics and Probability, pp. 281–297 (1967)
2. Zhang, T., Ramakrishnan, R., Livny, M.: BIRCH: An Efficient Data Clustering Method for Very Large Databases. In: Proceedings of The ACM SIGMOD International Conference on Management of Data, pp. 103–114 (1996)
3. Guha, S., Rastogi, R., Shim, K.: CURE: An Efficient Clustering Algorithm for Large Databases. In: Proceedings of The 1998 ACM SIGMOD International Conference on Management of Data, vol. 27(2), pp. 73–84 (1998)
4. Guha, S., Rastogi, R., Shim, K.: ROCK: A Robust Clustering Algorithm for Categorical Attributes. In: Proceedings of 15th International Conference on Data Engineering, pp. 512–521 (1999)
5. Ester, M., Kriegel, H.P., Sander, J., Xu, X.: A Density-Based Algorithm for Discovering Clusters in Large Spatial Databases with Noise. In: Proceedings of the 2nd International Conference on Knowledge Discovery and Data Mining, pp. 226–231 (1996)
6. Borah, B., Bhattacharyya, D.K.: An Improved Sampling-based DBSCAN for Large Spatial Databases. In: Proceedings of ICISIP, pp. 92–96 (2004)
7. Wang, W., Yang, J., Muntz, R.: STING: A Statistical Information Grid Approach to Spatial Data Mining. In: Proceedings of 23rd International Conference on Very Large Data Bases, pp. 186–195 (1997)
8. Wang, W., Yang, J., Muntz, R.: STING+: An Approach to Active Spatial Data Mining. Technical report, UCLA CSD, No. 980031 (1998)
9. Tsai, C.F., Liu, C.W.: KIDBSCAN: A New Efficient Data Clustering Algorithm for Data Mining in Large Databases. In: Rutkowski, L., Tadeusiewicz, R., Zadeh, L.A., Żurada, J.M. (eds.) ICAISC 2006. LNCS, vol. 4029, pp. 702–711. Springer, Heidelberg (2006)
10. Tsai, C.F., Yen, C.C.: ANGEL: A New Effective and Efficient Hybrid Clustering Technique for Large Databases. In: Zhou, Z.-H., Li, H., Yang, Q. (eds.) PAKDD 2007. LNCS, vol. 4426, pp. 817–824. Springer, Heidelberg (2007)
11. Tsai, C.F., Yen, C.C.: G-TREACLE: A New Grid-based and Tree-alike Pattern Clustering Technique for Large Databases. In: Washio, T., Suzuki, E., Ting, K.M., Inokuchi, A. (eds.) PAKDD 2008. LNCS, vol. 5012, pp. 739–748. Springer, Heidelberg (2008)

Applying ELECTRE and Maximizing Deviation Method for Stock Portfolio Selection under Fuzzy Environment

Chen-Tung Chen and Wei-Zhan Hung

Abstract. The purpose of stock portfolio selection is how to allocate the capital to a large number of stocks in order to bring a most profitable return for investors. In most of past literature, expert considered portfolio problem only based on past data. It is very important for experts to use their experience and knowledge to predict the performance of each stock. In this paper, 2-tuple linguistic variables are used to express the opinions of experts to predict the performance of each stock with respect to each criterion. According to experts' linguistic evaluations, we use maximizing deviation method to derive the weight of each criterion. And then, the linguistic ELECTRE method is used to derive the credibility matrix and calculate the net credibility degree of each stock. Based on the outranking index and selection threshold, we can easily obtain portfolio set and decide the investment ratio of each stock. An example is implemented to demonstrate the practicability of proposed method.

Keywords: stock portfolio selection, 2-tuple linguistic variable, ELECTRE, maximizing deviation method.

1 Introduction

The purpose of stock portfolio selection is how to allocate the capital to a large number of stocks in order to bring a most profitable return for investors [8]. Markowitz proposed the mean–variance method for the portfolio selection problem in 1952 [10]. The capital asset pricing model (CAPM), Black model and two-factor model are derived from the mean–variance method. In 1980, Saaty proposed Analytic Hierarchy Process (AHP) to deal with portfolio selection problem [12]. Edirisinghe and Zhang selected the securities in the context of data envelopment analysis (DEA) [2]. In the aforementioned portfolio selection models, experts decide investment portfolio only based on past numerical data except AHP. However, AHP is a subjective method and has the consistent problem of judgment by experts. In real situation, expert can use his experience and knowledge to

Chen-Tung Chen
Department of Information Management, National United University, Miao-Li, Taiwan
e-mail: ctchen@nuu.edu.tw

Wei-Zhan Hung
Graduate Institute of Management, National United University, Miao-Li, Taiwan

B.-C. Chien and T.-P. Hong (Eds.): Opportunities and Challenges, SCI 214, pp. 85–91.
springerlink.com

predict the performance of each stock; it is risky to select stock to invest only based on past numerical data in more and more competitive environment. The Elimination et choice in Translating to Reality (ELECTRE) method is a highly developed multi-criteria analysis model which takes into account the uncertainty and vagueness in the decision process [11]. It is based on the axiom of partial comparability; it can simplify the evaluation procedure of stock selection. Due to imprecise and subjective information that often appears in stock selection process, crisp values are inadequate for solving the problems. A more realistic approach may be to use linguistic assessments instead of numerical values [3].

In fact, experts can apply 2-tuple linguistic variables to express their opinions and obtain the final evaluation result with appropriate linguistic variable. The 2-tuple linguistic representation model is based on the concept of symbolic translation [3, 16]. It is an effective method to reduce the mistakes of information translation and avoid information loss through computing with words [6].

The maximizing deviation method is proposed by Wang [14] to compute the weight of each criterion in multiple attribute decision making (MADM) problems with numerical information. If some criterion makes the performance values among all the stocks have obvious differences, such a criterion plays a more important role in choosing the best stock. The distinguish ability and objectivity of the maximizing deviation method is better than AHP which is based on expert's subjective opinion.

This paper is organized as follows. In section 2, we present the context of the 2-tuple linguistic variable. In section 3, we discuss the concept and formula of the maximizing deviation method. In section 4, we describe the detail of the proposed method, and then an example is implemented to demonstrate the procedure for the proposed method and compare with the method of Tiryaki and Ahlatcioglu [13]. Finally, the conclusion is discussed at the end of this paper.

2 The 2-Tuple Linguistic Representation

Let $S=\{s_0, s_1, s_2, \ldots, s_g\}$ be a finite and totally ordered linguistic term set. A 2-tuple linguistic variable can be expressed as (s_i, α_i), where s_i is the central value of i-th linguistic term in S and α_i is a numerical value representing the difference between calculated linguistic term and the closest index label in the initial linguistic term set. The symbolic translation function Δ is presented in [4] to translate a crisp value into a 2-tuple linguistic variable. The generalized translation function can be represented as [1] $\Delta:[0,1]\rightarrow S\times[-(1/2g),(1/2g))$, $\Delta(\beta)=(s_i,\alpha_i)$ where $i=round(\beta\times g)$, $\alpha_i=\beta-(i/g)$, $\alpha_i\in[-(1/2g),(1/2g))$ and $\beta\in[0,1]$. A reverse function Δ^{-1} is defined to return an equivalent numerical value β ($\beta\in[0,1]$) from 2-tuple linguistic variable (s_i, α_i). According to the symbolic translation, an equivalent numerical value β is obtained as $\Delta^{-1}(s_i,\alpha_i)=(i/g)+\alpha_i=\beta$ [1]. Let $x = \{(r_1, \alpha_1),\ldots,(r_n, \alpha_n)\}$ be a 2-tuple linguistic variable set. The arithmetic mean $\bar{x}$ is computed as $\bar{x}=\Delta\left(\frac{1}{n}\sum_{i=1}^{n}\Delta^{-1}(r_i,\alpha_i)\right)=(s_m,\alpha_m)$ [7]. In general, decision makers would use the different

Table 1 Different types of linguistic variables

Linguistic variable	Figure
Very Poor (s_0^5), Poor (s_1^5), Fair (s_2^5), Good (s_3^5), Very Good (s_4^5)	Fig. 1(type 1)
Very Poor (s_0^7), Poor (s_1^7), Medium Poor (s_2^7), Fair (s_3^7), Medium Good (s_4^7), Good (s_5^7), Very Good (s_6^7)	Fig. 2(type 2)

2-tuple linguistic variables based on their knowledge or experiences to express their opinions [5]. For example, the different types of linguistic variables show as Table 1. Each 2-tuple linguistic variable can be represented as a triangle fuzzy number [3].

3 The Maximum Deviation Method

If the performance values among all the alternatives are little differences with respect to criterion, it shows that the criterion plays a less important role in the decision-making procedure. Contrariwise, if one criterion makes the performance values among all the alternatives have obvious differences, such a criterion plays a more important role in choosing the best alternative. According to the concept, the maximizing deviation method [15] is applied to calculate the weight of each criterion.

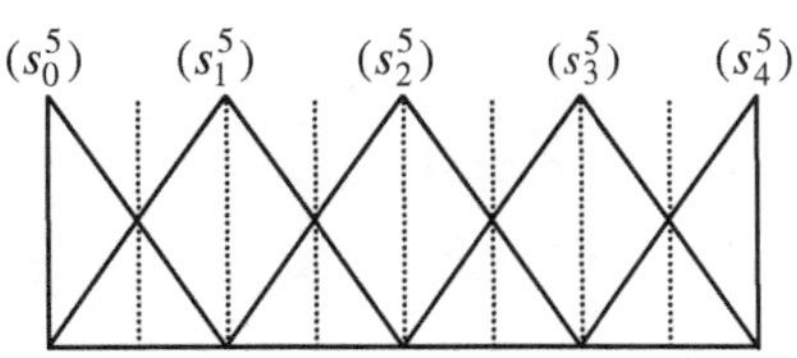

Fig. 1 Membership functions of linguistic variables at type 1 (t=1)

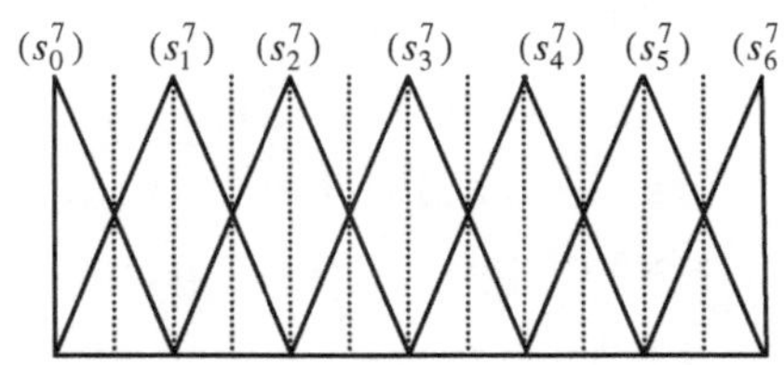

Fig. 2 Membership functions of linguistic variables at type 2 (t=2)

Assume that an expert group has K experts, and the fuzzy rating of alternative A_i respect to criterion c_j of each expert E_k (k = 1,2,...,K) can be represented as a 2-tuple linguistic variable $\tilde{x}_{ij}^k=\left(s_{ij}^k,\alpha_{ij}^k\right)$. The deviation method is used to compute the differences of the performance values of each alternative with respect to all criteria. For the expert E_k and the criterion C_j, the deviation of alternative A_i to all the other alternatives can be defined as $H_{ij}^k(w)=\sum_{l=1}^{n}\left(\Delta^{-1}\left(\tilde{x}_{ij}^k\right)-\Delta^{-1}\left(\tilde{x}_{lj}^k\right)\right)^2 w_j$ and

$$H_j^k(w)=\sum_{i=1}^{n}\sum_{l=1}^{n}\left(\Delta^{-1}\left(\tilde{x}_{ij}^k\right)-\Delta^{-1}\left(\tilde{x}_{lj}^k\right)\right)^2 w_j .$$

The $H_j^k(w)$ represents the deviation value of all alternatives to other alternatives with respect to the criterion c_j by the expert E_k. Based on the maximum deviation method, a non-linear programming model can be constructed as [15]

$$\max \quad H(w)=\sum_{k=1}^{K}\lambda_k \sum_{j=1}^{m}\sum_{i=1}^{n}\sum_{l=1}^{n}\left(\Delta^{-1}\left(\tilde{x}_{ij}^{k}\right)-\Delta^{-1}\left(\tilde{x}_{lj}^{k}\right)\right)^2 w_j \quad s.t. \quad w_j \geq 0, \quad \sum_{j=1}^{m} w_j^2 = 1 \tag{1}$$

where λ_k the represents the weight of expert E_k .The weight (w_j) of criterion C_j can be calculated as [15]

$$w_j^* = \frac{\sum_{k=1}^{K}\lambda_k \sum_{i=1}^{n}\sum_{l=1}^{n}\left(\Delta^{-1}\left(\tilde{x}_{ij}^{k}\right)-\Delta^{-1}\left(\tilde{x}_{lj}^{k}\right)\right)^2}{\sum_{j=1}^{m}\sum_{k=1}^{K}\lambda_k \sum_{i=1}^{n}\sum_{l=1}^{n}\left(\Delta^{-1}\left(\tilde{x}_{ij}^{k}\right)-\Delta^{-1}\left(\tilde{x}_{lj}^{k}\right)\right)^2} \tag{2}$$

4 Proposed Method

In general, stock selection problem may be described as a multiple criteria decision making (MCDM) problem with multiple experts. The fuzzy rating of each expert E_k $(k = 1,2,...,K)$ can be represented as a 2-tuple linguistic variable $\tilde{x}_{ij}^{k}=\left(s_{ij}^{k},\alpha_{ij}^{k}\right)$. The aggregated linguistic ratings ($\tilde{x}_{ij}$) of stocks with respect to each criterion can be calculated as $\tilde{x}_{ij} = \Delta \; (\frac{1}{n}\sum_{k=1}^{K}\Delta^{-1}(S_{ij}^{k}, \alpha_{ij}^{k})) = (S_{ij}, \alpha_{ij})$. A linguistic decision matrix can be concisely expressed as $\tilde{D}=[\tilde{x}_{ij}]_{m\times n}$ with $\tilde{x}_{ij}=(S_{ij},\alpha_{ij})$. According to the ELECTRE method, the concordance index $C_j(S_i,S_l)$ is calculated for each pair of stocks (S_i , S_l) with respect to each criterion as

$$C_j(S_i,S_l)=\begin{cases} 1 & ,\Delta^{-1}\left(\tilde{x}_{ij}\right)\geq\Delta^{-1}\left(\tilde{x}_{lj}\right)-q_j \\ \dfrac{\Delta^{-1}\left(\tilde{x}_{ij}\right)-\Delta^{-1}\left(\tilde{x}_{lj}\right)+p_j}{p_j-q_j} & ,\Delta^{-1}\left(\tilde{x}_{lj}\right)-q_j\geq\Delta^{-1}\left(\tilde{x}_{ij}\right)\geq\Delta^{-1}\left(\tilde{x}_{lj}\right)-p_j \\ 0 & ,\Delta^{-1}\left(\tilde{x}_{ij}\right)\leq\Delta^{-1}\left(\tilde{x}_{lj}\right)-p_j \end{cases} \tag{3}$$

where q_j and p_j are indifference and preference threshold values for criterion C_j, $p_j > q_j$ .The discordance index $D_j(S_i,S_l)$ is calculated for each pair of stocks with respect to each criterion as

$$D_j(S_i,S_l)=\begin{cases} 1 & ,\Delta^{-1}\left(\tilde{x}_{ij}\right)\leq\Delta^{-1}\left(\tilde{x}_{lj}\right)-v_j \\ \dfrac{\Delta^{-1}\left(\tilde{x}_{lj}\right)-p_j-\Delta^{-1}\left(\tilde{x}_{ij}\right)}{v_j-p_j} & ,\Delta^{-1}\left(\tilde{x}_{lj}\right)-p_j\geq\Delta^{-1}\left(\tilde{x}_{ij}\right)\geq\Delta^{-1}\left(\tilde{x}_{lj}\right)-v_j \\ 0 & ,\Delta^{-1}\left(\tilde{x}_{ij}\right)\geq\Delta^{-1}\left(\tilde{x}_{lj}\right)-p_j \end{cases} \tag{4}$$

where v_j is the veto threshold for criterion C_j, $v_j > p_j$.

Calculate the overall concordance index $C(S_i,S_l)$ as $C(S_i,S_l)=\sum_{j=1}^{n}W_j^{*}C_j(S_i,S_l)$. The credibility matrix $S(S_i,S_l)$ of each pair of the stocks is calculated as

$$S(S_i,S_l)=\begin{cases} C(S_i,S_l), & if\ D_j(S_i,S_l)\le C(S_i,S_l)\ \forall j \\ C(S_i,S_l)\prod\limits_{j\in J(S_i,S_l)}\dfrac{1-D_j(S_i,S_l)}{1-C(S_i,S_l)}, & otherwise \end{cases} \quad (5)$$

where $J(S_i,S_l)$ is the set of criteria for which $D_j(S_i,S_l)>C(S_i,S_l)$.

The concordance credibility and discordance credibility degrees are defined as $\phi^+(S_i)=\sum\limits_{S_l\in S} S(S_i,S_l)$ and $\phi^-(S_i)=\sum\limits_{S_l\in S} S(S_l,S_s)$ [9].

Then, the net credibility degree is defined as $\phi(S_i)=\phi^+(S_i)-\phi^-(S_i)$. If the net credibility degree is higher, then represents a higher attractiveness of stock. In order to determine the ranking order and the investment ratio, the outranking index of stock S_i can be defined as $OTI(S_i)=((\phi(S_i)/(m-1))+1)/2$. A portfolio set for investment can be determined based on threshold value β as $\Omega=\{S_i \mid OTI(S_i)\ge\beta\}$. Finally, investment ratio of stocks can be calculated as $P(S_i)=\left(OTI(S_i)/\left(\sum\limits_{S_i\in\Omega} OTI(S_i)\right)\right), S_i\in\Omega$ and $P(S_i)=0\ , S_i\notin\Omega$.

5 Numerical Example

In this paper, the data of Tiryaki and Ahlatcioglu [13] are used to implement in order to demonstrate the practicability of the proposed method. In their paper [13], three experts make the portfolio selection decision. They consider six criteria and 22 stocks. All of the experts use linguistic variables with 7 scale of linguistic term set to express their opinions (see Table 1). According to the proposed method, the computational procedures of the problem are summarized as follows.

Step 1. Each expert expresses his opinion about the performance of each stock refer to the data in [13].

Step 2. Assume that the importance of each expert is equal. We use maximizing deviation method to compute the weight of each criterion as 0.145, 0.230, 0.168, 0.098, 0.117, and 0.242.

Step 3. Calculate the aggregated linguistic ratings of each stock are shown in Table 2.

Step 4. The indifference threshold, preference threshold, and veto threshold values of each criterion can be determined in accordance with linguistic term set as $q_j=1/6,\ p_j=2/6,\ v_j=3/6,\ j=1,2,\ldots 6$.

Step 5. Calculate the concordance credibility degree, the discordance credibility degree, the net credibility degree, and the outranking index as Table 3.

Step 6. The investment ratio of each stock in accordance with different thresholds is shown as Table 4. For example, the portfolio set is $\{S_1,S_2,S_7,S_8,S_9,S_{10}\}$ in accordance with $\beta=0.7$. Compared with the method [13], the advantage of our method is that it provides a more flexible and reasonable tool to select the stock portfolio and investment ratio of each stock.

Table 2 The aggregated linguistic ratings of each stock

Stock	C_1	C_2	C_3	C_4	C_5	C_6	Stock	C_1	C_2	C_3	C_4	C_5	C_6
s_1	0.833	0.778	0.889	0.667	0.500	0.778	s_{12}	0.722	0.167	0.389	0.500	0.778	0.444
s_2	0.611	0.611	0.722	0.667	0.444	0.778	s_{13}	0.389	0.500	0.111	0.278	0.333	0.389
s_3	0.556	0.778	0.444	0.389	0.667	0.444	s_{14}	0.389	0.222	0.389	0.556	0.500	0.444
s_4	0.556	0.056	0.500	0.500	0.444	0.056	s_{15}	0.278	0.222	0.222	0.556	0.222	0.389
s_5	0.556	0.222	0.611	0.444	0.500	0.278	s_{16}	0.333	0.111	0.056	0.222	0.222	0.389
s_6	0.444	0.611	0.722	0.667	0.389	0.500	s_{17}	0.500	0.222	0.500	0.556	0.444	0.389
s_7	0.778	0.611	0.611	0.667	0.611	0.778	s_{18}	0.556	0.778	0.222	0.556	0.278	0.722
s_8	0.944	0.611	0.778	0.778	0.444	0.833	s_{19}	0.389	0.333	0.667	0.611	0.333	0.500
s_9	0.611	0.611	0.722	0.500	0.444	0.556	s_{20}	0.500	0.722	0.333	0.611	0.333	0.333
s_{10}	0.722	0.667	0.611	0.500	0.722	0.722	s_{21}	0.722	0.444	0.444	0.556	0.722	0.556
s_{11}	0.778	0.611	0.389	0.444	0.611	0.389	s_{22}	0.500	0.167	0.389	0.389	0.333	0.333

Table 3 The concordance credibility degree, the discordance credibility degree, the net credibility degree and the outranking index

Stock	$\phi^+(s_i)$	$\phi^-(s_i)$	$\phi\ (s_i)$	OTI	Stock	$\phi^+(s_i)$	$\phi^-(s_i)$	$\phi\ (s_i)$	OTI
s_1	21.845	6.422	15.423	0.867	s_{12}	11.420	14.059	-2.639	0.437
s_2	21.495	9.854	11.641	0.777	s_{13}	8.538	19.615	-11.077	0.236
s_3	18.688	11.790	6.898	0.664	s_{14}	11.717	20.868	-9.150	0.282
s_4	6.341	20.400	-14.059	0.165	s_{15}	6.989	21.595	-14.605	0.152
s_5	11.790	18.550	-6.761	0.339	s_{16}	4.362	21.758	-17.396	0.086
s_6	18.629	12.606	6.023	0.643	s_{17}	12.431	20.189	-7.758	0.315
s_7	21.888	10.012	11.876	0.783	s_{18}	11.809	11.559	0.250	0.506
s_8	21.689	6.224	15.465	0.868	s_{19}	14.034	16.882	-2.849	0.432
s_9	21.026	12.532	8.495	0.702	s_{20}	14.272	13.097	1.176	0.528
s_{10}	21.774	9.866	11.909	0.784	s_{21}	19.519	14.628	4.891	0.616
s_{11}	17.850	13.949	3.901	0.593	s_{22}	9.858	21.510	-11.653	0.223

Table 4 Investment ratio with different threshold and Comparison with Tiryaki's result

Rank	1	2	3	4	5	6
F.Tiryaki's result	S_1 ,0.216	S_8 ,0.196	S_2 ,0.157	S_7 ,0.157	S_{10} ,0.157	S_9 , 0.117
$\beta = 0.6$	(S_8 ,0.129),(S_1 ,0.129),(S_{10} ,0.117),(S_7 ,0.117),(S_2 ,0.116),(S_9 ,0.105) (S_3 ,0.099),(S_6 ,0.096),(S_{21} ,0.092)					
$\beta = 0.7$	S_8 ,0.182	S_1 ,0.181	S_{10} ,0.164	S_7 ,0.164	S_2 ,0.163	S_9 ,0.147
$\beta = 0.8$	S_8 ,0.500	S_1 ,0.500				

6 Conclusions

In general, the problem of stock selection and evaluation adhere to uncertain and imprecise data, and fuzzy set theory is adequate to deal with it. In this proposed model, 2-tuple linguistic variables are applied to express the subjective judgment of each expert. Expert can easily express his opinion by different types of 2-tuple linguistic variables. According to experts' opinions, the weight of each criterion can be determined by maximizing deviation method. The linguistic ELECTRE method is used to derive the credibility matrix and calculate the net credibility degree of each stock. Based on the outranking index and selection threshold, we can easily obtain portfolio set and decide the investment ratio of each stock. In the future, a decision support system will be developed based on the proposed method for dealing with the stock selection problems.

References

1. Chen, C.T., Tai, W.S.: Measuring the intellectual capital performance based on 2-tuple fuzzy linguistic information. In: The 10TH Annual Meeting of APDSI, Asia Pacific Region of Decision Sciences Institute, Taiwan, p. 20 (2005)
2. Edirisinghe, N.C.P., Zhang, X.: Generalized DEA model of fundamental analysis and its application to portfolio optimization. Journal of Banking & Finance 31, 3311–3335 (2007)
3. Herrera, F., Martinez, L.: A 2-tuple fuzzy linguistic representation model for computing with words. IEEE Transactions on Fuzzy Systems 8, 746–752 (2000)
4. Herrera, F., Martinez, L.: A model based on linguistic 2- tuples for dealing with multigranular hierarchical linguistic contexts in multi-expert decision-making. IEEE Transactions on Systems, Man, and Cybernetics Part B: Cybernetics 31, 227–234 (2001)
5. Herrera, F., Martinez, L., Sanchez, P.J.: Managing nonhomogeneous information in group decision making. European Journal of Operational Research 166, 115–132 (2005)
6. Herrera-Viedma, E., Cordón, O., Luque, M., Lopez, A.G., Muñoz, A.M.: A model of fuzzy linguistic IRS based on multigranular linguistic information. International Journal of Approximate Reasoning 34, 221–239 (2003)
7. Herrera-Viedma, E., Herrera, F., Martínez, L., Herrera, J.C., López, A.G.: Incorporating filtering techniques in a fuzzy linguistic multi-agent model for information gathering on the web. Fuzzy Sets and Systems 148, 61–83 (2004)
8. Huang, X.: Portfolio selection with a new definition of risk. European Journal of Operational Research 186, 351–357 (2008)
9. Li, H.F., Wang, J.J.: An Improved Ranking Method for ELECTRE III. In: International Conference on Wireless Communications. Networking and Mobile Computing, vol. 21-25, pp. 6659–6662 (2007)
10. Markowitz, H.: Portfolio selection. Journal of Finance, 77–91 (1952)
11. Papadopoulos, A., Karagiannidis, A.: Application of the multicriteria analysis method Electre III for the optimisation of decentralised energy systems. Omega 36, 766–776 (2008)
12. Saaty, T.L., Rogers, P.C., Bell, R.: Portfolio selection through hierarchies. Journal of Portfolio Manage, 16–21 (1980)
13. Tiryaki, F., Ahlatcioglu, M.: Fuzzy stock selection using a new fuzzy ranking and weighting algorithm. Applied Mathematics and Computation 170, 144–157 (2005)
14. Wang, Y.M.: Using the method of maximizing deviations to make decision for multi-indices. System Engineering and Electronics 7, 24–26 (1998)
15. Wu, Z., Chen, Y.: The maximizing deviation method for group multiple attribute decision making under linguistic environment. Fuzzy Sets and Systems 158, 1608–1617 (2007)
16. Xu, Z.S.: Deviation measures of linguistic preference relations in group decision making. Omega 33, 249–254 (2005)

Constraint Programming Approach for a University Timetabling Decision Support System with Hard and Soft Constraints

Li-Yen Shue, Pei-Chun Lin, and Chia-Yin Tsai

Abstract. This paper develops a university timetabling decision support system that considers both hard and soft constraints of the problem. A timetable solution must satisfy all hard constraints, and may be only capable of partially meeting soft constraints. We modeled the problem as a Constraint Satisfactory Problem and adapted lexicographic optimization approach to implement the solution procedure; where each soft constraint is treated as an objective with a priority. The solution approach utilizes the solution space reduction nature of constraint propagation to optimize objectives sequentially. Different value assignment strategies for constraint propagation are investigated to explore their robustness and effectiveness in performances. The system allows department management to indicate different combinations of preferences and parameters, and view the resulting timetable and related statistics in real time mode. The resulting timetabling contributes to a better teaching environment for both faculty and students.

1 Introduction

University timetabling [11][12] is about allocating courses to a weekly timetable, which are taught by staff and are taken by students. Thus, a good timetable must satisfy curriculum requirements and meet needs of both students and teachers at the same time. This problem has generally been regarded as one type of resource allocation problems in Operations Research,

Li-Yen Shue and Chia-Yin Tsai
Department of Information Management, National Kaohsiung
First University of Science and Technology,
No.2, Juoyue Road, Kaohsiung 802, Taiwan
e-mail: `ly_shue@ccms.nkfust.edu.tw`

Pei-Chun Lin
Department of Transportation and Communication Management Science,
National Cheng Kung University, No.1, University Road, Tainan 701, Taiwan
e-mail: `peichunl@mail.ncku.edu.tw`

B.-C. Chien and T.-P. Hong (Eds.): Opportunities and Challenges, SCI 214, pp. 93–98.
springerlink.com

where resources, such as timeslots, classrooms, personnel (students, teachers), and teaching facilities, are to be allocated to the same weekly timetable to achieve an objective function while subject to constraints of the problem. The heart of this problem lies in the inherent combinatorial nature, which has been classified as a NP-complete problem [7]. Various techniques have been proposed to solve timetabling problems [5][6]. These techniques include mathematics-based methods in operations research [4], interactive approaches of man-machine interaction, and various approaches in artificial intelligence [8]. However, most of these techniques require long period to solve a problem, and their models may not be easily reformulated to support changes of constraints in a decision support environment.

The objective of this research is to take into considerations both hard and soft constraints of a university timetabling problem and develop a decision support system, which could help department management to decide the composition of soft constraints and provide schedule solutions in real time. We treat the problem as a constraint satisfaction problem (CSP) and utilize the space reducing feature of the Constraint-Based Reasoning (CBR) [8][9] to develop a solution approach. The remaining of this paper is arranged as follows. Sect. 2 provides a brief description of CSP and the constraint framework of a university scheduling system. Sect. 3 presents the model formulation of constraints and objectives, which is followed by the solution approach in Sect. 4. The system description is given in Sect. 5, which is followed by the conclusions in Sect. 6.

2 Constraint Satisfaction Problem

Constraint Programming (CP) is the study of computational systems based on constraints; it solves problems by stating constraints that must be satisfied by the solution. Barták [1] stated that there are two branches of constraint programming; constraint satisfaction problem and constraint solving techniques. A CSP is characterized by a set of variables, together with a domain for each variable that contains possible finite values for that variable, and a list of constraints that govern the relationships between variables and thus restricting the values that the variables can simultaneously take [3]. A feasible solution to a CSP is an assignment of a value from its domain to every variable, in such a way that every constraint is satisfied. Currently, CSPs are solved by constraint logic programming languages, where constraints and constraint solving methods are incorporated in logic programming. CBR is a problem-solving technique that combines logic programming and constraint-solving technique, which utilize arc-consistency algorithms for the purpose of constraint propagation [8]. Arc consistency is defined relative to a specific binary constraint: a binary constraint is arc-consistent if every value of one variable has a value of the second variable such that they satisfy the constraint. Constraint propagation technique prunes the search space by preventing the variable instantiation that is not consistent with the constraints of the problem. One feature of CBR, which is different from traditional techniques, is that it allows users to find either feasible (approximate) solutions, or optimal (exact) solutions. The former could help users find a good solution with a reasonable computational time. This feature is particularly well suited to scheduling problems, which would otherwise require an exponential time if pursuing optimal solutions.

3 Model Formulation

The timetabling problem that is used as the case for this study is that of the information management department national Kaoshiung first university of science and technology in Taiwan. The department has 16 staff and is offering 42 courses in a semester. Out of the 42 classes, 11 must be split into two-hour and one-hour pattern (two-one pattern) and the remaining 31 are scheduled in three consecutive slots. Some important constraints for this case include:

1. Splitting every three-hour compulsory course into a two-one pattern on separate days; exceptions are given to those courses that use computer laboratory as part of teaching requirements.
2. No more than five teaching hours of the same day for a staff.
3. No crushes among courses of the same year.
4. No crushes between compulsory courses of a lower year and some of its higher year that could be taken by students of the lower year.
5. Faculty members with administration duty cannot schedule courses on Wednesday morning.

A survey was also conducted to solicit inputs from staff members that, in their opinions, may contribute toward a better teaching environment. A final analysis of the survey reveals the following preferences that are the wishes of most staff:

1. Avoiding teaching at the 1st (8:10-9:00) and 9th (17:30-18:20) slot.
2. Teaching the two-hour part before the one-hour part for each two-one course.
3. Finishing the two parts of a two-one course in less than n days.
4. Finishing weekly teachings in no more than n days. item Having at least n half days free from teaching.

These preferences are treated as soft constraints in developing the CSP problem model, and are formulated as objectives along with the system objectives that maximize unused first and ninth slot.

4 Solution Approach

This problem, as described above, is characterized by constraints of both hard and soft nature, which when considered simultaneously with traditional approaches may lead to over-constrained situations. Thus, we require that any feasible solution for this problem must meet all hard constraints, and may only satisfy soft constraints to the extent the solution approach is capable of. We treat the hard constraints as the traditional systems constraints, convert soft constraints into objectives with different priority, and formulated the problem into a lexicographic optimization problem. Hence, the objectives consist of the system objectives and objectives representing soft constraints. We then applied the proposed iterative solution method to find a solution.

4.1 The Iterative Solution Method

The proposed iterative solution method optimizes objectives iteratively; one objective at a time. With objectives being prioritized according to their importance, this method starts with the system objectives, then the less important ones. In order to ensure that objective values achieved in earlier iterations will not degenerate in later ones, this method utilizes the inherent nature of CBR in reducing solution space with additional constraints, that is to add the objective value found in an iteration as an additional new constraint for the subsequent iteration. A solution is found when either its optimal objective value is achieved, or when the stopping condition of 60 CPU seconds is reached. When the later is the case, we further apply Tabu search to the solution to see if further improvement is possible; this may prevent the backtracking process from being trapped in a thrashing situation. In this way, the solution process although starts each iteration as a new search process, it is in fact achieving almost the same effect as a continuous optimization process. The addition of new constraints will, as a result of constraint propagation in CBR, quickly reduce the search space, and will never repeat the search process of its earlier iterations.

4.2 Variable Selection Strategy

The problem solving process of CSP depends very much on constraint propagation, which maintains arc consistency [2] through the reduction of solution space. The strategy for variable selection and that for value assignment are the two essential elements in carrying out constraint propagation, both together will determine the computation efficiency of a solution approach. In this research, we apply dynamic fail-first heuristic as the strategy to select a variable from the pool of available variables. Fail-first heuristic, essentially, chooses the next variable with the most constraints, which will normally have the fewest solution alternatives. The dynamic fail-first heuristic reinitiates the variable selection process every time when a variable is to be selected.

4.3 Value Selection Strategy

Once a variable is selected, the system must assign to it a value to initiate constraint propagation. We adopted succeed-first as the value selection strategy, which selects a value that is more likely to lead to a solution than others. The maximum value space of any course is the entire timeslot of a week, which consists of nine slots a day for five days. The lunch break breaks the nine slots of each day into 4 slots in the morning and five slots in the afternoon. It is obvious that courses that require three consecutive slots are more difficult to fit into both sessions than those requiring only two or even one slot. We thus investigate how both vertical ordering and horizontal ordering of slots can affect the performance of value selection strategy; with two versions for each case.

5 The System Configuration

The system configuration for developing the decision support system to assist timetabling was designed with the four criteria [3] for system evaluation in mind; they are: ease of implementation, flexibility to handle a variety of constraints, computation time, and solution quality. There are three major components in the system: user interface, database, and timetabling DSS. The user interface was written using Borland C++ Builder, and is designed for department head to investigate the feasibility of considering different combinations of soft constraint(s) and "play" with different parameters. Under the "Constraints" tab, a department head can, from the list of four soft constraints, indicate which one to consider, its priority, parameter, and, for the last one, if 100% satisfaction is required. As explained in the solution method, differentials in priority will lead to different sequence of soft constraints during the optimization process, and parameters allows for experiments with varying degree of tightness of constraints. The database is built with Microsoft Access, and its component allows department secretary to input information of courses and staff for a particular semester. Each course contains: course code, course name, class year the course is for, hours required, two-one pattern or not, and staff responsible. The information for staff contains name and his/her scheduling constraints that may prevent him/her from being scheduled on a particular slot or day. The timetabling DSS has three modules: ILOG Solver (constraint solving techniques), ILOG Scheduler (scheduling techniques), and the problem representation and search strategy. We use ILOG Solver[10], which is a product by ILOG, to simplify the process and capture the specificities of problem representation, and facilitate the search processes for optimization. Under each setting, the system will return with a timetable and the related statistics that show the percentage for meeting each soft constraint. The average calculation time is 156.23 seconds to obtain a timetable.

6 Conclusions

This research applies constraint-based methodology to investigate strategies that could aid the development of a timetabling decision support system for a university environment. The system takes into consideration both hard and soft constraints. Hard constraints are the traditional system constraints that must be fully satisfied by all feasible solutions, and soft constraints represent staff's preferences in scheduling courses that may be partially satisfied. We treated the problem as a constraint satisfaction problem with soft constraints as additional objectives, and employ lexicographic optimization to develop an iterative solution procedure, which can take care of over-constrained situations. The solution procedure can preserve the objective values of earlier iterations, and avoid the solution degradation problem common to most iterative approaches.

References

1. Barták, R.: Modeling soft constraints: A Survey. Neural Network World 12(5), 421–431 (2002)
2. Bessiere, C.: Arc-consistency and arc-consistency again. Artificial Intelligence 65, 179–190 (1994)
3. Brailford, S.C., Potts, C.N., Smith, B.M.: Constraint satisfaction problems algorithm and applications. European Journal of Operational Research 119, 557–581 (1999)
4. Burke, E., Erben, W. (eds.): PATAT 2000. LNCS, vol. 2079. Springer, Heidelberg (2001)
5. Carter, M.W., Laporte, G.: Recent Developments in Practical Examination Timetabling. Interface 4, 3–21 (1996)
6. Carter, M.W., Laporte, G.: Recent Developments in Practical Course Timetabling. Interface 5, 3–19 (1998)
7. Cooper, T.B., Kingston, J.H.: The Complexity of Timetabling Construction Problem. Interfaces 17, 183–195 (1996)
8. Deris, S., Omatu, S., Ohta, H.: Timetable planning using the constraint-based reasoning. Computers & Operations Research 27, 819–840 (2000)
9. Fahrion, R., Dollansky, G.: Construction of University Faculty Timetables Using Logic Programming. Discrete Applied Mathematics 35(3), 221–236 (1992)
10. ILOG Inc. ILOG [Computer software]: Mountain View, California (2002)
11. Schaerf, A.: A Survey of Automated Timetabling. Artificial Intelligence Review 13, 87–127 (1995)
12. Werra, D.E.: An introduction to timetabling. European Journal of Operational Research 19, 151–162 (1985)

Behavioral Scoring Model for Bank Customers Using Data Envelopment Analysis

I-Fei Chen, Chi-Jie Lu, Tian-Shyug Lee, and Chung-Ta Lee

Abstract. This study proposes a behavior scoring model based on data envelopment analysis (DEA) to classify the customers into the high contribution and low contribution customers. Then, the low contribution customers are examined by using the slack analysis of DEA model to promote their contributions. The experiment results showed that the proposed method can provide indeed directions for bank to improve the contribution of the low contribution customers, and facilitates marketing strategy development.

1 Introduction

With more importance attached to risk management, the financial industry now has to take into account the credit risks of contributive customers. Accordingly, to achieve a balance between profit and risk, a customer's contribution is not only considered in terms of expenditure amount or purchase frequency but also the probability of credit risks. Behavioral scoring is a useful tool to assist risk management in the financial industry [1, 2]. Through analysis on the payment history and expenditure records, evaluation can be made on existing customer's credit risk and their consumption habit, so as to assess customer contribution. Behavioral scoring model can be employed for financial institutes to distinguish highly contributive customers from less contributive ones, and its ultimate goal is to reduce loss and increase profit via risk control.

I-Fei Chen
Tamkang University, Department of Management Science and Decision Making,
151 Ying-chuan Road, Tamsui, Taipei County, 25137 Taiwan
e-mail: enfa@mail.tku.edu.tw

Chi-Jie Lu
Ching Yun University, Department of Industrial Engineering and Management,
229 Chien-hsin Road, Jung Li, Taiwan, R.O.C
e-mail: jerrylu@cyu.edu.tw

Tian-Shyug Lee and Chung-Ta Lee
Fu-Jen Catholic University, Graduate Institute of Management,
510 Chung Cheng Rd , Hsin chuang , Taipei County 24205 Taiwan
e-mail: 036665@mail.fju.edu.tw

B.-C. Chien and T.-P. Hong (Eds.): Opportunities and Challenges, SCI 214, pp. 99–104.
springerlink.com

Nowadays, numerous approaches have been proposed to construct the behavioral scoring model to address relevant issues [3, 4, 5, 6, 7, 8, 9, 10, 11, 12, 13, 14]. However, most of the existing behavioral scoring models can facilitate discrimination process but not specify directions towards which improvements can be made on customers' performance. Such models can serve as reference on examining whether a customer has good credit, and deserves retaining efforts or whether his and/or her contributes are considerably to the enterprise. They may not be able to analyze less contributive customers, not to provide specific suggestions on marketing or managerial strategies for coping with customers of lower credit score or less contribution. In light of this, this study proposes a customer behavioral scoring model based on data envelopment analysis (DEA). This model is expected to assist banks understanding customer contribution and provide them with directions to improve earnings from less contributive customers.

DEA is an efficiency evaluation model proposed by [15]. As a multi-objective decision-making tool, it analyzes multiple input and output variables of DMUs (decision making unit) to figure out their relative efficiency [16]. Before assessing each DMU's efficiency, DEA dose not presume the relationship between each input and output variable but compares the relative efficiency among DMUs to decide their efficiency value. Also, for inefficient DMUs, specific suggestions can be provided so that the composition of input and output items can be properly adjusted to achieve higher efficiency. DEA has been widely applied to performance evaluation in many fields [17, 18, 19, 20, 21, 22, 23]. Nevertheless, very few studies have been dedicated to the application of DEA in issues related to customer behavioral scoring model.

This study examines the credit card holder database provided by a bank in Taiwan. Each card holder is regarded as a DMU, whose performance is evaluated in a DEA-based behavioral scoring model. Considering the impact of credit risks on banks, customer's payment behaviors (such as whether balance is paid in full each month or revolved) are set as input variables of DEA, while customer's expenditures (such as total amount spent and credit line usage rate) are set as output variables. The purpose of this design is to maximize corporate profit with risks considered. Based on each customer's efficiency score figured out by DEA, efficient customers can be distinguished from inefficient ones. Efficient customers are viewed as high-contribution customers because, in the trade-off relationship between customer's credit risk and bank's revenue, performance of such customers is better and more balanced than inefficient (or low-contribution) ones. Finally, the proposed model is applied to further analyze low-contribution customers, so as to derive appropriate marketing or administrative strategies with an aim of developing personalized marketing and customized management to reduce misuse or abuse of resources and improve customer contribution.

The remainder of this paper is organized as follows. Section 2 reviews literatures about behavioral scoring model and DEA; Session 3 gives a brief introduction to DEA; Session 4 presents a detailed discussion on the empirical results obtained from the proposed behavioral scoring model; Session 5 concludes the study with suggestions for further researches.

2 Data Envelopment Analysis

Based on the concept of replacing "predicted production functions" with "non-predicted production functions", DEA measures the relative efficiency of each DMU. Among various models of DEA hinging on their corresponding assumptions, CCR models, based on constant return to scale, are most often used in literature [19, 20].

The CCR model is a type of non-parameter estimation method that does not require assumption of production function and estimate function parameters. Weights in the model are determined by the input and output factors. And depending on perspectives used, CCR models can be either input-oriented or output-oriented. The input-oriented approach examines the current output level to understand which input method is the most efficient, while the output-oriented method compares the efficiency of different output methods under the same input level. In practice, the analysis approach is usually selected according to the control over either of input and output variables.

3 Empirical Study

This study examined a dataset of 1000 credit card holders (including customer data and credit records) provided by a bank in Taiwan. Each card holder's contribution was measured with the proposed DEA behavioral scoring model, and suggestions were also made on low contribution customers.

Credit risk not only has significant influence on credit granting but also reflects card holder's ability to pay debts. Thus, card holder's payment behavior is taken as an input variable in DEA. Considering data constraints, findings from literature review and expert suggestions, three variables were selected: the worst payment ranking of the month (X1), the worst payment ranking of the last 3 months (X2), and the worst payment ranking of the last 6 months (X3). Each input variable has two categories, namely "pay in full" and "revolve balance". Since DEA processes only numerical data, the two categories have to be coded as 1 and 2 respectively.

The output variables were also selected through literature review and expert interview. Factors that can reflect a card holder's purchasing power and consumption capacity should be used as output variables in DEA, namely "credit line usage rate of the month" (Y1), "average credit line usage rate of the last 3 months" (Y2), "average credit line usage rate of the last 6 months" (Y3), and "expenditure of the month" (Y4). Considering the convenience and practicality of customer management for banks, instead of using actual values, each variable was further divided into 5 categories as suggested by experts to make the analysis results more concise and comprehensive.

Besides, there should be significant and positive correlations between the output and input variables so that the isotropy between the variables and significant interactions between input resources and output performance can be presented. Therefore, correlation analysis was performed on the input and output variables and results shows that they do exhibit strongly positive relationship.

In this study, card holders were clustered via DEA. Those with efficiency value of 1 were considered as high-contribution customers while the rest were viewed as low-contribution ones. After computation, the average efficiency score of all the card holders was 0.558, with a standard deviation of 0.225. Among all the 1000 samples, 128 card holders were rated as high-contribution customers (12.8%) and the remaining 872 card holders were judged as low-contribution customers whose expenditure or payment behavior could be further improved.

Sample ID 561 and 871 are examples of a detailed DEA analysis on the high-contribution customers. Both high-contribution card holders paid the balance in full each month, indicating that they had very low credit risks. In view of the output variables (consumption), their credit line usage rates were 5% - 10% and 10% - 50% respectively, with total expenditure of each month over NT$20,000, implying that their consumption capacities were good. DEA's recommended value was the same as the actual value and there was no need to improve these customers' efficiencies. Hence, from the standpoint of bank, they were regarded to as efficient customers or "high-contribution credit card holders".

Further on the analysis of low-contribution card holders, three examples were employed, including Sample ID 477, 134 and 703. Sample ID 477 had an efficiency value of 0.5, relied on revolving balance in all the three variables for payment behavior, and reached the highest expenditure level. It can be inferred that this customer had very good consumption capacity but also high credit risks. Hence, DEA suggested that, in order to improve this customer's contribution, the emphasis should be put on his/her payment behavior. If the reliance on revolving balance can be replaced by paying balance in full each month, the bank can have one more efficient customer. For this customer, DEA prescribed directed improvements on inputs from the perspective of risk control.

Furthermore, DEA results also indicated that the efficiency value of Sample ID 134 was 0.75 and this customer tended to pay balance in full each month, implying a very low credit risk. However, this customer's consumption should be encouraged. In terms of credit line usage rates of the last 3 months and the last 6 months, the customer performed well and could remain in the current level, yet credit line usage rate of the month should be raised from below 5% to 5% ~ 10%. Moreover, the total expenditure of the month should be increased from below NT$5,000 to the level of NT$20,000 ~ NT$60,000. For this customer, DEA recommended that the output, i.e. consumption of this customer, can be improved.

Finally, the efficiency value of Sample ID 703 was 0.62. DEA gives suggestions on improvements for each card holder in terms of risk management and expenditure. To reduce risk, "payment ranking of the latest 6 months" should move from "revolve balance" to "pay in full". To encourage consumption of this customer, the credit line usage rate of the month should be raised from below 5% (Category 2) to the level of 5% ~ 10% (Category 3) and total expenditure of the month should increase from below NT$5,000 (Category 2) to NT$5,000 ~ NT$20,000 (Category 3). Therefore, DEA suggests that in the case of this customer, improvements should be made for both risk and the expenditure aspects.

4 Conclusions and Suggestions

The empirical results revealed that the proposed DEA-based customer behavioral scoring model was able to provide specific suggestions on customer categorization. Of the 1000 card holders in the sample, 128 were rated as high-contribution customers, and the rest 872 were judged as low-contribution customers. For these low-contribution customers, specific suggestions in the aspects of payment behavior or consumption were made to enhance their contribution levels. Through the proposed system, banks can provide individual customers with personalized marketing or customized management. Benefits brought about include better control over resources input, comprehensive consideration on risk and revenue, and effective improvement of customer contribution. Due to the constraint on data access, only payment behavior and expenditure amount were used as the input variables in this study. In future studies, practicability of the model can be further enhanced by integrating other factors associated with customer behavior scoring or contribution, such as total amount paid, the purchase of financial instruments, and consumption frequency.

References

1. Banasiak, M., O'Hare, E.: Behavior Scoring. Business Credit 103(3), 52–55 (2001)
2. Connors, M., Bona, S.: Scoring the Customer Lifecycle. Business Credit 105(2), 32–33 (2003)
3. Fritz, S., Hosemann, D.: Restructuring the Credit Process: Behaviour Scoring for German Corporates. Intelligent Systems in Accounting, Finance & Management 9(1), 9–21 (2000)
4. Thomas, L.C., Ho, J., Scherer, W.T.: Time Will Tell: Behavioural Scoring and the Dynamics of Consumer Credit Assessment. IMA Journal of Management Mathematics 12(1), 89–103 (2001)
5. Lin, Y.: Improvement on Behavior Scores by Dual-Model Scoring System. International Journal of Information Technology and Decision 1(1), 153–164 (2002)
6. He, J., Liu, X., Shi, Y., Xu, W., Yan, N.: Classifications of Credit Cardholder Behavior by Using Fuzzy Linear Programming. International Journal of Information Technology and Decision Making 3(4), 633–650 (2004)
7. Hsieh, N.C.: An Integrated Data Mining and Behavioral Scoring Model for Analyzing Bank Customers. Expert Systems with Applications 27(4), 623–633 (2004)
8. Hsieh, N.C.: Hybrid Mining Approach in the Design of Credit Scoring Models. Expert Systems with Applications 28(4), 655–665 (2005)
9. Frias-Martinez, E., Magoulas, G., Chen, S., Macredie, R.: Modeling Human Behavior in User-Adaptive Systems: Recent Advances Using Soft Computing Techniques. Expert Systems with Applications 29(2), 320–329 (2005)
10. Kou, G., Peng, Y., Shi, Y., Wise, M., Xu, W.: Discovering Credit Cardholders' Behavior by Multiple Criteria Linear Programming. Annals of Operations Research 135(1), 261–274 (2005)
11. Larivičre, B., Van den Poel, D.: Predicting Customer Retention and Profitability by Using Random Forests and Regression Forests Techniques. Expert Systems with Applications 29(2), 472–484 (2005)

12. Crook, J.N., Edelman, D.B., Thomas, L.C.: Recent Developments in Consumer Credit Risk Assessment. European Journal of Operational Research 183(3), 1447–1465 (2007)
13. Hadden, J., Tiwari, A., Roy, R., Ruta, D.: Computer Assisted Customer Churn Management: State-of-the-Art and Future Trends. Computers and Operations Research 34(10), 2902–2917 (2007)
14. Lim, M.K., Sohn, S.Y.: Cluster-Based Dynamic Scoring Model. Expert Systems with Applications 32(2), 427–431 (2007)
15. Chanes, A., Cooper, W.W., Rhodes, E.: Measuring the Efficiency of Decision Making Units. European Journal of Operational Research 2, 429–444 (1978)
16. Cooper, W.W., Seiford, L.M., Zhu, J.: Handbook on Data Envelopment Analysis. Kluwer Academic, Boston (2004)
17. Seiford, L.M., Thrall, R.M.: Recent Developments in DEA: The Mathematical Programming Approach to Frontier Analysis. Journal of Econometrics 46, 7–38 (1990)
18. Seiford, L.M.: Data Envelopment Analysis: The Evolution of the State of the Art (1978–1995). Journal of Productivity Analysis 7, 99–137 (1996)
19. Cherchye, L., Post, T.: Methodological Advances in DEA: A Survey and an Application for the Dutch Electricity Sector. Statistica Neerlandica 57(4), 410–438 (2003)
20. Emrouznejad, A., Parker, B.R., Tavares, G.: Evaluation of Research in Efficiency and Productivity: A Survey and Analysis of the First 30 Years of Scholarly Lliterature in DEA. Socio-Economic Planning Sciences 42(3), 151–157 (2007)
21. Cielen, A., Peters, L., Vanhoof, K.: Bankruptcy Prediction Using a Data Envelopment Analysis. European Journal of Operational Research 154(2), 526–532 (2004)
22. Chang, T.C., Chiu, Y.H.: Affecting Factors on Risk-Adjusted Efficiency in Taiwan's Banking Industry. Contemporary Economic Policy 24(4), 634–648 (2006)
23. Al-Tamimi, H.A.H., Lootah, A.M.: Evaluating the Operational and Profitability Efficiency of a UAE-Based Commercial Bank. Journal of Financial Services Marketing 11(4), 333–348 (2007)

Decision Support for Combinatorial Reverse Auction with Multiple Buyers and Sellers

Fu-Shiung Hsieh and Cheng Chung Hua

Abstract. We consider a multiple buyers/sellers combinatorial reverse auction problem in which multiple buyers want to acquire items from a set of sellers to process the task on hand. Each seller owns a set of items to bid for the required items requested by the buyers. The problem is to determine the winners to minimize the total cost to perform acquire the required items. The main results include: (1) a problem formulation for the winner determination problem; (2) a solution methodology based on Lagrangian relaxation; (3) analysis of numerical results obtained by our algorithms.

1 Introduction

Auctions are popular, distributed and autonomy preserving ways of allocating items or tasks among multiple agents to maximize revenue or minimize cost. Combinatorial auctions [1, 3] are beneficial, when complementarities exist between the items to be auctioned. Allowing bids for bundles of items is the foundation of combinatorial auctions. However, combinatorial auctions have been notoriously difficult to solve from a computational point of view [4, 8, 10, 11]. Many algorithms have been developed for combinatorial auction [2, 8, 4, 7, 12, 13]. However, in real world, there are usually multiple buyers and sellers involved in auction. Motivated by the deficiency of the existing methods, we consider a multiple buyers/sellers combinatorial reverse auction problem (MBSCRA) in which multiple buyers want to acquire items from a set of sellers. Each buyer requests a minimum bundle of items that can be provided by a set of bidders. The problem is to determine the winners to minimize the total cost. The remainder of this paper is organized as follows. In Section 2, we present the problem formulation. In Section 3, we propose the solution algorithms. We analyze the performance of our algorithm in Section 4. We conclude in Section 5.

Fu-Shiung Hsieh and Cheng Chung Hua
Department of Computer Science and Information Engineering,
Chaoyang University of Technology, 168 Jifong E. Rd.,Wufong Township,
Taichung County, 41349 Taiwan
e-mail: {fshsieh, s9627622}@cyut.edu.tw

B.-C. Chien and T.-P. Hong (Eds.): Opportunities and Challenges, SCI 214, pp. 105–110.
springerlink.com

2 Multiple Buyers/Sellers Combinatorial Reverse Auction

Fig. 1 illustrates a scenario in which buyers request to purchase three different bundles of items from the sellers. Suppose there are two buyers and five sellers. Buyer 1 requests to purchase 2A, 1B and 2C while Buyer 2 requests to purchase 1A and 3B. There are five sellers, Seller 1, Seller 2, Seller 3, Seller 4 and Seller 5, who place bids. Suppose Seller 1 places the bid: (1A, 1B, 1C, p11, p21), Seller 2 places the bid: (2A, 1B, 0C, p12, p22), Seller 3 places the bid: (0A, 1B, 2C, p13, p23), Seller 4 places the bid: (0A, 3B, 0C, p14, p24) and Seller 5 places the bid: (1A, 2B, 0C, p15, p25), where p_{in} denotes the prices of the bid placed by Seller n to Buyer i. We assume that all the bids entered the auction are recorded.

Let's formulate the multiple buyers/sellers combinatorial reverse auction (MBSCRA) problem. Let I denote the number of buyers in MBSCRA. Each $i \in \{1,2,3,....,I\}$ represents a buyer. Let N denote the number of sellers that place bids in MBSCRA. Each $n \in \{1,2,3,....,N\}$ represents a seller. Let K denote the number of items requested. Let d_{ik} denote the desired units of the $k-th$ items requested by Buyer $i \in \{1,2,3,....,I\}$, where $k \in \{1,2,3,....,K\}$. In MBSCRA, we use a vector $b_n = (q_{in1}, q_{in2}, q_{in3},..., q_{inK}, p_{in},)$ to represent the bid submitted by bidder n, where q_{ink} denotes the quantity of the $k-th$ items and p_{in} denotes the price of the bundle. Bid b_n is an offer to deliver q_{ink} units of the $k-th$ items a total price of p_{in}. We use the variable x_{in} to indicate the bid placed by bidder n is active (x_{in} =1) or inactive (x_{in} =0). We formulate the problem as follows.

Multiple Buyers/Sellers Combinatorial Reverse Auction Problem

$$\min \sum_{i=1}^{I} \sum_{n=1}^{N} x_{in} p_{in}$$

$$s.t. \sum_{n=1}^{N} x_{in} q_{ink} \geq d_{ik} \ \mathit{for\ all}\ k \in \{1,2,...K\}, i \in \{1,2,...I\}$$

$$\sum_{i=1}^{I} x_{in} \leq 1 \quad \mathit{for\ all}\ n \in \{1,2,..., N\}, x_{in} \in \{0,1\}$$

We form a Lagrangian function by applying Lagrangian relaxation.

$$L(\lambda) = \min \sum_{i=1}^{I} \sum_{n=1}^{N} x_{in} p_{in} + \sum_{i=1}^{I} \sum_{k=1}^{K} \lambda_{ik} (d_{ik} - (\sum_{n=1}^{N} x_{in} q_{ink}))$$

$$\text{st.} \sum_{i=1}^{I} x_{in} \leq 1 \ \mathit{for\ all}\ n, x_{in} \in \{0,1\}$$

$$L(\lambda) = \sum_{i=1}^{I} \sum_{k=1}^{K} \lambda_{ik} d_{ik} + \sum_{n=1}^{N} L_n(\lambda),$$

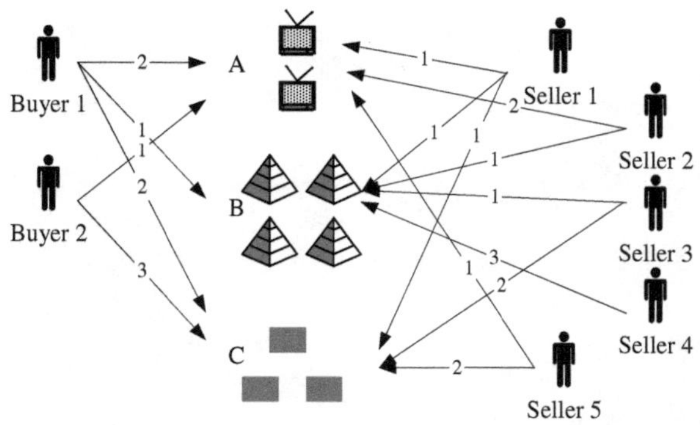

Fig. 1 Multiple Buyers/Sellers Combinatorial Auction

$$where\ L_n(\lambda) = \min \sum_{i=1}^{I} x_{in}(p_{in} - \sum_{k=1}^{K} \lambda_{ik} q_{ink})$$

$$s.t. \sum_{i=1}^{I} x_{in} \le 1 \quad for\ all\ n \in \{1,2,...,N\},\ x_{ij} \in \{0,1\}$$

Lagrangian relaxation of constraints decomposes the original problem into a number of subproblems that can be solved independently. Lanrange multipliers are determined by solving the dual problem: $\max_{\lambda \ge 0} L(\lambda)$.

3 Solution Algorithms

Our algorithms developed based on Lagrangian relaxation consists of three parts: (1) an algorithm for solving subproblems; (2) a subgradient method for solving the dual problem; (3) a heuristic algorithm for finding a near-optimal feasible solution.

(1) An algorithm for solving subproblems: Given Lagrange multiplier λ, the optimal solution to SS subproblem $L_n(\lambda)$ can be solved as follows. Let $i^* = \arg\min \sum_{i=1}^{I} x_{in}(p_{in} - \sum_{k=1}^{K} \lambda_{ik} q_{ink})$. The optimal solution to $L_i(\lambda)$ is as follows.

$$x_{in} = \begin{cases} 0\ \forall i \in \{1,2,...,I\} \setminus \{i^*\} \\ 1\ if\ i = i^*\ and\ p_{i^*n} - \sum_{k=1}^{K} \lambda_{i^*k} q_{i^*nk} < 0 \\ 0\ if\ p_{i^*n} - \sum_{k=1}^{K} \lambda_{i^*k} q_{i^*nk} \ge 0 \end{cases}$$

(2) A subgradient method for solving the dual problem $\max_{\lambda \geq 0} L(\lambda)$: Let x^l be the optimal solution to the subproblems for given Lagrange multipliers λ^l of iteration l. We define the subgradient of $L(\lambda)$ as $g_{ik}^l = \frac{\partial L(\lambda)}{\partial \lambda_{ik}} |\lambda_{ik}^l = d_{ik} - \sum_{n=1}^{N} x_{in} q_{ink}$, where, $i \in \{1,2,...I\}$ and $k \in \{1,2,..., K\}$. The subgradient method proposed by Polak [9] is adopted to update λ by $\lambda_{ik}^{l+1} = \begin{cases} \lambda_{ik}^l + \alpha^l g_{ik}^l & if\ \lambda_{ik}^l + \alpha^l \lambda_{ik}^l \geq 0; \\ 0 & otherwise. \end{cases}$ where $\alpha^l = c \frac{\overline{L} - L(\lambda)}{\sum_k (g_k^l)^2}$, $0 \leq c \leq 2$ and $\overline{L}$ is an estimate of the optimal dual cost. The iteration step terminates if α^l is smaller than a threshold. Polyak proved that this method has a linear convergence rate and iterative application converges to an optimal dual solution (x^*, λ^*).

(3) A heuristic algorithm for finding a near-optimal, feasible solution based on the solution of the relaxed problem: The solution (x^*, λ^*) may result in one type of constraint violation due to relaxation: assignment of the quantity of items less than the demand of the items. Our heuristic scheme first find $K^0 = \{k | k \in \{1,2,3,...., K\}, \sum_{n=1}^{N} x_{in} q_{ink} < d_{ik}\}$ to identify the demand constraints that have not been satisfied. Let $N^0 = \{n | n \in \{1,2,3,...., N\}, x_{in}^* = 0\}$ be the set of bidders that is not a winner in solution x^*. To make the set of constraints K^0 satisfied, we first pick $k \in K^0$ with $k = \arg \min_{k \in K^0} d_{ik} - \sum_{n=1}^{N} x_{in}^* q_{ink}$. Select $n \in N^0$ and $j \in \{1,2,..., n_i\}$ with $n = arg \min_{n \in N^0, q_{ink} > 0} p_{in}$ and set $x_{in}^* = 1$. Then we set $N^0 \leftarrow N^0 \setminus \{n\}$. If the violation of the k-th constraint cannot be completely resolved, the same procedure repeats. If all the constraints are satisfied after applying the aforementioned procedure, a solution is obtained.

4 Numerical Results

The effectiveness of the solution algorithms can be evaluated based on the duality gap, which is the difference between primal and dual objective values. That is, duality gap is defined by $f(x^*) - L(\lambda^*)$. Based on the proposed algorithms for combinatorial reverse auction, we conduct several examples to illustrate the validity of our method.

Table 1 Buyers' Requirements

	Item1	Item2	Item3
Buyer 1	2	1	2
Buyer 2	1	3	0

Table 2 Sellers' Bids

	Item1	Item2	Item3
Seller 1	1	1	1
Seller 2	2	1	0
Seller 3	0	1	2
Seller 4	0	3	0
Seller 5	1	2	0

Example 1: Consider two buyers who will purchase a set of items as specified in Table 1. Five potential sellers' bids as shown in Table 2. For this example, we have $I = 2, \quad N = 5, \quad K = 3, d_{11} = 2, d_{12} = 1, d_{13} = 2, d_{21} = 1,$ $d_{22} = 3, d_{23} = 0$. According to Table 2, we have:

$q_{111} = q_{211} = 1, q_{112} = q_{212} = 1, q_{113} = q_{213} = 1, q_{121} = q_{221} = 2, q_{122} = q_{222} = 1,$
$q_{123} = q_{223} = 0, q_{131} = q_{231} = 0, q_{132} = q_{232} = 1, q_{133} = q_{233} = 2, q_{141} = q_{241} = 0,$
$q_{142} = q_{242} = 3, q_{143} = q_{243} = 0, q_{151} = q_{251} = 1, q_{152} = q_{252} = 2, q_{153} = q_{253} = 0,$

Suppose the prices of the bids are: $p_{11} = 38, p_{12} = 30, p_{13} = 30, p_{14} = 25, p_{15} = 25,$

$p_{21} = 35, p_{22} = 22, p_{23} = 21, p_{24} = 24, \quad p_{25} = 25.$

Suppose we initialize the Lagrange multipliers as follows.
$\lambda(1) = 10.0, \lambda(2) = 10.0, \lambda(3) = 10.0, \lambda(4) = 10.0, \lambda(5) = 10.0$.
Our algorithm the subgradient algorithm converges to the following solution: $x_{13}^* = 1, \ x_{24}^* = 1$ and $x_{in}^* = 0$ for all the other (i, n). As the above solution is a feasible one, the heuristic algorithm needs not be applied. Therefore, $\bar{x}_{13} = 1, \ \bar{x}_{12} = 1,$ $\bar{x}_{24} = 1, \ \bar{x}_{22} = 1$. The solution x^* is also an optimal solution. The duality gap of the solution is 3.75%. The duality gap is within 5%. This means the solution methodology generates near optimal solution.

5 Conclusion

We study multiple buyers/sellers combinatorial reverse auction problem. By applying Lagrangian relaxation technique, the original optimization can be decomposed into a number of sellers' subproblems. Numerical results indicate that our proposed algorithms yield near optimal solutions for small problems. Our future research directions are to study the optimality of the near optimal solutions obtained from our algorithms for large problems and compare our algorithms with existing methods.

Acknowledgements. This paper is supported in part by National Science Council of Taiwan under Grant NSC97-2410-H-324-017-MY3.

References

1. de Vries, S., Vohra, R.V.: Combinatorial Auctions:A Survey. INFORMS Journal on Computing (3), 284–309 (2003)
2. Guo, Y., Lim, A., Rodrigues, B., Tang, J.: Using a Lagrangian heuristic for a combinatorial auction problem. In: Proceedings of the 17th IEEE International Conference on Tools with Artificial Intelligence (2005)
3. Peke , A., Rothkopf, M.H.: Combinatorial auction design. Management Science 49, 1485–1503 (2003)
4. Andersson, A., Tenhunen, M., Ygge, F.: Integer programming for combinatorial auction winner determination. In: Proceedings of the Seventeenth National Conference on Artificial Intelligence, pp. 39–46 (2000)
5. Fujishima, Y., Leyton-Brown, K., Shoham, Y.: Taming the computational complexity of combinatorial auctions:Optimal and approximate approaches. In: Sixteenth International Joint Conference on Artificial Intelligence, pp. 548–553 (1999)
6. Hoos, H.H., Boutilier, C.: Solving combinatorial auctions using stochastic local search. In: Proceedings of the Seventeenth National Conference on Artificial Intelligence, pp. 22–29 (2000)
7. Sandholm, T.: Algorithm for optimal winner determination in combinatorial auctions. Artificial Intelligence 135(1-2), 1–54 (2002)
8. Sandholm, T., Suri, S., Gilpin, A., Levine, D.: CABOB: A fast optimal algorithm for combinatorial auctions. In: IJCAI, pp. 1102–1108 (2001)
9. Polyak, B.T.: Minimization of Unsmooth Functionals. USSR Computational Math. and Math. Physics 9, 14–29 (1969)
10. Rothkopf, M., Peke , A., Harstad, R.: Computationally manageable combinational auctions. Management Science 44, 1131–1147 (1998)
11. Vemuganti, R.R.: Applications of set covering, set packing and set partitioning models: a survey. In: Du, D.-Z. (ed.) Handbook of Combinatorial Optimization, vol. 1, pp. 573–746. Kluwer Academic Publishers, Netherlands (1998)
12. Gonen, R., Lehmann, D.: Optimal solutions for multi-unit combinatorial auctions: branch and bound heuristics. In: The Proceedings of the Second ACM Conference on Electronic Commerce (EC 2000), pp. 13–20 (2000)
13. Jones, J.L., Koehler, G.J.: Combinatorial auctions using rule-based bids. Decision Support Systems 34, 59–74 (2002)

Visualization-Based Approaches to Support Context Sharing towards Public Involvement Support System

Shun Shiramatsu, Yuji Kubota, Kazunori Komatani, Tetsuya Ogata, Toru Takahashi, and Hiroshi G. Okuno

Abstract. In order to facilitate public involvement in the consensus building process needed for community development, a lot of time and effort needs to be spent on assessing and sharing public concerns. This paper presents new approaches for support for context sharing that involve visualizing public meeting records. The first approach is to visualize the transition of topics to enable the user to grasp an overview and to find specific arguments. The second is to visualize topic-related information to enable the user to understand background. The third is to visualize the auditory scene to enable the user to find and to listen to paralinguistic (prosodic) information contained in audio recordings. These approaches are designed on the basis of Visual Information-Seeking Mantra, "Overview first, zoom and filter, then details on demand." These approaches support citizens and stakeholders to find, to track, and to understand target arguments from the records of a public meeting.

1 Introduction

Public involvement (PI), citizen participation process in the decision-making of public policy, is characterized as an interactive communication process among citizens and stakeholders [1]. Events that improve levels of PI such as public debates, town meetings, and workshops have been getting popular in Japan recently. A benefit of PI, consensus building, involves aggregating and sharing public concerns [2, 3]. The decision-making through PI processes, however, requires a lot of time and effort

Shun Shiramatsu, Yuji Kubota, Kazunori Komatani, Tetsuya Ogata,
Toru Takahashi, and Hiroshi G. Okuno
Graduate School of Informatics, Kyoto University
e-mail: {siramatu,ykubota,komatani,ogata}@kuis.kyoto-u.ac.jp,
{tall,okuno}@kuis.kyoto-u.ac.jp

B.-C. Chien and T.-P. Hong (Eds.): Opportunities and Challenges, SCI 214, pp. 111–117.
springerlink.com

because public concerns need to be assessed. For instance, readers of the transcription of a long meeting find overviewing the contextual flow and identifying the target section difficult. This problem is an obstacle to the consensus-building process because citizens and stakeholders have various concerns or backgrounds. We aim to develop PI support methodologies that help stakeholders to track, to find, and to understand each other's concerns and the context.

Since a record of a public meeting tends to be lengthy, effective visualizations for browsing meeting records are needed to support context sharing. Visualizing arguments prevents a conflict about the agenda between stakeholders. We presume that support for sharing the following contextual information is required.

(1) An overview of a meeting record to show how a topic evolved and flowed. A lengthy meeting record is hard use to understand the overview of topic dynamics.
(2) Topic-related information to help users to understand the background to the arguments put forward. This is because important background knowledge is sometimes omitted from utterances.
(3) Paralinguistic information contained in audio recordings. Understanding emotional prosody is important to detect conflicting arguments because the meaning of utterances can differ on the basis of prosody contained in auditory information.

We designed visualizers on the basis of *Visual Information-Seeking Mantra*, "Overview first, zoom and filter, then details on demand." [4] Such overviewing work may be alleviated if tools for visualizing the arguments are available to identify and to track participants' concerns sentence by sentence. In addition, such tools may speed up the process of building consensus among stakeholders.

2 SalienceGraph: Topic Transition Visualizer

We reported the *SalienceGraph* which is a graph shaving the temporal change of joint attention to the major latent topics with discourse salience [5]. The GUI of a prototype system is shown in Figure 1. It contains GUI components designed on the basis of the Visual Information-Seeking Mantra. When a user wants to read arguments related to specific terms, they choose the terms they wish to find from the menus and drag the slide bar to the position in which the terms have high salience. The transcription window then scrolls to the section which contains the terms, and the user can read the intended arguments.

To visualize the transition of topics, we designed a metric for discourse salience, *reference probability* [5], on the basis of the assumption that a salient entity tends to be referred to in the subsequent utterance unit. This assumption is consistent with Centering Theory [6]. Since calculating the reference probability requires to extract linguistic features, the transcription should be analyzed by CaboCha [7], a Japanese dependency parser. The analysis result is annotated with Global Document Annotation (GDA) [8], a XML tagset for linguistic information. This method for visualization does not need text segmentation because the reference probability can deal with the transition on a sentence-by-sentence basis.

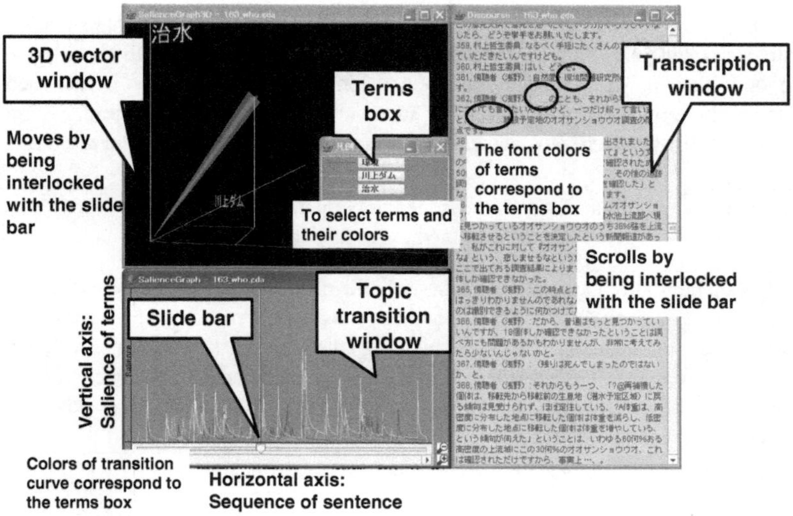

Fig. 1 SalienceGraph: Visualization of transition of topics

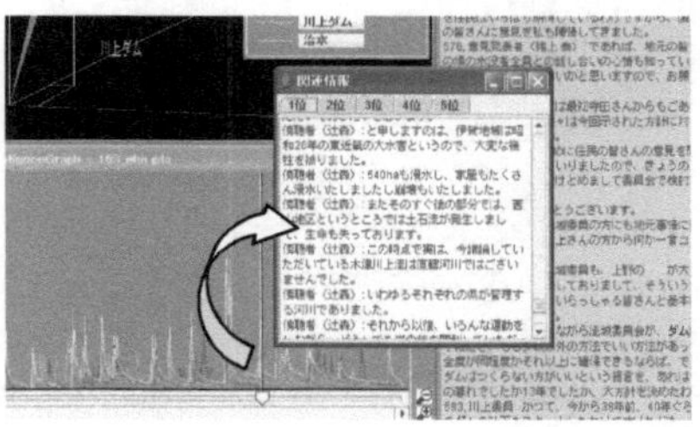

Fig. 2 Visualization of topic-related information: search and provide information related to the transient topic at slide bar position specified by user

3 Topic-Related Information Visualizer

To provide background information that may have been omitted in an argument, we developed a visualizer of finer-grain information related to the transient topic at the selected point in SalienceGraph (Figure 2). The topic-related information is retrieved by a query vector automatically generated from the transient topic. The query vector is generated as a vector comprizing the salience of terms. It is a natural expansion of the method for visualizing topic transition, which can deal with topic transition and influences from the preceding context. In order to reduce the processing load, we use probabilistic latent semantic analysis (pLSA) [9] to compress the dimensions of a query vector and approximate nearest neighbor (ANN) [10] to search for the nearest neighbors because the vocabulary size is in the tens of thousands. The *n*-best candidates as topic-related information are provided by a tabbed window in order to easily choose an effective candidate.

An instance of an experimental result on how to retrieve information related to a point specified by a user is shown in Figure 3. We used 226 meeting records (176,971 sentences) from the minutes of a meeting of the Yodo River Basin

... And so I think now we have various problems. We had been burdened by that situation for 38 years, and eight years have passed since we had left our hometown with tears. In the meantime, our village people were split into pro and con camps and were at odds with each other. When looking objectively, you may say such situation happens often in many dam plans. **But in actuality, we citizens are greatly affected by the national enterprise like that.**

An example of an argument assumed to be in a browsed meeting
(Bold-faced sentence is assumed to be the slide-bar position)

... It is truly excellent alternative plan to simply build rivers elsewhere, but it will never materialize without destroying rice paddies and anything and everything. If they do that, many dams can be built. From that viewpoint, although the brochure keep hammering away at citizen participation, I can't help feeling that they don't actually hear from citizens. Many related parties have desperate problems. For example, all Yogo-cho's plans will go haywire. **And there is also a matter about local revitalization.**

The nearest candidate (Bold-faced sentence has the nearest vector)

Fig. 3 Example of provided information related to transient topic (translated from Japanese)

Committee who were discussing a dam, which are published as lengthy PDF files on the Web [11]. The retrieval-target documents were 225 records of debates other than the browse-targeted record, 774 newspaper articles (9,989 sentences), and 2,277 Web pages (106,094 sentences). The top excerpt shown in the figure is assumed to be a browsed record and the bold-faced sentence is assumed to be a slide bar position specified by a user. The lower excerpt of the figure shows the result of the nearest neighbor candidate.

In the browsed record, the speaker mentions the magnitude of the effect of the dam construction. On the other hand, the retrieved candidate mentions a negative effect of the project and shows a citizen's concern. This example is regarded as a good way to support the user's understanding the argument.

4 Auditory Scene Visualizer

Listening to emotional prosody of utterances is essential to be aware of the conflicting arguments that are expressed in meetings. For this requirement, we developed a 3D auditory scene visualizer [12] on the basis of the Visual Information-Seeking Mantra to provide a view of sound sources in a user-friendly manner. Overview first, zoom and filter, then details on demand functions are switched by the user's face moving in three ways: approach, back away, and look inside for an auditory scene.

The GUI we developed is shown in Figure 4. A control panel has buttons for five ordinary audio functions: Play, Pause, Stop, Fast-forward, and Record. A user can operate these buttons not only with a mouse but also by using the above face movements. The viewer component contains five contents: SalienceGraph, timeline viewer, 3D viewer of directions, sound playback component, and closed-captioned

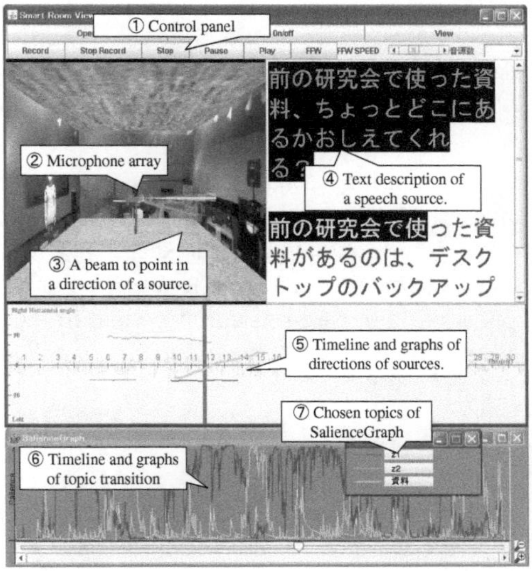

Fig. 4 Integration of the SalienceGraph and the Auditory Scene Visualizer

speech recognition in a karaoke-like manner. The contents work in synchronization with the playing back sound. In the online mode, captured sounds are played and the sound source directions are displayed by a timeline-viewer and a 3D-viewer. In the offline mode, SalienceGraph and speech recognition results are also provided.

The 3D auditory scene visualization system consists of the following subsystems.

1. CASA system, HARK open source software: (a) audio signal recording module, (b) sound source localization module, (c) sound source separation module, and (d) automatic speech recognition module
2. Face tracking client system
3. SalienceGraph client system
4. 3D visualizer server system

These subsystems enable a user to choose the intended utterance and to listen to emotional prosody from mixture of utterances and noises.

5 Related Work

The SalienceGraph is a visualizer of an argument. Several argument visualization tools currently exist [13]. Typically, these tools produce "box and arrow" diagrams in which premises and conclusions are formulated as statements [14]. However, SalienceGraph employs a different approach from these conventional works because to help the users to find arguments, it provides an overview of the whole transition of a long meeting rather than local diagramming. Moreover, overviewing the whole transition is a function suitable to be integrated with a multimodal visualizer, e.g., our 3D auditory scene visualizer. We presume that our apporoach can also be integrated with other multimodal meeting browsers [15, 16].

6 Conclusion

We developed tools that visualize the flow of a meeting record towards a system to support PI on the basis of the Visual Information-Seeking Mantra. SalienceGraph, a visualizer of the topic dynamics of discourse, helps stakeholders to browse the minutes of a long debate because it can be used to identify a particular point in the minutes. The search engine based on SalienceGraph helps stakeholders to refer to related information in order to share background knowledge. An auditory scene visualizer helps users to listen to a particular utterance to listen to emotional prosody contained in the spoken utterance and to thus assess the extent of any conflict between participants. We are planning to apply our system to improve PI processes.

References

1. Jeong, H., Hatori, T., Kobayashi, K.: Discourse analysis of public debates: A corpus-based approach. In: Proceedings of 2007 IEEE International Conference on Systems, Man and Cybernetics (SMC 2007), pp. 1782–1793 (2007)
2. Renn, O., Webler, T., Rakel, H., Dienel, P., Johnson, B.: Public participation in decision making: A three-step procedure. Policy Sciences 26(3), 189–214 (1993)
3. Rowe, G., Frewer, L.: A typology of public engagement mechanisms. Science Technology Human Values 30(2), 251–290 (2005)
4. Shneiderman, B.: Designing the User Interface: Strategies for Effective Human-Computer Interaction. Pearson Addison Wesley (1998)
5. Shiramatsu, S., Komatani, K., Ogata, T., Okuno, H.G.: SalienceGraph: Visualizing Salience Dynamics of Written Discourse by Using Reference Probability and PLSA. In: Proc. of PRICAI 2008, pp. 890–902. Springer, Heidelberg (2008)
6. Grosz, B., Joshi, A., Weinstein, S.: Centering: A Framework for Modeling the Local Coherence of Discourse. Computational Linguistics 21(2), 203–226 (1995)
7. Kudo, T., Matsumoto, Y.: Japanese dependency analysis using cascaded chunking. In: Proc. of CoNLL-2002, COLING 2002 Post-Conference Workshops, pp. 1–7 (2002)
8. Hasida, K.: Global Document Annotation (GDA) (2004), `http://i-content.org/GDA/`
9. Hofmann, T.: Probabilistic latent semantic analysis. In: Proc. of UAI 1999, pp. 289–296 (1999)
10. Arya, S., Mount, D.M.: Approximate Nearest Neighbor Queries in Fixed Dimensions. In: Proc. of the 4th Annual ACM-SIAM Symposium on Discrete Algorithms, pp. 271–280 (1993)
11. Yodo River Basin Committee: List of meeting minutes (2007) (in Japanese), `http://www.yodoriver.org/doc_list/gijiroku.html`
12. Kubota, Y., Komatani, K., Ogata, T., Okuno, H.: Design and implementation of 3d auditory scene visualizer towards auditory awareness with face tracking. In: Proc. of IEEE ISM 2008, pp. 468–476 (2008)
13. Kirschner, P., Shum, S., Carr, C.: Visualizing Argumentation: Software Tools for Collaborative and Educational Sense-Making. Springer, Heidelberg (2003)
14. van den Braak, S.W., van Oostendorp, H., Prakken, H., Vreeswijk, G.A.W.: A critical review of argument visualization tools: Do users become better reasoners? In: Workshop Notes of the ECAI-2006 Workshop on CMNA, pp. 67–75 (2006)

15. Bouamrane, M.M., Luz, S.: Navigating multimodal meeting recordings with the meeting miner. In: Larsen, H.L., Pasi, G., Ortiz-Arroyo, D., Andreasen, T., Christiansen, H. (eds.) FQAS 2006. LNCS, vol. 4027, pp. 356–367. Springer, Heidelberg (2006)
16. Bouamrane, M.M., Luz, S.: Meeting browsing: State-of-the-art review. Multimedia Systems 12(4-5), 439–457 (2007)

A Case Study of Genetic Algorithms for Quay Crane Scheduling

Yi Wang, Yun Chen, and Kesheng Wang

Abstract. In the operations of container terminals, a proper organized quay-crane-scheduling is critical to the operational efficiency. The aim of this paper is to develop a two-quay-crane schedule with non-interference constraints for the port container terminal of Narvik. First, a mathematical formulation of the problem is provided, and then a Genetic Algorithm (GA) approach is developed to obtain near optimal solutions. Finally, computational experiments on GA approach with different parameters are conducted.

1 Introduction

A container terminal is a facility where cargo containers are transshipped between different transport vehicles, for onward transportation. Containerization has enabled globalization [6] and allowed for economy of scale driving the vessels to be bigger and bigger. However, this has led to new challenges concerning terminals management and container handling operations for ports in order to keep up the pace with the supply and demand. The handling speed at the container terminal is a prerequisite for ships to achieve this economy of scale [6]. Hence the port competitiveness relies heavily on minimizing the *makespan* or transshipment time of a vessel at the terminal [5, 6]. *Makespan* means the latest completion time of all handling tasks concerning a given container vessel.

Kim and Park [4] discussed the quay crane scheduling problem with non-interference constraints in which only single container vessel was considered.

Yi Wang
Nottingham Business School, NTU, UK
e-mail: yi.wang@ntu.ac.uk

Yun Chen
School of Public Economics & Administration, SHUFE, China
e-mail: chenyun@mail.shufe.edu.cn

Kesheng Wang
Department of Production and Quality Engineering, NTNU, Norway
e-mail: kesheng.wang@ntnu.no

B.-C. Chien and T.-P. Hong (Eds.): Opportunities and Challenges, SCI 214, pp. 119–125.
springerlink.com

They established a mixed integer programming model for the problem and proposed a branch and bound method and a heuristic algorithm called "greedy randomized adaptive search procedure (GRASP)" for the solution of the quay crane scheduling problem. Park and Kim [7] also studied the problem using two stages approaches. Lee et al. [5] provided a mixed integer programming model for quay scheduling problem with non-interference constraints. Hybrid intelligent system was discussed by Sun [8] for solving quay crane scheduling problem.

The objective of this project is to focus on the quay crane scheduling with non-interface constraints for any one single container vessel in the port of Narvik . This work was stimulated from Kim and Park [4) and Lee el al. [5], which will be adapted and simplified for the practical case of Narvik contain terminal berth. A Genetic Algorithm was used to find near optimization solution for the problem.

2 Problem Description

Because both cranes are on the same rail, the Port-Authority is faced with the problem of scheduling them so that they avoid interference, respect the physical constraints, let at least one hold free between them at any given time, and yield the minimum total *makespan* for any given ship at the berth. The vessel's stability is not taken into consideration here. As shown in Figure 1, the container vessel is divided into up to ten holds and the goal of this paper is to provide a schedule that determines a handling sequence for these two cranes while avoiding interferences between them. A hold is assumed to be served by only one crane at any given time. The transition time from one hold to another is around 1 minute. Both the speed of the cranes and the vessel's capacity can be updated thus obtaining new solutions without influencing the robustness of the algorithm.

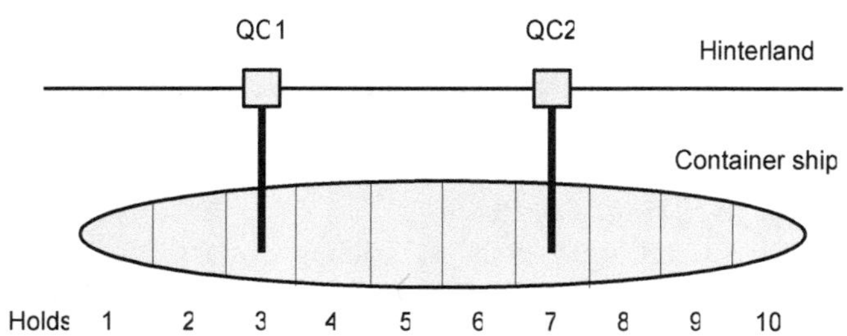

Fig. 1 A schema of the container vessel and the two quay cranes

3 Problem Formulation

For Narvik crane terminal berth, a model of this problem is described as a mixed integer programming and is given below, which is adapted from Kim and Park [4] and Lee et al [5] to the case of two cranes:

Parameters: H the number of holds (= 10); p_h the processing time of hold h by a quay crane ($1 \leq h \leq H$); M a sufficiently large positive and constant number.

Decision variables: $X_{h,1}$ 1, if hold h is handled by QC 1; 0, otherwise; $X_{h,2}$ 1, if hold h is handled by QC 2; 0, otherwise; $Y_{h,h'}$ 1, if hold h finishes no later than hold h'; C_h the completion time of hold h

Objective function:

(1.1) Minimize $[\underset{h}{Max}\, C_h]$ / minimize the *makespan* of handling one single container ship

Subject to:

(1.2) $C_h - p_h \geq 0$; $(\forall\ 1 \leq h \leq H)$/ property of C_h

(1.3) $X_{h,1} + X_{h,2} = 1$; $(\forall\ 1 \leq h \leq H)$ / every hold must be handled by only one QC

(1.4) $C_h - (C_{h'} - p_{h'}) + Y_{h,h'}M > 0$; $(\forall\ 1 \leq h, h' \leq H)$/ property of $Y_{h,h'}$

(1.5) $C_h - (C_{h'} - p_{h'}) - (1 - Y_{h,h'})M \leq 0$; $(\forall\ 1 \leq h, h' \leq H)$/ property of $Y_{h,h'}$

(1.6) $M(Y_{h,h'} + Y_{h',h}) \geq X_{h,1} - (X_{h',1} + 2X_{h',2}) + 1$; $1 \leq h < h' \leq H$ / it must avoid interference between QCs

(1.7) $X_{h,1}, X_{h,2}, Y_{h,h'} = 0 \text{ or } 1$; $(\forall\ 1 \leq h, h' \leq H)$ / binary decision variables

For the general version of this formulation, please refer to Lee et al [5] where a proof of NP-completeness is given. This means that there is no polynomial time algorithm for the exact solution. Therefore, a Genetic Algorithm is adopted to obtain near optimal solutions. The proposed Genetic Algorithm approach is effective, efficient and robust in solving the considered quay cranes scheduling problem.

4 Methodology

The concept of GAs was developed by Holland and his colleagues in the 1960s and 1970s [3]. A GA is a stochastic optimization method based on the mechanisms of natural selection and evolution. In GAs, searches are performed based on a population of chromosomes representing solutions to the problem. A population starts from random values and then evolves through succeeding generations. During each generation a new population is generated by propagating a good solution to replace a bad one and by crossing over or mutating existing solutions to construct new solutions. GAs have been theoretically and empirically proven robust for identifying solutions to combinatorial optimization problems. This is due to GAs ability to conduct an efficient parallel exploration of the search space, while only requiring minimum information on the function to be optimized. [9]

As a development tool, GeneHunter [2], is used as an add-in (solver) in Microsoft Excel. The population size is the size of the genetic breeding pool and is initialized to 50 individuals. The fitness function is the same as the objective function and the goal is to minimize total completion time (equation 1.1). The *generation gap* is set at 0.98 meaning that only 2 percent of individuals will go directly into the next generation without having to go through crossover and mutation. For the population of size 50, one individual will be sent directly to the next generation. As *elitism* is the chosen selection operator, then this one individual sent directly to the next generation is the fittest one. The *crossover* and *mutation rates* are set to 0.90 and 0.01 respectively.

Quay Crane Scheduling is a combinatorial problem where a sequence of holds is needed to be found so as it yields a near optimum solution. This sequence should be constituted of integer and unique values from 1 to 10. Thus, *enumerated chromosomes* with *unique genes* will be used. With enumerated chromosome, genes can have more allele values (integers) than just 0 and 1 while the unique genes property does not allow the chromosome to contain duplicate genes, or else the hold may be handled more than once which is absurd. The GA will strive to find the optimum order of these values so that the total completion time is minimized. The GA will go through the selection, crossover and mutation creating generations and seeking the optimum until a stopping criterion is met. Here, the GA will stop once the best fitness value remains unchanged for 50 generations. These GA parameter values are the one that give the best results from a practical point of view. Below are summarized the main feature of this GA: (1) Tool: GeneHunter v 2.4 embedded into Microsoft Excel 2003, (2) Genetic Algorithm (Fitness (objective) function: *(1.1)*; Population size: 50; Chromosome type: Enumerated with unique genes; Crossover: crossover with probability: 0.90; Mutation rate: 0.01; Generation gap: 0.98; Selection strategy: Elitist; Stopping criterion: Best fitness value unchanged after 50 generations).

QC1				QC2			
OS	HO	PT	CT	OS	HO	PT	CT
1	11	0,0	0,0	1	12	0,0	0,0
2	8	266,7	267,7	2	9	300,0	301,0
3	6	533,3	802,0	3	7	600,0	902,0
4	1	133,3	936,3	4	4	300,0	1203,0
5	3	266,7	1204,0	5	10	200,0	1404,0
6	2	133,3	1338,3	6	14	0,0	1405,0
7	5	400,0	1739,3	7	13	0,0	1406,0

Total Time	Unit
1739,3	Min.
28,92	Hour

Fig. 2 Implementation of interference check, two-list chromosome, and dummy holds. (OS: Operation Sequence; HO: Hold Order; PT: Processing Time; CT: Completion Time)

A chromosome is modeled as a list of adjustable cells (variables) corresponding to genes with 1 as a minimum value and 10 as maximum mapping obviously the ten holds of the container vessel. The order of these genes is important as it represents the sequence in which the holds will be handled by the two quay cranes. For implementation purposes, having one list of ten adjustable cells proved to be a poor modeling technique because the resulting chromosome has to be assigned to the two QCs offline, i.e. after the optimization. This means that the interference between QCs is not resolved during optimization time. As a good alternative, first, the chromosome is split into two lists with 5 cells each and corresponding to the two QCs. This allows for dynamically checking the interference between the cranes for the considered potential solution. Second, the interference check is implemented in the business logic of MS Excel as tests that return 0 if no interference and 1 otherwise. Then, in GeneHunter, constraints are forcing these cells to always return 0, thus avoiding interference in the adopted solutions. A problem arising from this design choice is how to split these two lists, is it logical to have five for each list when on crane is much faster than the other and might handle more holds than five? To overcome this, the original chromosome is extended from 10 to 14, hence introducing four dummy holds, from 11 to 14, with zero volume. The choice of 4 dummy holds stems from comparing real world QC throughputs. These dummy holds let the optimization be much flexible in finding the near optimal solution and assigning holds to cranes. At the end, these dummy holds can be discarded from the solution. The total *makespan* will be adjusted accordingly (by subtracting four times the transfer time) as shown in Figure 2. The solution here will be holds 8, 6, 1, 3, 2 and 5 for QC1 while holds 9, 7, 4 and 10 will be assigned to QC2 with a total *makespan* of 28.92 hours

5 Computational Results

GAs with different paremeters are tried on different QC throughputs. It is clear that the strategy adopted in the previous section, is the most suitable one as it gives the best results. It could be argued that the population size could be decreased to 40 or perhaps 30, this will decrease the processing time, but a population of 50 is good enough. Another observation is that when the elitist strategy is not used, the search worsens and can not stabilize around a solution and thus the quality of the final solution is not as good as with other strategies.

In solutions given by GeneHunter, knowledge about the distribution of Holds' volume is not exploited. This distribution tends to be of the same nature for different container vessels and this could be used to the operator's advantage when using either GeneHunter or developing a manual heuristic. For example, considering the distribution given in the Excel sheet (200, 200, 400, 300, 600, 800, 600, 400, 300, 200), one could proceed as follows:

- take the maximum *(800 at the 6th position)* assign it to the fastest crane *(QC1)*,
- take the second greatest volume *(600 at the 7th position which is physically possible)* assign it to the other crane *(QC1)*,

- with QC1, go left assigning as many as possible
- with QC2, go right assigning as many as possible
- cross (go beyond the 7th position for *QC1* and beyond the 6th position for *QC2*) when the steps 3 and 4 are not feasible (or not optimal) anymore.
- for the manual operator, make sure that hold 1 is handled by *QC1* and hold 10 by *QC2*

This heuristic was tried and proved to yield good results. For instance, for throughputs *(QC1, QC2) = (90, 60)* with Strategy C in GeneHunter, this heuristic gives the following result:

Makespan of *20 hr 02 min* obtained at generation *99* with a processing time of *18 sec. QC1* will handle holds *6, 1,(11), 3, 4, 2, and 8* while holds *7,10, (13), 9, 5, (14), and 12* will be assigned to *QC2.*

It should be noticed that the dummy holds introduce a problem of having an additional transfer time of one minute (4 min in total) in the original solution, this allows for some feasible solutions that are not possible otherwise. One has only to wait one or two minutes to reach better solutions. This can be regarded both as problem and as an additional flexibility in the design of the solution.

6 Conclusion

In this paper, a methodology based on GA is developed for solving the two quay-crane scheduling problem for the port of Narvik. This work is applied to the two quay cranes of Narvik Container Terminal. A GA has been developed through GeneHunter tool interfaced with MS Excel. Advantages of this methodology is the fact that near optimal solutions are obtained while the software is user-friendly as the user interacts with a known program (MS Excel) and the input can be changed easily (vessel's capacity; holds' capacity, cranes' throughput, and transfer time). GeneHunter can be then rerun and new solutions will be available. Furthermore, with some modeling changes, the number of quay cranes can be extended to three cranes but for a higher number of cranes, interference check will become difficult to carry on this way and developing one's own code (VBA or C++) will be more appropriate. As a result, this software could easily be used by operators in Narvik Container terminal. The limitations of this work, which constitute further research alleys, are neither the stability of the boat nor the handling priorities of the holds are taken into consideration here. Specific knowledge about the holds' volume distribution should be exploited to obtain minimal *makespan*..

References

1. Cheng, J.L., You, F.H., Yang, Y.: Research on quay crane scheduling problem by hybrid genetic algorithm. In: Proceedings of the IEEE International Conference on Automation and Logistics, ICAL 2008, pp. 2131–2135 (2008)
2. GeneHunter v 2.4 Getting Started Manual, `http://www.wardsystem.com`

3. Holland, J.H.: Adaptation in natural and artificial systems. The University of Michigan Press (1975)
4. Kim, K.H., Park, Y.M.: A crane scheduling methods for port container terminal. European Journal of Operational Research 156, 752–768 (2004)
5. Lee, D.H., Wang, H.Q., Miao, L.: Quay crane scheduling with non-interference constraints in port container terminals. Transportation Research, Part E 44, 124–135 (2008)
6. Meersmans, P.J.M., Dekker, R.: Operation research supports container handling, Econometric Institute Report EI, pp. 2001–2022 (2001)
7. Park, Y.M., Kim, K.H.: OR Spectrum 25, 1–23 (2003)
8. Sun, J.Q., Li, P., Han, M.: The crane scheduling problem and the hybrid intelligent optimization algorithm GASA. In: Proceedings of the 26th Chinese Control Conference, CCC 2007, pp. 92–96 (2007)
9. Wang, K.: Applied Computational Intelligence in Intelligent Manufacturing Systems, Advanced Knowledge International Pty Ltd., Australia (2005)

Genetic Algorithm and Time-Scale Separation on the Reduced-Order Observer-Based Control Design for a Class of Discrete Systems

Shing-Tai Pan and Ching-Fa Chen

Abstract. The design of the control for discrete multiple time-delay systems subject to input constraint is considered in this paper. Genetic algorithms will be applied to adjust the control gain and then a observer-based feedback control will be designed to stabilize the closed-loop system with input constraint. Moreover, using the two time-scale property of the system, the slow and fast subsystem of the discrete systems will be derived. Then the controls are design for the two subsystems such that the two subsystems are both stabile. The control of the original full-order system is then derived from the two controls. Finally, an application example will also be given in this paper to illustrate the results of this paper.

1 Introduction

The most significant advantage of singularly perturbed systems is the two-time-scale property which permits separate design of feedback controls for the slow and fast subsystems and then a composite feedback control can be synthesized from these feedback controls. This simplifies the procedure of controller design. A key to the analysis of singularly perturbed systems thus lies in the construction of the slow and fast subsystems. It is noted that the approximation of the original singularly perturbed system via its corresponding slow and fast subsystems is valid only when the singular perturbation parameters of this system are sufficiently small. Therefore, it is important to find the upper bound of singular perturbation parameters such that stability of the original system can be investigated by establishing that of its corresponding slow and fast subsystems, provided that the singular

Shing-Tai Pan
Department of Computer Science and Information Engineering,
National University of Kaohsiung, Kaohsiung, Taiwan 811, R.O.C.
e-mail: stpan@nuk.edu.tw

Ching-Fa Chen
Department of Electronic Engineering, Kao Yuan University,
Kaohsiung County, Taiwan 821, R.O.C.
e-mail: cfchen@cc.kyu.edu.tw

B.-C. Chien and T.-P. Hong (Eds.): Opportunities and Challenges, SCI 214, pp. 127–132.
springerlink.com

perturbation parameters are within this bound. Numerous reports in regard to this subject without control have been published [1-6].

In this study, the observer-based controllers for the slow and the fast subsystems are then separately designed and a composite observer-based controller for the original system is subsequently synthesized from these observer-based controllers. An illustrative example is given to demonstrate that the upper bound of the singular perturbation parameter ε can be obtained by examining this criterion.

2 Problem Formulation

Consider the following discrete singularly perturbed system which is referred to as the C-model [7]

$$x_1(k+1)=\sum_{i=0}^{n} A_{1i}x_1(k-i)+\varepsilon\sum_{i=0}^{n}\tilde{A}_{1i}x_2(k-1)+B_1u(k) \tag{2.1a}$$

$$x_2(k+1)=\sum_{i=0}^{n} A_{2i}x_1(k-i)+\varepsilon\sum_{i=0}^{n}\tilde{A}_{2i}x_2(k-i)+B_2u(k) \tag{2.1b}$$

$$y(k)=C_1x_1(k)+c_2x_2(k) \tag{2.1c}$$

where A_{1i}, A_{2i}, $\tilde{A}_{1i}$, $\tilde{A}_{2i}$ (for i=0,1,2,...,n), B_1, B_2, C_1, and C_2 are constant matrices with appropriate dimensions. The small positive scalar ε is a singular perturbation parameter.

Before proceeding to the main result, a lemma is given in the following.

Lemma 2.1 [8]. For any matrix $A\in R^{m\times n}$, if $\rho[A]<1$ then $|\det(I\pm A)>0|$.

The notation ρ[A] denotes the spectral radius of the matrix *A*. Now, according to the quasi-steady-state approach as that in([9], p. 130) and ([10], p. 276) the slow and fast subsystems of the original system (2.1) can then be derived as follows.

The slow subsystem of the original system (2.1) can be expressed as

$$x_s(k+1)=\sum_{i=0}^{n} A_{si}x_s(k-i)+B_su_s(k), \tag{2.2a}$$

$$y_s(k)=\sum_{i=0}^{n} C_{si}x_s(k-i)+D_su_s(k), \tag{2.2b}$$

where

$$A_{si}=A_{1i}+\varepsilon(\sum_{j=0}^{n}\tilde{A}_{1j})(I-\varepsilon\sum_{j=0}^{n}\tilde{A}_{2j})^{-1}A_{2i},\ B_s=B_1+\varepsilon(\sum_{i=0}^{n}\tilde{A}_{1i})(I-\varepsilon\sum_{i=0}^{n}\tilde{A}_{2i})^{-1}B_2,$$

$$C_{s0}=C_1+C_2(I-\varepsilon\sum_{i=0}^{n}\tilde{A}_{2i})^{-1}A_{20},\ C_{si}=C_2(I-\varepsilon\sum_{j=0}^{n}\tilde{A}_{2j})^{-1}A_{2i},$$

$$D_s=C_2(I-\varepsilon\sum_{i=0}^{n}\tilde{A}_{2i})^{-1}B_2.$$

The fast subsystem of the original system (2.1) is derived as follows:

$$x_f(k+1) = \varepsilon \sum_{i=0}^{n} \tilde{A}_{2i} x_f(k-i) + B_2 u_f(k) \tag{2.3a}$$

$$y_f(k) = C_2 x_f(k) \tag{2.3b}$$

3 Observer-Based Controller Design for the Slow and the Fast Subsystems

The design of observer-based controller for stabilizing discrete two-time-scale systems was discussed in [11–13]. In this section, the observer-based controllers for the slow subsystem (2.2) and for the fast subsystem (2.3) are separately designed such that both subsystems are stable.

3.1 *Controller Design for the Slow Subsystem*

By defining a new slow state vector $X_s(k) = [x_s(k), x_s(k-1),....,x_s(k-n)]^T$, the slow subsystem (2.2) can be rewritten as

$$X_s(k+1) = \overline{A}_s X_s(k) + \overline{B}_s u_s(k), \tag{3.1}$$
$$y_s(k) = \overline{C}_s X_s(k) + \overline{D}_s u_s(k),$$

where

$$\overline{A}_s = \begin{bmatrix} A_{s0} & A_{s1} & \cdots & A_{s(n-1)} & A_{sn} \\ I & 0 & \cdots & 0 & 0 \\ 0 & I & \cdots & 0 & 0 \\ \vdots & \vdots & \ddots & \vdots & \vdots \\ 0 & 0 & \cdots & I & 0 \end{bmatrix}, \overline{B}_s = \begin{bmatrix} B_s \\ 0 \\ 0 \\ \vdots \\ 0 \end{bmatrix}, \overline{C}_s = [C_{s0} \quad C_{s1} \quad \cdots \quad C_{s(n-1)} \quad C_{sn}], \overline{D}_s = D_s.$$

The observer-based controller for the slow subsystem (3.1) is given by

$$\hat{X}_s(k+1) = (\overline{A}_s - \overline{B}_s K_s - F_s \overline{C}_s)\overline{X}_s(k) + F_s \hat{u}_s(k) - F_s \overline{D}_s u_s(k), \tag{3.2}$$
$$\hat{y}_s(k) = K_s \hat{X}_s(k),$$

where the two constant matrices K_s and F_s are chosen such that $\overline{A}_s - \overline{B}_s K_s$ and $\overline{A}_s - F_s \overline{C}_s$ are both Hurwitz. Letting $e_s(k) \equiv X_s(k) - \hat{X}_s(k)$, we have

$$\begin{bmatrix} X_s(k+1) \\ e_s(k+1) \end{bmatrix} = M_s \begin{bmatrix} X_s(k) \\ e_s(k) \end{bmatrix}. \tag{3.3}$$

where $M_s \equiv \begin{bmatrix} \overline{A}_s - \overline{B}_s K_s & \overline{B}_s K_s \\ 0 & \overline{A}_s - F_s \overline{C}_s \end{bmatrix}$. Since the matrices $\overline{A}_s - \overline{B}_s K_s$ and $\overline{A}_s - F_s \overline{C}_s$ are both Hurwitz, the closed-loop system (3.3) is thus stable.

3.2 Controller Design for the Fast Subsystem

By defining a new fast state vector $X_f(k)=[x_f(k), x_f(k-1),\ldots, x_f(k-n)]^T$, the fast subsystem (2.3) can be rewritten as

$$X_1(k+1) = \overline{A}_f X_f(k) + \overline{B}_f u_f(k), \quad (3.4)$$

$$y_f = \overline{C}_f X_f(k),$$

where
$$\overline{A_f} = \begin{bmatrix} \varepsilon\tilde{A}_{20} & \varepsilon\tilde{A}_{21} & \cdots & \varepsilon\tilde{A}_{(2n-1)} & \varepsilon\tilde{A}_{2n} \\ 1 & 0 & \cdots & 0 & 0 \\ 0 & 1 & \cdots & 0 & 0 \\ \vdots & \vdots & \ddots & \vdots & \vdots \\ 0 & 0 & \cdots & 1 & 0 \end{bmatrix}, \quad \overline{B_f} = \begin{bmatrix} B_2 \\ 0 \\ 0 \\ \vdots \\ 0 \end{bmatrix}, \quad \overline{C_f} = [C_2 \ \ 0 \ \ 0 \ \ \cdots \ \ 0]$$

The observer-based controller for the fast subsystem (3.4) is given by

$$\hat{X}_f(k+1) = (\overline{A_f} - \overline{B_f}K_f - F_f\overline{C_f})\hat{X}_f(k) + F_f\hat{u}_f(k), \quad (3.5)$$

$$\hat{y}_f(k) = K_f\hat{X}_f(k),$$

where the two constant matrices K_f and F_f are chosen such that $\overline{A_f} - \overline{B_f}K_f$ and $\overline{A_f} - F_f\overline{C_f}$ are both Hurwitz. Letting $e_f(k) \equiv X_f(k) - \hat{X}_f(k)$, we have

$$\begin{bmatrix} x_f(k+1) \\ e_f(k+1) \end{bmatrix} = M_f \begin{bmatrix} x_f(k) \\ e_f(k) \end{bmatrix}, \quad (3.6)$$

where $M_f = \begin{bmatrix} \overline{A_f} - \overline{B_f}K_f & \overline{B_f}K_f \\ 0 & \overline{A_f} - F_f\overline{C_f} \end{bmatrix}$. Since the matrices $\overline{A_f} - \overline{B_f}K_f$ and $\overline{A_f} - F_f\overline{C_f}$ are both Hurwitz, the closed-loop system (3.6) is thus stable.

4 Composite Observer-Based Controller for the Original System

By defining the new state vectors $X_1(k) = [x_1(k), x_1(k-1),\ldots, x_1(k-n)]^T$ and $X_2(k) = [x_2(k), x_2(k-1),\ldots, x_2(k-n)]^T$, the original system (2.1) can then be transformed into the following form:

$$X(k+1) = A_c X(k) + B_c u(k)$$
$$y(k) = C_c X(k), \quad (4.1)$$

in which $X(k) = \begin{bmatrix} X_1(k) \\ X_2(k) \end{bmatrix}, A_c = \begin{bmatrix} \overline{A_1} & \overline{A_2} \\ \overline{A_3} & \overline{A_4} \end{bmatrix}, B_c = \begin{bmatrix} \overline{B_1} \\ \overline{B_2} \end{bmatrix}, C_c = [\overline{C_1} \ \ \overline{C_2}]$ and

$$\overline{A_1}=\begin{bmatrix} A_{10} & A_{11} & \cdots & A_{1(n-1)} & A_{1n} \\ I & 0 & \cdots & 0 & 0 \\ \vdots & \vdots & \cdots & 0 & 0 \\ 0 & 0 & \ddots & \vdots & \vdots \\ 0 & 0 & \cdots & I & 0 \end{bmatrix} \qquad \overline{A_2}=\begin{bmatrix} \varepsilon\tilde{A}_{10} & \varepsilon\tilde{A}_{11} & \cdots & \varepsilon\tilde{A}_{1(n-1)} & \varepsilon\tilde{A}_{1n} \\ 0 & 0 & \cdots & 0 & 0 \\ 0 & 0 & \cdots & 0 & 0 \\ \vdots & \vdots & \ddots & \vdots & \vdots \\ 0 & 0 & \cdots & 0 & 0 \end{bmatrix},$$

$$\overline{A_3}=\begin{bmatrix} A_{20} & A_{21} & \cdots & A_{2(n-1)} & A_{2n} \\ 0 & 0 & \cdots & 0 & 0 \\ 0 & 0 & \cdots & 0 & 0 \\ \vdots & \vdots & \ddots & \vdots & \vdots \\ 0 & 0 & \cdots & 0 & 0 \end{bmatrix} \qquad \overline{A_4}=\begin{bmatrix} \varepsilon\tilde{A}_{20} & \varepsilon\tilde{A}_{21} & \cdots & \varepsilon\tilde{A}_{2(n-1)} & \varepsilon\tilde{A}_{2n} \\ I & 0 & \cdots & 0 & 0 \\ 0 & I & \cdots & 0 & 0 \\ \vdots & \vdots & \ddots & \vdots & \vdots \\ 0 & 0 & \cdots & I & 0 \end{bmatrix}$$

$$\overline{B_1}=\begin{bmatrix} B_1 \\ 0 \\ 0 \\ \vdots \\ 0 \end{bmatrix}, \quad \overline{B_2}=\begin{bmatrix} B_2 \\ 0 \\ 0 \\ \vdots \\ 0 \end{bmatrix} \qquad \overline{C_1}=\begin{bmatrix} C_1 & 0 & 0 & \cdots & 0 \end{bmatrix}, \overline{C_2}=\begin{bmatrix} C_2 & 0 & 0 & \cdots & 0 \end{bmatrix}.$$

Lemma 4.1[13,14]. If the gains (K_s , F_s) and (K_f , F_f) are designed such that the systems (3.3) and (3.6) are both stable, then the following composite observer-based controller (4.2) can stabilize the original system (4.1), provided that ε is sufficiently small.

$$\hat{X}(k+1)=(A_c-B_cK_c-F_cC_c)\hat{X}(k)+F_c\hat{u}(k) \tag{4.2}$$
$$\hat{y}(k)=K_c\hat{X}(k),$$

where

$$\hat{u}(k)=y(k), u(k)=-\hat{y}(k), \hat{X}(k)=\begin{bmatrix}\hat{X}_1(k)\\ \hat{X}_2(k)\end{bmatrix}, K_c(k)=\begin{bmatrix}K_{c1}(k)\\ K_{c2}(k)\end{bmatrix}, F_c(k)=\begin{bmatrix}F_{c1}(k)\\ F_{c2}(k)\end{bmatrix}$$

and
$$K_{c1}=\left[I+K_f(I-\overline{A_4})^{-1}\overline{B_2}\right]K_s-K_f(I-\overline{A_4})^{-1}\overline{A_3}, \quad K_{c2}=K_f \tag{4.3a}$$

and
$$F_{c1}=F_s\left[I+\overline{C_2}(I-\overline{A_4})^{-1}F_f\right]-\overline{A}2(I-\overline{A_4})^{-1}F_f, \quad F_{c2}=F_f \tag{4.3b}$$

Based on above discussions, the design of controller is then concluded as follows.

Design procedure

Step 1. Use genetic algorithm to adjust K_s such that $|\lambda(\overline{A_S}-\overline{B_S}K_S)|$ as small as possible. *Step 2*. Use genetic algorithm to adjust F_s such that $|\lambda(\overline{A_S}-F_S\overline{C_S})|$ as small as possible. *Step 3*. Use genetic algorithm to adjust K_f such that $|\lambda(\overline{A_f}-\overline{B_f}K_f)|$ as small as possible. *Step 4*. Use genetic algorithm to adjust F_s such that $|\lambda(\overline{A_S}-F_S\overline{C_S})|$ as small as possible. *Step 5*. Find $K_{C1}=\left[I+K_f(I-\overline{A_4})^{-1}\overline{B_2}\right]K_S-K_f(I-\overline{A_4})^{-1}\overline{A_3}$. and $K_{C2}=K_f$. *Step 6*. Find

$F_{C1} = F_S\lfloor I + \overline{C_2}(I - \overline{A_4})^{-1} F_f \rfloor - \overline{A_2}(I - \overline{A_4})^{-1} F_f$ and $F_{c2} = F_f^{\ T}$ *Step 7*. Form the observer-based controller (4.3) *Step 8*. Simulate the closed-loop system (4.1) and (4.3).

References

1. Li, T.H.S., Li, J.H.: Stabilization Bound of Discrete Two-Time-Scale Systems. Syst. Control Lett. 18, 479–489 (1992)
2. Feng, W.: Characterization and Computation for The Bound e* in Linear Time-invariant Singularly Perturbed Systems. Syst. Control Lett. 11, 195–202 (1988)
3. Chen, B.S., Lin, C.L.: On The Stability Bounds of Singularly Perturbed Systems. IEEE Trans. Autom. Control. 35, 1265–1270 (1990)
4. Sen, S., Datta, K.B.: Stability Bounds of Singularly Perturbed Systems. IEEE Trans. Autom. Control. 38, 302–304 (1993)
5. Pan, S.T., Hsiao, F.H., Teng, C.C.: Stability Bound of Multiple Time-delay Singularly Perturbed Systems. Electron. Lett. 32, 1327–1328 (1996)
6. Hsiao, F.H., Pan, S.T., Teng, C.C.: D-Stabilization Bound Analysis for Discrete Multi-parameter Singularly Perturbed Systems. IEEE Trans. Circuits Syst. (I) 44, 347–351 (1997)
7. Hsiao, F.H., Hwang, J.D., Pan, S.T.: Stabilization of discrete Singularly Perturbed Systems Under Composite Observer-Based Control. ASME Journal of Dynamic Systems, Measurement, and Control 123, 132–139 (2001)
8. Chou, J.H., Chen, B.S.: New Approach for The Stability Analysis of In-terval Matrices. Control-Theory and Advanced Technology 6, 725–730 (1990)
9. Mahmoud, M.S.: Order Reduction and Control of Discrete Systems. IEE Proceeding–Control Theory Appl. 129, 129–135 (1982)
10. Saksena, V.R., O'Reilly, J., Kokotovic, P.V.: Singular Perturbations and Time Scale Methods in Control Theory—Survey 1976–1983. Automatica 20, 273–293 (1984)
11. Oloomi, H., Sawan, M.E.: The Observer-Based Controller Design of Discrete-Time Singularly Perturbed Systems. IEEE Trans. Autom. Control. 32, 246–248 (1987)
12. Li, J.H., Li, T.H.S.: On the Composite and Reduced Observer–based Control of Discrete Two-Time-Scale Systems. J. Franklin Inst. 332b, 47–66 (1995)
13. Wang, M.S., Li, T.H.S., Sun, Y.Y.: Design of Near-Optimal Observer-Based Controllers for Singularly Perturbed Discrete Systems. JSME International Journal: Series C 39, 234–241 (1996)

AKDB–Tree: An Adjustable KDB-Tree for Efficiently Supporting Nearest Neighbor Queries in P2P Systems

Ye-In Chang, Lee-Wen Huang, and Hung-Ze Liu

Abstract. A P2P system is a system in which peers can directly communicate with other peers and share resources. In this paper, we propose an Adjustable KDB–tree (AKDB–tree) to answer the nearest neighbor queries for spatial data in the P2P environment. The AKDB–tree has five properties: reducing load unbalance, low cost of the tree construction, storing the data in the internal nodes and leaf nodes, high accuracy and low search cost of the NN query. Besides, in order to combine AKDB–tree with the Chord system, we design the IDs of the nodes in the AKDB–tree. From our simulation results, for the NN query, our AKDB-tree can provide the higher accuracy and lower search cost than the P2P MX–CIF quadtree.

1 Introduction

P2P networks have become a powerful means for online data exchange. Currently, users are primarily utilizing these networks to perform exact–match queries and retrieve complete files. However, future more data intensive applications, such as P2P auction networks, P2P job–search networks, and P2P multi–player games, will require the capability to respond to more complex queries such as the nearest neighbor queries involving numerous data types including those that have a spatial component. Although the P2P MX–CIF quadtree [5] can do the nearest neighbor query efficiently, the accuracy of the nearest neighbor query is not perfect. Besides the difficulties, there are another two problems of the nearest neighbor query in the P2P MX–CIF quadtree. One is that some control points contain no data, and the other one is that some control points contain a lot of data items.

In fact, the index structures for the region data can also work for the point data which can be considered as the degenerated case of the region data. Many applications in the P2P systems really work on the point data, for example, Web GIS

Ye-In Chang, Lee-Wen Huang, and Hung-Ze Liu
Dept. of Computer Science and Engineering, National Sun Yat-Sen University, Kaohsiung, Taiwan
e-mail: changyi@cse.nsysu.edu.tw

B.-C. Chien and T.-P. Hong (Eds.): Opportunities and Challenges, SCI 214, pp. 133–138.
springerlink.com

(geographic information systems) and the systems of combination of GIS and GPS. For the point data, the KDB–tree is one of well–know tree structures [3]. The KDB–tree can reduce the problem of load unbalance, because it can control the amount of the data which is stored in the node. But it has the same problem as the quadtree, the data is stored only in the leaf nodes of KDB–tree. Therefore, the KDB–tree is unsuited for the P2P systems.

In the P2P systems, in order to reduce the problem of load unbalance and keeps the property of storing the data in the leaf nodes and internal nodes at the same time, in this paper, we propose the Adjustable KDB–tree (AKDB–tree) to achieve these two goals. Besides, the IDs (bit patterns) of the nodes are used to hash to the locations in the Chord system. Therefore, we do not need to trace the surrounding edges or regions from the root in the 2D space for the property of ID. We can know where the surrounding edges or regions are in the Chord system immediately, and connect the location directly without going through the root, resulting in reducing the search cost of the nearest neighbor query by a wide margin. From our simulation results, for the nearest neighbor query, our approach can provide the higher accuracy and lower search cost than the P2P MX–CIF quadtree.

The rest of the paper is organized as follows. In Section 2, we give a survey of spatial index structure in the P2P system. In Section 3, we present the AKDB–tree. In Section 4, we study the performance of the proposed approach, and make a comparison with P2P system with MX–CIF quadtree by simulation. Finally, we give the conclusion.

2 A Survey of Spatial Index Structures

Spatial indexes are used by spatial databases to optimize spatial queries. Indexes used by non–spatial databases cannot effectively handle features such as how far two points differ and whether points fall within a spatial area of interest. The problem of retrieving multi–key records via range queries from a large, dynamic index is considered. The KDB–tree has been proposed as a solution to this problem [3]. In case of the P2P R–tree, the universe is first divided *statically* into a set of *blocks*. The static decomposition of space has an important advantage from the perspective of the P2P systems [1]. There are many variants of the quadtree data structure, with the region quadtree being the most common. We choose MX–CIF quadtrees to exhibit our P2P index and associated algorithms although other quadtree types could have also been utilized [5].

3 An AKDB–Tree Approach

In this Section, we first describe the data structure of our proposed strategy, *i.e.*, AKDB–tree. Then, we present the strategies of the nearest neighbor query, and the hashing functions between the AKDB–tree and the Chord system.

3.1 Data Structure

Basically, the tree structure of the AKDB–tree is the revised version of the KDB–tree. The AKDB–tree is a complete binary tree. The data in the AKDB–tree can be stored in the internal nodes or the leaf nodes, but the data in the KDB–tree can only be stored in the leaf nodes. Therefore, the height of a tree can be reduced by a wide margin. Each node records some information. The ID of the node is calculated based on the Hamming Code. For example, each node with ID = X of the AKDB–tree has two child nodes, the ID of its left child node is "$X0$", and its ID of the right child node is "$X1$". Then, the ID of the right child node of the node with ID "0" is "01". The nodes of the AKDB–tree have two types: *edge nodes* and *region nodes*. *Edge nodes* are internal nodes, and they represent an edge in the 2D space.

3.2 The Nearest Neighbor Query

In this section, we present the algorithms for answering nearest neighbor query. Based on the above process for tree construction, there are two important properties about the IDs of the nodes in the AKDB–tree as follows.

1. **Difference between the odd and the even length of IDs:** If the length of the ID of the edge node is an odd number, this edge is a horizontal edge in 2D space. On the contrary, if the length of ID of the edge node is an even number, this edge is a vertical edge in 2D space.
2. **Relationship between the parent and the children:** For the node α with an odd length ID, if the last bit of the ID of the child node is "1", the child node is in the upper side of node α in the 2D space. On the other hand, if it is "0", the child node is in the lower side of node α in the 2D space.

These two properties help us to reduce the cost of the query processing for the nearest neighbor query. When the nearest neighbor of a query point *Qpoint* is issued in the 2D space, in Step 1, we try to find which node *Qpoint* is located in the AKDB–tree (*i.e.*, finding that *Qpoint* is located in a region or on an edge in the 2D space). Then, *Qpoint* is found to be located at the node *QNode* (*i.e.*, a region or an edge in the 2D space) in the AKDB–tree. In Step 2, the temporary nearest neighbor in the data items of *QNode* is found. *QNode* and the temporary nearest neighbor *TempNN* are obtained. *Qpoint* may be located in any position in the 2D space. But only when *Qpoint* is located in the region, instead of on the edge in the 2D space, we will expand the tracing scope. Therefore, in Step 3, we check whether *QNode* is an edge node or a region node in the AKDB–tree.

3.3 Hashing Functions between the AKDB–Tree and the Chord System

In this section, we present how a AKDB–tree is combined with the Chord system [4]. Three hashing functions are used in these cases: *the hashing function of data insert into the AKDB–tree*, *the hashing function of peer insertion*, and *the hashing function of data insert into the Chord system*. The hashing function of data insert into the AKDB–tree (DH_{AKDB}) is used to calculate the coordinate of each data item in the AKDB–tree. The hashing function of peer insertion (PH) is used to calculate the position of each peer in the Chord system. The hashing function of data insert into the Chord system (DH_{Chord}) is used to calculate the position of each node of a AKDB–tree in the Chord system. Among these hashing functions, DH_{Chord} is the most important hashing function as follows.

1. $DH_{Chord}(L,V) = 0$, if $L = 0$,
2. $DH_{Chord}(L,V) = (2^L - 1 + V) \bmod n$, if $L > 0$.

L is the length of the ID of the node. But, if the node is a root of the AKDB–tree, L is 0. V is the value of the ID of the node in the decimal system. n is the Chord ring size. For example, If the ID of a node is "01" and n = 36, we have L = 2 and V = 1. Then, $DH_{Chord}(2,1) = (2^2 - 1 + 1) \bmod 36 = 4$. If the ID of a node is "001" and n = 36, we have L = 3 and V = 1. Then, $DH_{Chord}(3,1) = (2^3 - 1 + 1) \bmod 36 = 8$. Therefor, the values of the hashing function DH_{Chord} of "01" and "001" will be different even though their decimal value is the same. Furthermore, every node in the AKDB–tree can be located on different position in the Chord system by DH_{Chord}.

4 Performance

In this section, we compare the performance of nearest neighbor finding strategies based on the P2P AKDB–tree and the P2P MX–CIF quadtree. We consider the two–dimensional space, so that the selected points constitute a straight (horizontal or vertical) line through the grid. We focus on the search cost of the nearest neighbor query and the accuracy of the nearest neighbor query in the simulation model. Here, we define that the search cost in the P2P system is the number of visited peers [2].

Given that the data space is 1000*1000. There are 1000, 2000, 3000, 4000, 5000, 6000, 7000, 8000, and 9000 data items. For the region data, the data items with the average sizes 0.0025% and 0.0001% are random distributed (with overlap) on the whole space. The minimal level of the MX–CIF quadtree containing the data, f_{min}, was assigned to be 7 and 0 that are the best cases for the nearest neighbor query and the construction in the P2P MX–CIF quadtree. The maximal level of the MX–CIF quadtree containing the data, f_{max}, was assigned to be 9 that is the normal

Table 1 Four different situations of the P2P MX–CIF quadtree

Case	f_{min}	*avg_size*
F7A25	7	25
F7A1	7	1
F0A25	0	25
F0A1	0	1

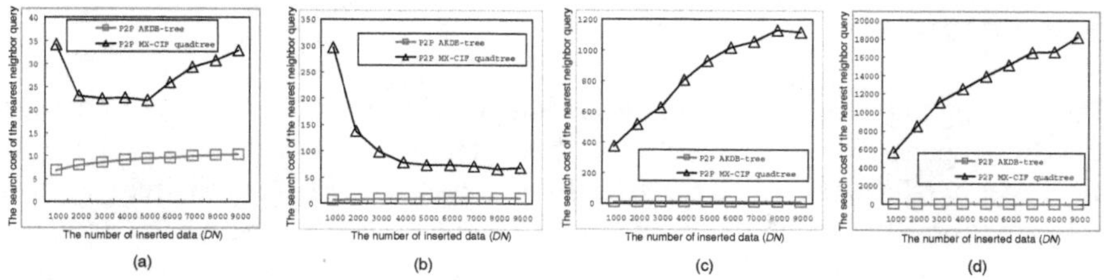

Fig. 1 A comparison of the search cost of the nearest neighbor query: (a) F7A25; (b) F7A1; (c) F0A25; (d) F0A1

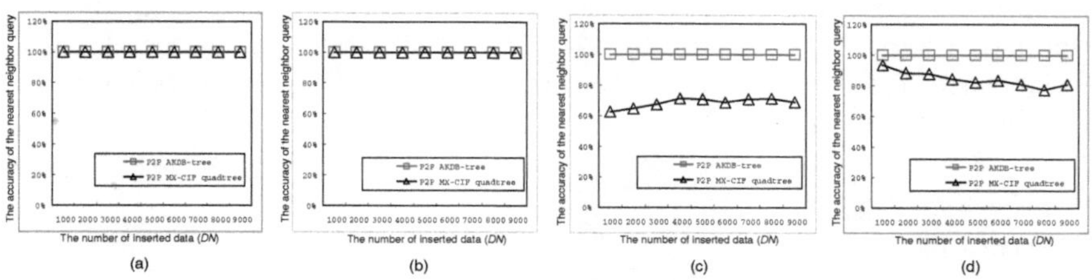

Fig. 2 A comparison of the accuracy of the nearest neighbor query: (a) F7A25; (b) F7A1; (c) F0A25; (d) F0A1

case of the nearest neighbor query and the tree construction in the P2P MX–CIF quadtree. We make four comparisons with four different situations of the MX–CIF quadtree. We let m_p of the AKDB–tree be 50, which is a normal case of the nearest neighbor query and the tree construction in the P2P AKDB–tree. Then, we consider four performance measures under four different situations as shown in Table 1, of the P2P MX–CIF quadtree.

- **The search cost of the nearest neighbor query in the P2P system:** Figure 1 shows the comparisons of the search cost of the nearest neighbor query between the P2P AKDB–tree (with $m_p = 50$) and the P2P MX–CIF quadtree (with f_{min} = 7 or 0) under the cases of avg_size = 0.0025% and 0.0001%, respectively. When the size of the spatial region data decreases or f_{min} decreases, the search cost of the nearest neighbor query in the P2P MX–CIF quadtree also increases. The search cost of our approach is always less than that in the P2P MX–CIF quadtree.
- **The accuracy of the nearest neighbor query in the P2P system:** Figure 2 shows the comparisons of the accuracy of the nearest neighbor query between the P2P AKDB–tree (with $m_p = 50$) and the P2P MX–CIF quadtree (with f_{min} = 7 or 0) under the cases of avg_size = 0.0025% and 0.0001%, respectively. The accuracy of our strategy is always 100%, but the accuracy of the P2P MX–CIF quadtree might be 60%. With f_{min} decreases, the accuracy of the nearest neighbor query in the P2P MX–CIF quadtree also decreases. Obviously, in the nearest neighbor query, the accuracy of P2P AKDB–tree is higher than that of the P2P MX–CIF quadtree.

5 Conclusion

The growing importance and ever–increasing popularity of peer–to–peer (P2P) systems have opened new and exciting possibilities for global sharing of spatial data. In this paper, we have presented the AKDB–tree. Based on the AKDB–tree, we have presented some strategies about the nearest neighbor query of the AKDB–tree, and the combination of the AKDB–tree and the Chord systems. From our simulation results, for the nearest neighbor query, our approach can provide the higher accuracy and lower search cost than the P2P MX–CIF quadtree.

Acknowledgements. This research was supported in part by the National Science Council of Republic of China under Grant No. NSC-97-2221-E-110-058-MY3. The authors also like to thank "Aim for Top University Plan" project of NSYSU and Ministry of Education, Taiwan, for partially supporting the research.

References

1. Mondal, A., Lifu, Y., Kitsuregawa, M.: P2PR-tree: An R-tree-based Spatial Index for Peer-to-Peer Environments. In: Lindner, W., Mesiti, M., Türker, C., Tzitzikas, Y., Vakali, A.I. (eds.) EDBT 2004. LNCS, vol. 3268, pp. 516–525. Springer, Heidelberg (2004)
2. Kwon, O., Moon, J.W., Li, K.J.: DisTIN – A Distributed Spatial Index for P2P Environment. In: Proc. of Data Engineering Workshop, pp. 11–17 (2006)
3. Robinson, J.T.: The KDB-tree: A Search Structure for Large Multidimensional Dynamic Indexes. In: Proc. ACM SIGMOD Int. Conf. on Management of Data, pp. 10–18 (1981)
4. Stoica, I., Morris, R., Karger, D., Kaashoek, M.F., Dabek, F., Balakrishnan, H.: Chord: A Scalable Peer-to-Peer Lookup Service for Internet Applications. In: Proc. of the 9th Int. Conf. on Applications, Technologies, Architectures, and Protocols for Computer Communications, pp. 149–160 (2001)
5. Tanin, E., Harwood, A., Samet, H.: Using A Distributed Quadtree Index in Peer-to-Peer Networks. The Int. Journal on Very Large Data Bases 16(2), 165–178 (2007)

Optimal Lot Sizing and Algorithm for Dynamic Pricing under Random Yield and Demand

Guo Li, Shihua Ma, and Wenya Chou

Abstract. The literature under random component yield has focused on coordination of supply chain at the determined price, where decision maker chooses its optimal production quantities. We consider a centralized system when the price is not determined under both random yield and demand. Type A with perfect quality and type B with imperfect quality are produced due to the random yield. We prove the unique concavity of expected profit in centralized system at determined price. Then dynamic pricing is considered and algorithm is put forward for dynamic pricing. Errors can be sufficiently small as long as some parameters can be set suitably. Apart from lot sizing and dynamic pricing, we also provide qualitative insights based on numerical illustration of centralized and decentralized solutions.

1 Introduction

The intense competition in semiconductor and electronics industry pose great challenge for manufacturers to reduce cost. Many manufacturers try to reduce sales' representatives and adopt the direct marketing. So the manufacturers have to control the order quantity and pricing dynamically to get maximum profit and incur minimum cost.

On the other side, supply process of the component is also random since the suppliers have uncertain production processes leading to yield losses [1-2]. For example, in the LCD manufacturing industry, it is quite common to get production yield of less than 50%. So in these industries, the manufacturers have to face the random yields besides random demand.

Yano and Lee (1995) give through review about single item single stage, multi items multi stages in the assembly system with lot sizing [3]. Gurnani (2000) study a centralized assembly system facing random demand and random yield[4].

Guo Li and Shihua Ma
School of Management, Huazhong University of Science and Technology, Wuhan, China
e-mail: lg4229682@163.com, Stenvenmai@126.com

Wenya Chou
Changsha Telecommunications and Technology Vocational College, Changsha, China
e-mail: chouwenya@vip.sina.com

B.-C. Chien and T.-P. Hong (Eds.): Opportunities and Challenges, SCI 214, pp. 139–145.
springerlink.com

Gerchak and Wang (2004) studied coordination in decentralized assembly systems having random demand. But they do not consider dynamic pricing and random yield[5]. Gurnani and Gerchak (2007) study coordination in decentralized assembly systems with two suppliers and one manufacturer under uncertain component yield and determined demand[6]. Güray Güler (2008) considered a decentralized assembly system with multi suppliers and one manufacturer under uncertain yield and demand[7]. As to dynamic pricing under random yield, Li (2006) studied the joint inventory replenishment and pricing problem for production systems with random demand and yield [8]. Ismail (2006) considered the effects of recovery yield rate on pricing decisions in reverse supply chains and determined the optimal acquisition price for the end-of-life products [9]. Tomlin (2008) studied the production, pricing, down conversion and allocation decisions in a two-class, stochastic-yield co production system. [10].

To the best of our knowledge, most literature under random component yield has focused on coordination of supply chain at determined price[11,12,13]. Some have studied establishing properties of the profit function of the chain and finding the optimal order quantity. Few have concentrated on dynamic pricing under random yield[8,10] but they studied different aspects from ours. Since lot sizing with uncertain yields is an important area of production systems[3], we will consider optimal lot sizing and dynamic pricing at type A with perfect quality and type B with imperfect quality under random yield and demand.

The remainder of the paper is organized as follows. In section 2, we give the description and then construct the model of expected profit. In section 3, we analyze the basic model. In section 4, we consider dynamic pricing. In section 5, numerical analysis is given. Finally in section 6, we summarize the results and discuss extensions.

2 Basic Centralized Model

Consider a centralized system with a single supplier and a single retailer who sells one item in one period. The retailer places an order of size Q from its supplier. Due to random yield of the supplier, the retailer receives an amount of αQ of perfect quality (Type A) and $(1-\alpha)Q$ of imperfect quality (Type B), where α is a random yield rate with support [0, 1]. These two types of item are price sensitive and the selling amount $y_i(p_i)$ is the function of selling price $p_i (i = A, B)$.Assuming the demand is random during the selling period, and the demand for item i is $y_i(p_i)\cdot\varepsilon_i$, where ε_i is the random variable.

The parameters are defined as follows:

x_i :the yield amount of type i; c : unit production cost; p_i: unit selling price of type i;

h_i : unit holding cost of type i; s_i :unit salvage cost of type i; π_i :unit penalty cost of type i;

$f(.)$, $F(.)$: the probability density function and cumulative density function of ε_i;

$g(.), G(.)$: the probability density function and cumulative density function of α;

Then the yield amount of type i is defined as

$$x_A = \alpha Q \; (1)\; ; x_B = (1-\alpha)Q \tag{1}$$

And the basic model of the centralized system's profit is given by

$$\Pi_c(Q) = \sum_{i=A}^{B} [p_i \cdot \min(x_i, y_i\varepsilon_i) - (h_i - s_i)\cdot(\alpha Q - y_i\varepsilon_i)^+ - \pi_i \cdot (y_i\varepsilon_i - x_i)^+] - cQ \tag{2}$$

The expected profit is different between expected sales revenue, inventory holding and shortage cost. Inventory holding and salvage cost of type A item will occur when the yield of type A exceeds the random demand, i.e., $\alpha Q > y_A \cdot \varepsilon_A$, equivalently $\alpha > y_A\varepsilon_A / Q$. Otherwise, if $\alpha < y_A \varepsilon / Q$, then the shortage cost will occur. Similarly, inventory holding and salvage cost of type B item occur if $(1-\alpha)Q > y_B \cdot \varepsilon_B$, equivalently $\alpha < 1 - y_B\varepsilon_B / Q$, and the shortage cost is incurred, otherwise. Then the expected profit can be given as follows:

$$\begin{aligned} E[\Pi_c(Q)] = & \int_0^\infty \int_{y_A\varepsilon_A/Q}^1 [p_A y_A \varepsilon_A - (h_A - s_A)(\alpha Q - y_A\varepsilon_A)] f_A(\varepsilon_A) g(\alpha) d\alpha d\varepsilon_A \\ & + \int_0^\infty \int_0^{y_A\varepsilon_A/Q} [p_A \alpha Q - \pi_A(y_A\varepsilon_A - \alpha Q)] f_A(\varepsilon_A) g(\alpha) d\alpha d\varepsilon_A \\ & + \int_0^\infty \int_0^{1-y_B\varepsilon_B/Q} \{p_B y_B \varepsilon_B - (h_B - s_B)[(1-\alpha)Q - y_B\varepsilon_B)] f_B(\varepsilon_B) g(\alpha) d\alpha d\varepsilon_B \\ & + \int_0^\infty \int_{1-y_B\varepsilon_B/Q}^1 \{p_B(1-\alpha)Q - \pi_B[y_B\varepsilon_B - (1-\alpha)Q]\} f_B(\varepsilon_B) g(\alpha) d\alpha d\varepsilon_B - cQ \end{aligned} \tag{3}$$

While the first term is the condition that random yield of type A exceeds the random demand for Type A; the second term shows that random yield of type A is less than random demand for Type A. And the third term and forth term are similar to the first term and second term respectively for type B. The fifth term is the supplier's production cost.

3 Analysis of This Basic Centralized Model

The expected profit of centralized model is a function of Q when the price of type A and B is determined. Then we have to get the optimal Q^*, namely $Q^* = \arg\max E[\Pi_c(Q)]$.

Theorem 1. The expected profit $E[\Pi_c(Q)]$ is strictly concave in Q. And the optimal order quantity is the unique solution to the following equation:

$$\int_0^\infty \int_{y_A\varepsilon_A/Q^*}^1 \alpha g(\alpha) f_A(\varepsilon_A) d\alpha d\varepsilon_A + \frac{p_B + \pi_B + h_B - s_B}{p_A + \pi_A + h_A - s_A} \int_0^\infty \int_0^{1-y_B\varepsilon_B/Q^*} (1-\alpha) g(\alpha) f_B(\varepsilon_B) d\alpha d\varepsilon_B$$
$$= \frac{(p_A + \pi_A)\cdot \mu_\alpha + (p_B + \pi_B)\cdot(1-\mu_\alpha) - c}{p_A + \pi_A + h_A - s_A} \tag{4}$$

Proof. Differentiating $E[\Pi_c(Q)]$ with respect to Q

$$\frac{\partial E[\Pi_c(Q)]}{\partial Q}=\int_0^\infty \frac{\partial}{\partial Q}\{\int_{y_A\varepsilon_A/Q}^1 [p_A y_A\varepsilon_A-(h_A-s_A)(\alpha Q-y_A\varepsilon_A)]g(\alpha)d\alpha\}f_A(\varepsilon_A)d\varepsilon_A$$
$$+\int_0^\infty \frac{\partial}{\partial Q}\{\int_0^{y_A\varepsilon_A/Q}[p_A\alpha Q-\pi_A(y_A\varepsilon_A-\alpha Q)]g(\alpha)d\alpha\}f_A(\varepsilon_A)d\varepsilon_A$$
$$+\int_0^\infty \frac{\partial}{\partial Q}\{\int_0^{1-y_B\varepsilon_B/Q}\{p_B y_B\varepsilon_B-(h_B-s_B)[(1-\alpha)Q-y_B\varepsilon_B)]f_B(\varepsilon_B)d\varepsilon_B\}g(\alpha)d\alpha$$
$$+\int_0^\infty \frac{\partial}{\partial Q}\{\int_{1-y_B\varepsilon_B/Q}^1\{p_B(1-\alpha)Q-\pi_B[y_B\varepsilon_B-(1-\alpha)Q]\}g(\alpha)d\alpha\}f_B(\varepsilon_B)d\varepsilon_B-c$$

Therefore,

$$\frac{\partial E[\Pi_c(Q)]}{\partial Q}=-(h_A-s_A)\int_0^\infty\int_{y_A\varepsilon_A/Q}^1 \alpha g(\alpha)f_A(\varepsilon_A)d\alpha d\varepsilon_A$$
$$+(p_A+\pi_A)[\int_0^\infty\int_0^1 \alpha g(\alpha)f_A(\varepsilon_A)d\alpha d\varepsilon_A-\int_0^\infty\int_{y_A\varepsilon_A/Q}^1 \alpha g(\alpha)f_A(\varepsilon_A)d\alpha d\varepsilon_A]$$
$$-(h_B-s_B)\int_0^\infty\int_0^{1-y_B\varepsilon_B/Q}(1-\alpha)g(\alpha)f_B(\varepsilon_B)d\alpha d\varepsilon_B$$
$$+(p_B+\pi_B)[\int_0^\infty\int_0^1(1-\alpha)g(\alpha)f_B(\varepsilon_B)d\alpha d\varepsilon_B-\int_0^\infty\int_0^{1-y_B\varepsilon_B/Q}(1-\alpha)g(\alpha)f_B(\varepsilon_B)d\alpha d\varepsilon_B]-c$$
$$=(p_A+\pi_A)\cdot\mu_\alpha-(p_A+\pi_A+h_A-s_A)\int_0^\infty\int_{y_A\varepsilon_A/Q}^1 \alpha g(\alpha)f_A(\varepsilon_A)d\alpha d\varepsilon_A$$
$$+(p_B+\pi_B)\cdot(1-\mu_\alpha)-(p_B+\pi_B+h_B-s_B)\int_0^\infty\int_0^{1-y_B\varepsilon_B/Q}(1-\alpha)g(\alpha)f_B(\varepsilon_B)d\alpha d\varepsilon_B-c$$

Let $\frac{\partial E[\Pi_c(Q)]}{\partial Q}=0$, then equation (5) is got. To prove concavity of $E[\Pi_c(Q)]$,then

$$\frac{\partial^2 E[\Pi_c(Q)]}{\partial Q^2}=-(p_A+\pi_A+h_A-s_A)\int_0^\infty \frac{(y_A\varepsilon_A)^2}{Q^3}g(\tfrac{y_A\varepsilon_A}{Q})f_A(\varepsilon_A)d\varepsilon_A$$
$$-(p_B+\pi_B+h_B-s_B)\int_0^\infty \frac{(y_B\varepsilon_B)^2}{Q^3}g(1-\tfrac{y_B\varepsilon_B}{Q})f_B(\varepsilon_B)d\varepsilon_B<0$$

The expected profit $E[\Pi_c(Q)]$ is maximized when the equation (4) is satisfied. As we can see the left tern of equation (5) is increasing in Q. The best Q^* is got when the left term equals the right term.

4 Algorithm for Dynamic Pricing

Since the expected profit $E[\Pi_c(Q)]$ is strictly concave in Q under determined price of Type A and B, we can find the best Q^* to maximize the expected profit $E[\Pi_c(Q)]$. Under different price of Type A and B, different maximized expected profit $E[\Pi_c(Q)]$ is got. Then algorithm for dynamic Pricing is put forward to get the best price of Type A and B.

The step of this algorithm is as follows (see figure 1):

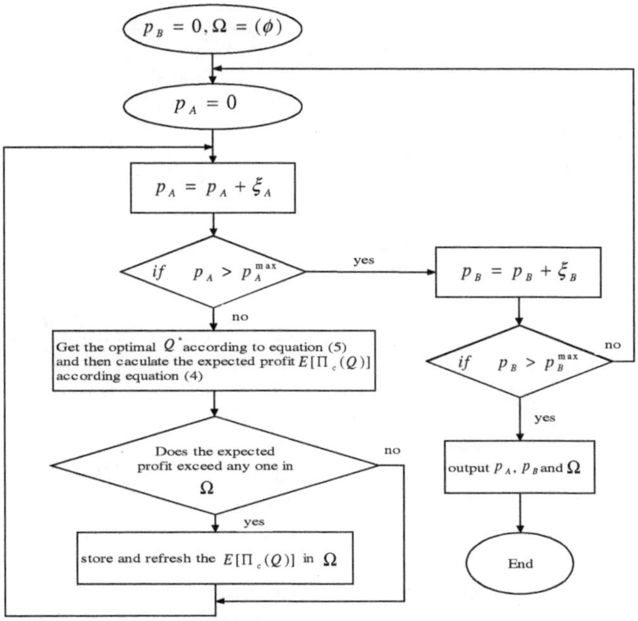

Fig. 1 Algorithm process for dynamic pricing

Step 1. let $p_B=0,\Omega=(\phi)$; Step 2 let $p_A=0$:Step 3 set $p_A=p_A+\xi_A(\xi_A$ is sufficiently small);

Step 4 if $p_A > p_A^{max}$,then let $p_B = p_B+\xi_B(\xi_B$ is sufficiently small) and go to step 2; if not , get the optimal Q^* according the equation (5), then calculate the expected profit $E[\Pi_c(Q)]$ according to equation (4).

Step 5 if the new the expected profit $E[\Pi_c(Q)]$ exceeds the former $E[\Pi_c(Q)]$, store and refresh the $E[\Pi_c(Q)]$ in Ω, then go to step 3; if not, return step 3 direct.

Step 6 output the best $E[\Pi_c(Q)]$, and the corresponding p_A and p_B.

As we can see, that the optimal p_A^* should be larger than p_B^* and the precision of p_A^* and p_B^* can be ensured as long as the ξ_A and ξ_B are sufficiently small.

5 Numerical Analysis

In this section, we assume that random variable ε_A and ε_B of demand obey the normal distribution with $\mu_A=1$, $\sigma_A=0.25$ and $\mu_B=1$, $\sigma_B=0.15$, respectively. The random variable α of supply has a uniform distribution of yield taking values in $(0,1]$. Then the demand function can be assumed as $y_i(p_i)=a_i \cdot p_i^{-b_i}$, where $b_i>1$. That means the demand for type A and Type B are both elastic. According to

Table 1 Centralized solution for different cases under determined price

case	p_A	p_B	h_A	h_B	s_A	s_B	π_A	π_B	C	Q^*	$E[\Pi_c(Q^*)]$
1	1.5	1	0.5	0.3	1	0.5	1	0.8	1	375.28	-295.3
2	2	1.5	0.5	0.3	1	0.5	1	0.8	1	430.69	-63.59
3	2.5	2	0.5	0.3	1	0.5	1	0.8	1	484.16	216.84
4	3	2.5	0.5	0.3	1	0.5	1	0.8	1	498.57	489.71
5	3.5	3	0.5	0.3	1	0.5	1	0.8	1	518.62	523.85
6	4	3.5	0.5	0.3	1	0.5	1	0.8	1	623.47	637.15
7	4.5	4	0.5	0.3	1	0.5	1	0.8	1	684.36	613.93
8	5	4.5	0.5	0.3	1	0.5	1	0.8	1	719.92	572.46
9	5.5	5	0.5	0.3	1	0.5	1	0.8	1	733.55	412.17
10	6	5.5	0.5	0.3	1	0.5	1	0.8	1	758.62	230.96
11	6.5	6	0.5	0.3	1	0.5	1	0.8	1	781.43	76.49
12	7	6.5	0.5	0.3	1	0.5	1	0.8	1	809.44	-193.61

equation (4) and (5), the optimal quantities and expected centralized supply chain profit are depicted in table 1 under determined price.

Table 1 shows that under determined price, there exist the optimal order quantities and expected profit of supply chain. One might wonder what the optimal price and expected profit of supply chain are when the price is not determined. So the algorithm of section 4 can be programmed in Matlab 7. As in this specific case, the optimal expected profit can be calculated, that is 677.02, and the precise of expected profit can be estimated. Then the optimal price for type A and B can be calculated under different ξ_A and ξ_B.As ξ_A and ξ_B become smaller in our cases , the errors also become smaller from the whole view. So the precise can be ensured as long as the ξ_A and ξ_B are sufficiently small. Then the profit changed with the price also is considered in the figure 2, while $E[\Pi_c]$ stands for the expected profit of the centralized supply chain, $E[\Pi_A]$ and $E[\Pi_B]$ are the expected profit of type A and type B, respectively.

Figure 3 shows that the expected centralized profit $E[\Pi_c]$ are changing with price of Type A and B. And our algorithm can achieve the optimal price of Type A and B, and the $E[\Pi_c^*(Q^{**})]$ by ensuring the ξ_A and ξ_B sufficiently small.

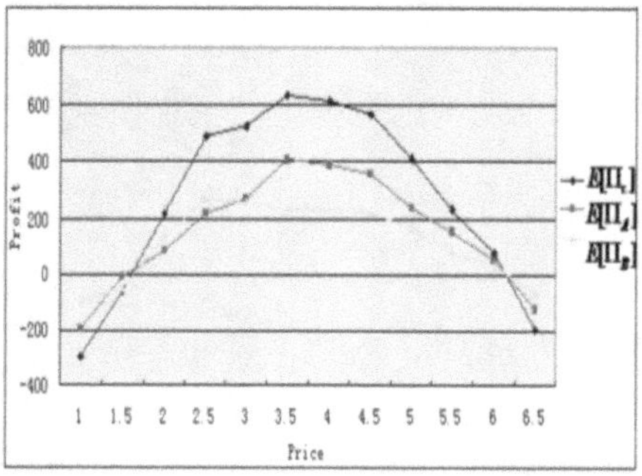

Fig. 2 Profit changing with the price

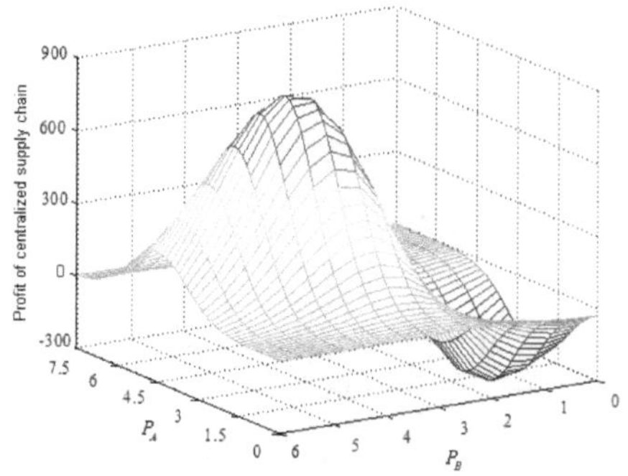

Fig. 3 Centralized profit changing with the price

6 Conclusion

In this paper, we considered optimal lot sizing and dynamic pricing under random yield and demand. Some observations are as follows: (1)The expected profit $E[\Pi_c(Q)]$ is strictly concave in Q when the prices of these two types are determined. So there exists the unique optimal expected profit $E[\Pi_c(Q)]$ when the prices of these two types are given. (2)Our algorithm for dynamic pricing is effective and the error can be decreased when ξ_A and ξ_B go smaller. So this algorithm for dynamic pricing can help centralized system to achieve the optimal expected profit $E[\Pi_c(Q)]$ as long as ξ_A and ξ_B are sufficiently small. From numerical examples, the expected profit curve of Type A ($E[\Pi_A]$) and Type B ($E[\Pi_B]$) can be observed with its corresponding price. But $E[\Pi_A]$ and $E[\Pi_B]$ don't achieve the maximized value with the same price. Our algorithm can help getting the optimal expected profit $E[\Pi_c(Q)]$ while the errors can be controlled sufficiently small.

References

1. Maddah, B., Salameh, M.K.G., Karame, M.: Lot sizing with random yield and different qualities.Applied Mathematical Modeling (article in press) (2008)
2. Gurnani, H.: Optimal lot-sizing policy with incentives for yield improvement. IEEE Transactions on Semiconductor Manufacturing 18(2), 304–308 (2005)
3. Yano, C.A., Lee, H.L.: Lot sizing with random yields: a review. Operation Research 43(2), 311–344 (1995)
4. Gurnani, H., Akella, R., Lehoczky, J.: Supply management in assembly systems with random yield and random demand. IIE Transactions 32, 701–714 (2000)
5. Gerchak, Y., Wang, Y.: Revenue sharing and wholesale-price contracts in assembly systems with random demand. Production and Operations Management 13, 23–33 (2004)
6. Gurnani, H., Gerchak, Y.: Coordination in decentralized assembly systems with uncertain component yields. European Journal of Operational Research 176, 1559–1576 (2007)
7. Güler, M.G., Bilgic, T.: On coordinating an assembly system under random yield and random demand. European Journal of Operational Research (article in press) (2008)
8. Li, Q., Zheng, S.: Joint inventory replenishment and pricing control for systems with uncertain yield and demand. Operations Research 54(4), 696–705 (2006)
9. Ismail, S.B., Elif, A.: Effects of random yield in remanufacturing with price-sensitive supply and demand. Production and operations management 15(3), 407–420 (2006)
10. Tomlin, B., Wang, Y.M.: Pricing and operational resource in co production systems. Management Science 54(3), 522–537 (2008)
11. Singh, M.R., Abraham, C.T., Akella, R.: A wafer design problem in semiconductor manufacturing for reliable customer service. IEEE Transactions on Components, Hybrids and Manufacturing Technology 13, 103–108 (1990)
12. Gerchak, Y., Wang, Y., Yano, C.A.: Lot sizing in assembly systems with random component yields. IIE Transactions 26(2), 19–24 (1994)
13. Gurnani, H., Akella, R., Lehoczky, J.: Optimal order policies in assembly systems with random demand and random supplier delivery. IIE Transactions 28, 865–878 (1996)

Collaborative Production Machine Choice System Based on Hierarchical Case Based Reasoning

Shian-Shyong Tseng, Fengming M. Chang, Huan-Yu Lin, Chi-Chun Hsu, Chung-Chao Ku, Jui-Pin Tsai, and Jun-Ming Su

Abstract. New cell phone styles and technologies developments are changing quickly. To improve the ability of competition for a company, a new designed cell phone has to be available in the market in the short time. Therefore, production processes including manufacturing machine choice for a new cell phone product have to be determined quickly. Facing the lack of information, it is difficult to choose manufacturing machines for a production process for a new product. Thus, this study proposed a Collaborative Production Machine Choice System (CPMCS) based on a hierarchical case based reasoning approach to solve the problem of machine choice for new cell phone production. A real interaction system is built up also to help users for machine choice in this study. Using non-fixed number of new product features, this system can advise suitable machines for users by case based reasoning approach and help users to find out the right machines for production in the short time.

Keywords: Case based reasoning, customization, cell phone, production process, machine.

1 Introduction

In modem globalization environment, to deal with uncertainty of multiple items that have short production life cycle (PLC) or have seasonal demands is an

Shian-Shyong Tseng, Huan-Yu Lin, Chi-Chun Hsu, and Jun-Ming Su
Department of Computer Science, National Chiao Tung University, Taiwan, ROC
e-mail: sstseng@cs.nctu.edu.tw, huan.cis89@nctu.edu.tw, justmaker@msn.com, jmsu@csie.nctu.edu.tw

Fengming M. Chang
Department of Information Science and Applications, Asia University, Taiwan, ROC
e-mail: paperss@gmail.com

Chung-Chao Ku and Jui-Pin Tsai
Mechanical and Systems Research Laboratories,
Industrial Technology Research Institute, Taiwan, ROC
e-mail: ccku@itri.org.tw, JuiPinTsai@itri.org.tw

B.-C. Chien and T.-P. Hong (Eds.): Opportunities and Challenges, SCI 214, pp. 147–152.
springerlink.com

important topic [1] and customization becomes a challenge for enterprises [2]. To satisfy customers, new product processes and production machines have to be determined quickly under unknown conditions. Machines selection shall be reached in the short time to satisfy the short PLC. Although production machines are unknown for new products, they can be adapted and re-designed from the similar products before. Therefore, a case based reasoning approach is useful for the machines choice. Using previous experiments for similar products and machine characteristics from equipment catalogs, a case based reasoning approach can help to seek suitable machines for a new product soon. To choose machines for a new product is difficult because of the paucity of experiments. The most difficulty is that a new product has to be available in the market in the short time. To increase the ability of the competition for a cell phone company, shorting developing time for a new product is very important. Although the knowledge of a new cell phone is lack, a new product is often similar to some old products. Hence, a hierarchical case based reasoning approach that retrieves similar knowledge can be applied. From the rough to the detailed features of a new product, a hierarchical case based reasoning approach can seek suitable machines and suitable machine attribute value setup soon. Therefore, using cell phone production as an example, this study proposes a Collaborative Production Machine Choice System (CPMCS) based on hierarchical case based approach as an assistant to help production machine chosen.

2 Case Based Reasoning

Case based reasoning (CBR) is a method to obtain preferential information from selected cases [3]. Many CBR researches have been proposed. For example, Nikolaychuk et al. applied CBR to computer-aided identification technical state for mechanical system [4]. Still, Chang et al. proposed a fuzzy CBR model to predict the demands of circuit board industries in Taiwan [5]. Recently, Lin et al. [6] proposed a Collaborative Interpretative Service Assisted Design System (CISAD) to help design museum interpretation services. In their study, a new service requirement can be created by partially reusing the previous applications to compose of a new application. The results indicate that CISAD is successful and the same concept can probably apply to the production process design in this study.

3 The Proposed Method

3.1 The Hierarchical Case Based Reasoning Approach

With the defined hierarchical case structure, the coarse-grained and fine-grained case can be retained in the case base. Besides, the designers can input their desired case, represented as a query feature list, and the proposed Intelligent Query Processor can generate the feature list of quires to represent the required

Fig. 1 Hierarchical case-based reasoning system architecture

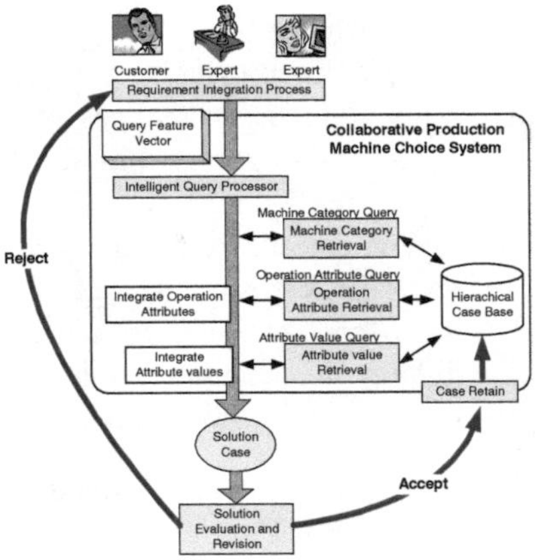

machines. Therefore, the CPMCS is proposed to retrieve and adapt the case hierarchically to fulfill the desired features and constraints. As shown in Fig. 1, when customers and machine experts input the required use case as query feature values into the CPMCS, the Intelligent Query Processor (IQP) firstly generates the query for machine category cases. Next, the IQP further generates the operation attribute level and attribute value level. In the case retrieval of different level, the objective functions and predicate function are used to evaluate the satisfaction of the solution cases, and these retrieved solution cases can be integrated into the original application case to generate a new application. Thus, the adapted solution case can be returned to designer, and after the designer check and revise the solution case, it can be retained to the case base.

3.2 *Hierarchical Case Machine Choice*

In our CPMCS, first, customers describe the new product characteristics, including the outlook, functions, materials, and so on. These characteristics are named as features of the product. Next, manufacturing experts provide additional features for this new product, most of them are about production needed. Then, the requirement Integration Process integrates these features provided by customers and experts to be a Query Feature Vector (QFV) as Fig. 1 presents. This QFV is sent to the proposed CPMCS via IQP.

The suitable machines can be sought by three levels of hierarchical structure. The first level is machine category level. In this level, according to the features from customers and experts, CPMCS seeks the suitable production processes and necessary machine categories for the new product, and the possible machines for each category are found also. Next in the operation attribute level, for each category, a right machine which has the most suitable operation attributes or features for the product is chosen from all the possible machines. Each attribute, spindle override or cutter radius compensation and so on for example, for this right machine has its operation domain or range. Hence, in the attribute value level of

CPMCS, each attribute value is adapted to the optimal one for the new product. For example, three machine categories, A, B, and C, and their possible machines are sought. In them, the possible machines for machine category A are A1, A2, ..., An. Category B has suitable machines B1, B2, ..., Bn. And machine category C has machines C1, C2, ..., C3. In the operation attribute level, if machine A7 is chosen from machine category A, and attributes needed for the new product are listed as a1, a2, ..., an. Also, B2 are chosen from machine category B, and C9 are chosen from machine category C with relevant attributes in this case. Finally in attribute value level, optimal value for each attribute is determined from the range of attribute values.

Example 1: The machine category level for house assembly for N73 cell phone
In the top of FIG 2, 5 features for N73 cell phone house assembly is listed. Using these 5 features, three machine categories are sought for production process. The possible machines can be applied for each category are listed in the machine category level box. As well as the features that are used to determine each category are listed also. In this example, blow molding machine, injection machine, and machinery center are advised by CPMCS for house assembly of N73. Three possible blow molding machines from the case database are sought, as well as four injection machines and three machinery centers. In Fig. 2, the possible blow molding machines are listed as VP2012, SP2016, and HVM4025. The advised injection machines are Ve-110, Ve120, VR-250, and VR-350. Besides, the proposed possible machinery centers are VC-H630, FANUC 0, and B130s.

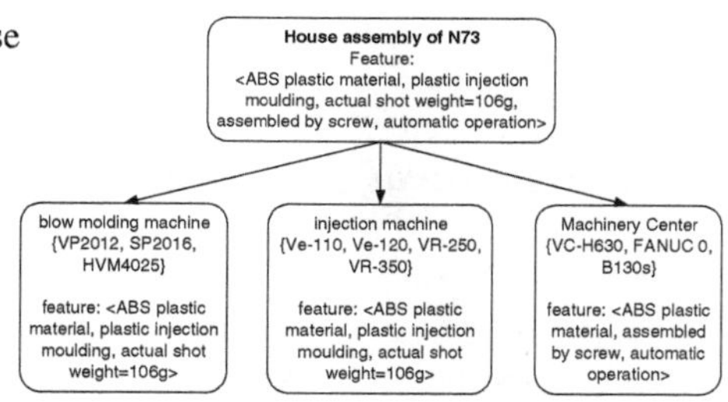

Fig. 2 The machine category level for house assembly for N73 cell phone

Example 2: The operation attribute level for Ve-110 injection machine in the house assembly of N73
Following Example 1 about the part of machine category, injection machine, a Ve-110 injection machine is chosen by the features and by the operation attribute range of itself as shown in the operation attribute level in Fig. 3. In the figure, five attributes are considered, they are actual shot weight, injection pressure Max., screw diameter, min/max mould thickness, and machine dimension. In them, four attributes have a range of attribute values, and one attribute, machine dimension, has fixed value that can not be adapted. That is to say, these five attributes and their attribute value ranges are suitable for the N73 plastic part production.

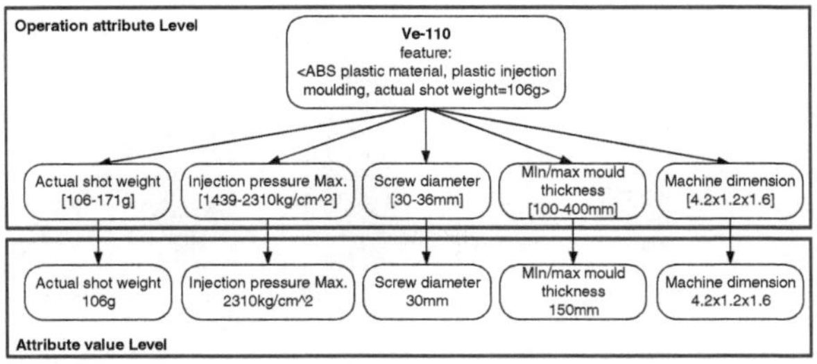

Fig. 3 The operation attribute level and attribute value for Ve-110 injection machine

Fig. 4 The N73 case

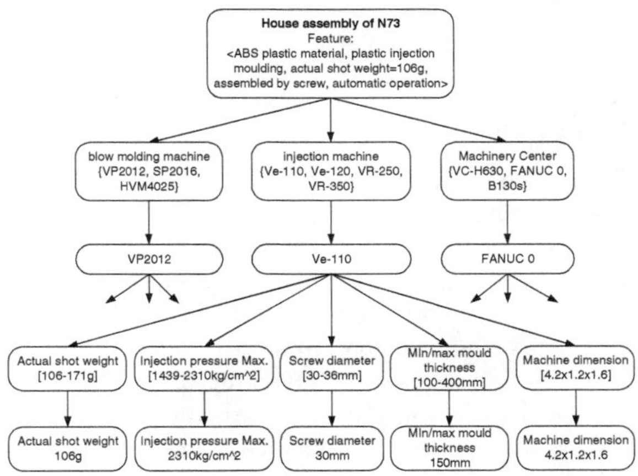

Example 3: The attribute value level for Ve-110 injection machine

After Ve-110 injection machine is chosen, its attribute values have to be adapted to optimal values to fit the production for N73 cell phone. These values are determined by QFV and are presented on the lower row in the Fig. 4 as an attribute value level. As actual shot weight is determined to be 106 grams from the range of 106 to 171 grams, injection pressure Max. is setup as 2310kg/cm^2 from the value of 1439 to 2310 kg/cm^2, screw diameter is 30 mm from the size of 30 to 36 mm, and mould thickness is designed as 150 mm. Finally, machine dimension is fixed as 4.2*1.2*1.6 that is considered by the factory environment.

3.3 A Real System for CPMCS

In this study, a real system is built up for CPMCS performance. There are two main functions in this system: one is to offer a function to key in case base data as Fig. 5(a) illustrates, and the other is to search for suitable machines by CPMCS as shown in Fig. 5(b).

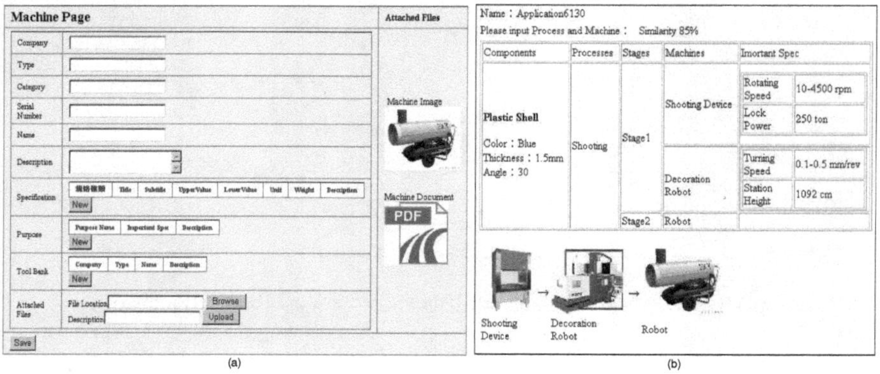

Fig. 5 A real system for CPMCS

4 Conclusions

The way to counsel enterprises to find out and choose the right machines for new product manufacture in the short time is very important. This study, using cell phone product as an example, develops a CPMCS hierarchical case base solution to reach the goal of seeking suitable production machines in the short time for a new product. From rough to detailed, CPMCS uses features by customers and experts to find out appropriate machine categories including possible machine solutions first to save the searching time. Then it further seeks the most suitable machine by machine attributes and features. The detailed attributes are adapted for new product. Finally, an interaction system is developed in this study. The results indicate that the proposed method indeed help for the production machines choice.

Acknowledgments. Thanks are due to the support in part by the Industrial Technology Research Institute of Taiwan under Grant No. ITRI 7353C23100B1.

References

1. Chung, C.-S., Flynn, J., Kirca, Ö.: A multi-item newsvendor problem with preseason production and capacitated reactive production. European Journal of Operational Research 188(3), 775–792 (2008)
2. Ko, E., Kim, S.H., Kim, M., Woo, J.Y.: Organizational characteristics and the CRM adoption process. Journal of Business Research 61, 65–74 (2008)
3. Chen, Y., Marc Kilgour, D., Hipel, K.W.: A case-based distance method for screening in multiple-criteria decision aid. Omega 36(3), 373–383 (2008)
4. Nikolaychuk, O.A., Yurin, A.Y.: Computer-aided identification of mechanical system's technical state with the aid of case-based reasoning. Expert Systems With Applications 34(1), 635–642 (2008)
5. Chang, P.-C., Liu, C.-H., Lai, R.K.: A fuzzy case-based reasoning model for sales forecasting in print circuit board industries. Expert Systems With Applications 34(3), 2049–2058 (2008)
6. Lin, H.-Y., Tseng, S.-S., Weng, J.-F., Su, J.-M.: Collaborative Interpretative Service Assisted Design System Based on Hierarchical Case Based Approach. In: Proc. of 2008 IEEE International Conference on Sensor Networks, Ubiquitous, and Trustworthy Computing (2008)

Passive Analog Filter Design Using GP Population Control Strategies

Mariano Chouza, Claudio Rancan, Osvaldo Clua, and Ramón García-Martínez

Abstract. This paper presents the use of two different strategies for genetic programming (GP) population growth control: decreasing the computational effort by plagues and dynamic adjustment of fitness; applied to passive analog filters design based on general topologies. Obtained experimental results show that proposed strategies improve the design process performance.

1 Introduction

Conventional techniques for filter design use a particular circuit topology. The selection of a topology introduces some limitations on what designers can do, constraining the optimality of obtained results. The use of genetic programming (GP) for solving the design of this problem [5], allows exploring different component values and possible topologies. Automatically exploring different topologies by GP has competitive results compared to those ones obtained by human design. One arising problem of using GP for filter design is the code growth [13]. Some

Mariano Chouza and Ramón García-Martínez
University of Buenos Aires, School of Engineering, Intelligent Systems Laboratory
Paseo Colon 850, (C1063ACV) Buenos Aires, Argentina
e-mail: ecalot@fi.uba.ar

Claudio Rancan
Buenos Aires Institute of Technology, Master Program on Software Engineering
Av. Madero 399, (C1063ACV) Buenos Aires, Argentina
e-mail: crancan@itba.edu.ar

Claudio Rancan and Ramón García-Martínez
University of La Plata, Computer Science School, PhD Program
Calles 50 y 120, (B1900) La Plata, Buenos Aires, Argentina
e-mail: rgarciamar@fi.uba.ar

Osvaldo Clua
University of Buenos Aires, School of Engineering, Distributed Systems Laboratory
Paseo Colon 850, (C1063ACV) Buenos Aires, Argentina
e-mail: oclua@acm.org

B.-C. Chien and T.-P. Hong (Eds.): Opportunities and Challenges, SCI 214, pp. 153–158.
springerlink.com

techniques developed to solve this have been explored in restricted domains [12] but the scalability of them remains as an open issue. Passive analog filter design is a particular field of filter design domain. Conventional design techniques do not deal correctly well with the non-linear interaction of the passive analog filter component values in general topologies. In that context, this paper explores the use of some GP population control techniques applied to passive analog filter design. State of the art is presented in section 2, describing passive analog filter design state (section 2.1) and GP for filter design uses (section 2.2); problems addressed in this paper are presented in section 3; the techniques proposed to solved the addressed problems are presented in section 4; the experiment results are shown in section 5 and some conclusions and future research work are drawn in section 6.

2 State of the Art

A filter is commonly defined as a signal processing device oriented to modify the signal spectrum. This modification is generally described by the filter transfer function. This research is particularly interested in analog passive filters. The term "analog" is used because this type of filters deals with analog inputs. This type of inputs is needed in a wide set of fields; being one of the most important human-computer interaction in medicine uses. The term "passive" is used in the sense that in this type of filters only resistors, capacitors and inductors are used [9]. Classical procedures use a fixed set of transfer functions and assume a *"ladder"* circuit topology. This type of procedures is based on component value table sets where the inputs are the approximation form and order of the designed analog circuit; and the outputs are the normalized component values for those circuits. The introduction of automated design tools has decreased the use of this kind of design procedures [4, 9]. In general, classical automated design tools are software derived from the "historical" design procedures [4, 8] in which inputs are: approximation form, approximation order and topology ("ladder" is one of them); and the circuit design is obtained as an output. A contribution of the use of automated design tools is that they allow analog passive circuit sensibility analysis that may be used to detect unsatisfactory circuit designs. A disadvantage of these tools is that they do not generate new (original) topologies [5]. This research continues the exploration of the use of genetic programming [6] to analog filter design [5]. The technique consists on using an evolving procedure similar to genetic algorithms [7] instead of using conventional approaches (see section 2.1.2). A genetic programming based design procedure has as inputs: desire transfer function and a convergence metric (which compares the obtained transfer function with the desired one); and as outputs: an analog filter design (which describes topology and component values). Genetic programs may take different forms, but in general two ways are used: tree structure and lineal structure. Tree structure is used since earlier research steps in the area [6] and its philosophy is based on Lisp programming language. Linear structure philosophy is based on imperative programming languages. The commonly used structure is the tree one [1].

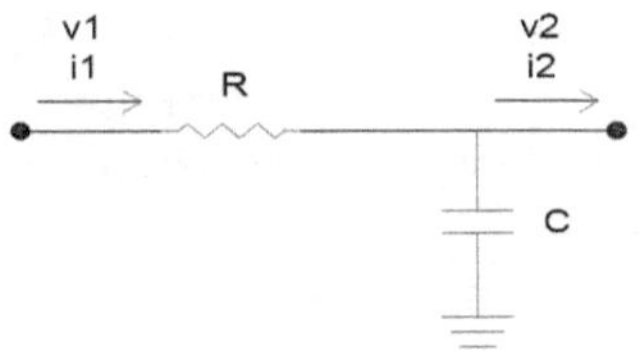

Fig. 1 Conventional representation of a low-pass filter

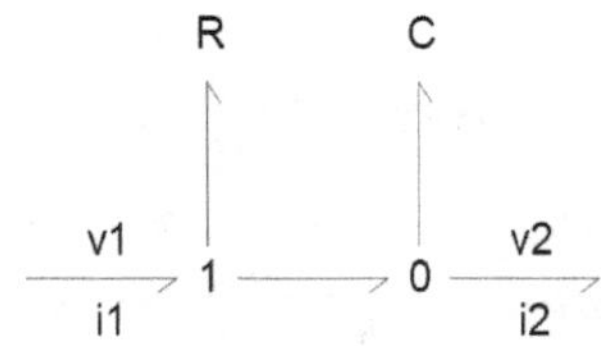

Fig. 2 Bond-graph representation of a low-pass filter

One decision in genetic programming applied to analog filter design is how the circuit is going to be represented. The current representations are: conventional [5] and bond graph based [3]. Conventional representation uses graphs where edges represent components (different type of arcs to represent different type of components) and nodes to represent the interconnection among components (edges). This type of representation takes advantage of the existence of different tools for circuit simulation *i.e.* SPICE tool [11].

3 The Approached Problems

When using genetic programming, *individual size growth* may occur [13]. This size growth becomes with two inconvenients: [a] higher memory consumption, more amount or more complex individuals implies more memory for their storage, [b] higher evaluation time, when complexity of individuals grow, the time for evaluating them grow. Both aspects limit the scalability of genetic programming because the use of computational resources (remember that part of them are used to evaluate and store individuals) cannot be used completely to improve the whole process performance. The problem of *premature convergence* occurs when the population variability decreases without having reached an acceptable solution, leaving the system stuck in a suboptimal solution. As the other mention problem it decreases the amount of computational resources applicable to improve the process performance.

4 Proposed Solution

One workaround for the identified problems may be to penalize individuals with high complexity. But this solution disturbs the development of the evolution process, because complex individuals are required to act as bridges among different solution space zones configurations. If they are eliminated the population may be stuck in several local minima. To solve these problems many strategies have been proposed [12]. We explore two of them in this paper: decreasing the computational effort by plagues, and dynamic adjustment of the fitness function. We will compare their performance solving the analog filter design problem.

Decreasing the computational effort by plagues. The excessive growth of the complexity individuals may acquire during the evolution process is called bloat.

This derives from the increasing of each generation processing time, and the amount of memory needed for the process. If the individuals with the worst fitness are eliminated best results of the evolution process are obtained during the same processing time. This improving process is called plague [2].

Dynamic adjustment of fitness function. One way of controlling individual size is to penalize them because of their size. However, this has undesirable consequences in the evolution process. Poli proposes the creation of "holes" in the fitness function to solve this problem in a dynamic way [10]. The "hole" creation process consists of the elimination of randomly selected individuals which size is over a certain number. This process put a brake to the population growth.

5 Experiments

All the comparisons use a population of 1000 individuals, 100 generations and 10 independent runs for each method. An elitist rank selection was done with an exponential probability distribution over all individuals not automatically selected in a way that the probability of selecting the individual *i* is $P(i) \approx \lambda e^{-\lambda i}$. A $\lambda = 0.002$ equal to $\lambda = 2$ over the normalized ranks in the interval $[0, 1)$. The probabilities of cross mutation and value modification (a mutation variant that only affects numerical values) was empirically optimized obtaining the values 0.1, 0.2 and 0.2 respectively. The probability of eliminating the individuals with over-average size in the dynamic adjustment of fitness function based method was fixed to 0.2. This probability value of eliminating individuals, was experimentally selected. The chosen problem for experimentation was the design of a low-pass filter with a cut frequency of 10 kHz and 1 kΩ for input and output impedance. The selected evaluation function was based on the sum of the square differences among the real and ideal transfer functions over 50 points, logarithmically distributed between 1 Hz and 100 kHz.

Relation among individuals size and evaluation time. First statistical analysis looks for establishing relation among individuals size and evaluation time. Its is shown in figure 3. It was developed over the described problem without applying any population control techniques. Fitting data with the expression $t_{eval} = A \cdot (size_{individual})^B$ $B = 2.03$ was obtained as result, indicating that the evaluation time depends on individuals size in a quadratic way (approximately).

Best individual score for each generation. The experimental results show three curves. Two of them correspond to the proposed strategies presented in this paper; the third (called reference) is related to the reference case which shows results of solving the analog filter design problem without applying any proposed strategy. Figures 4 shows minimum score average, and figure 5 shows minimum score among 10 runs using the proposed two strategies.

Relation among individuals score and time. Figure 6 shows results of minimum average score through time for both strategies, representing the computational cost in a more precise way than the results showed in figure 4.

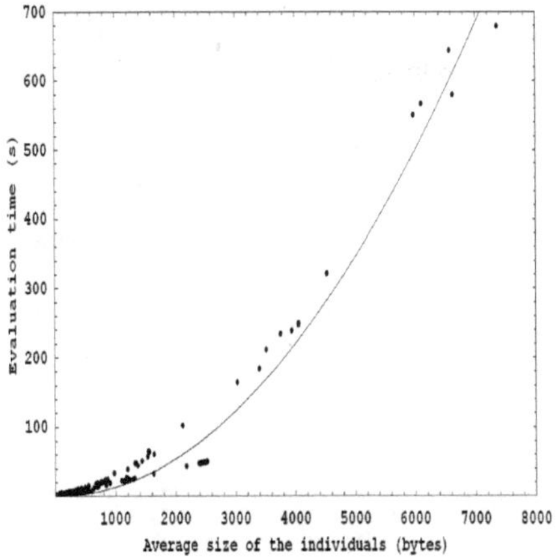

Fig. 3 Relation among individuals size and evaluation time

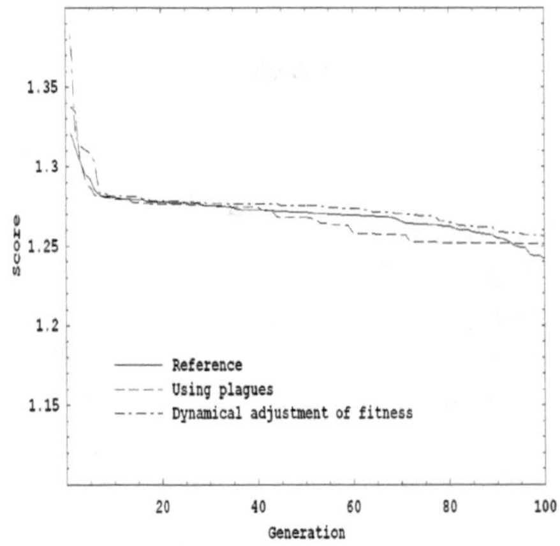

Fig. 4 Minimum score average among 10 runs

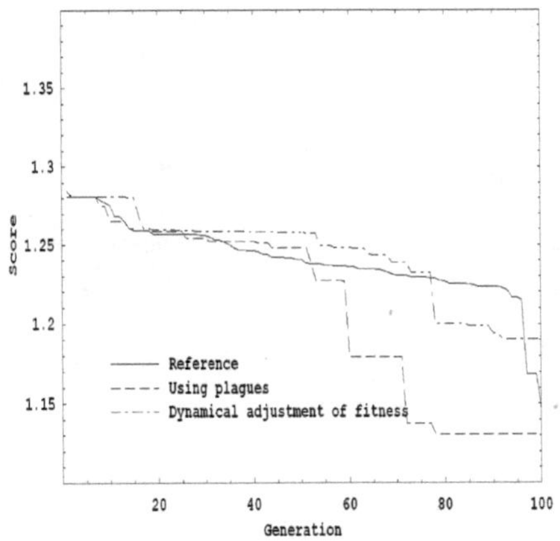

Fig. 5 Minimum score among 10 runs

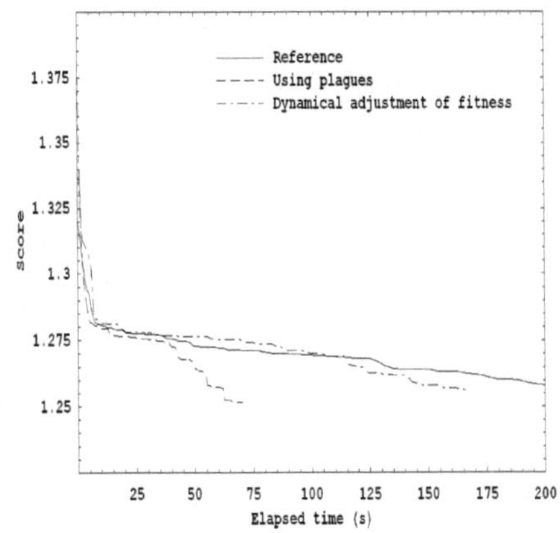

Fig. 6 Minimum average score through time

It is experimentally proven that the two proposed strategies based on GP population control strategies improve the analog filter design process performance (same solution in less time). Results shows that the strategy based on decreasing the computational effort by plagues produces better results than strategy based on dynamic adjustment of fitness function.

6 Conclusions

One limitation of the application of GP to real problems deals with that when problem complexity increases, it is necessary to increase the amount of memory to storage and the amount of time to process the population associated in order to find the solution. This paper presents a possible approach to this problem proposing the use of two different strategies for population growth control: decreasing the computational effort by plagues, and dynamic adjustment of fitness. As conventional design techniques do not deal correctly well with the non-linear interaction of the passive analog filters component values in general topologies; we select this problem to prove the two selected strategies. We found that that the strategy based on decreasing the computational effort by plagues produces better results than strategy based on dynamic adjustment of fitness function, but both proposed strategies based on GP population control improve the design process performance.

References

1. Brameier, M.: On Linear Genetic Programming. PhD Thesis. Universität Dortmund. Germany (2004)
2. Fernandez, F., Vanneschi, L., Tomassini, M.: The Effect of Plagues. In: Ryan, C., Soule, T., Keijzer, M., Tsang, E.P.K., Poli, R., Costa, E. (eds.) EuroGP 2003. LNCS, vol. 2610, pp. 317–326. Springer, Heidelberg (2003)
3. Hu, J., Zhong, X., Goodman, E.: Open Ended Robust Design of Analog Filters Using Genetic Programming. In: Proc. 2005 ACM Conf. & and Evolutionary Computation, pp. 1619–1626 (2005)
4. Koller, R., Wilamowski, B.: LADDER. A Microcomputer Tool for Passive Filter Design and Simulation. IEEE Transactions on Education 39(4), 478–487 (1996)
5. Koza, J., Bennett, F., Andre, D., Keane, M., Dunlap, F.: Automated Synthesis of Analog Electrical Circuits by Means of Genetic Programming. IEEE Transactions on Evolutionary Computations 1(2), 109–128 (1997)
6. Koza, J.: Genetic Programming: On the Programming of Computers by Means of Natural Selection. MIT Press, Cambridge (1992)
7. Mitchell, M.: An Introduction to Genetic Algorithms. MIT Press, Cambridge (1998)
8. Nuhertz Technologies, Passive Filters Solutions (2008), `http://www.filter-solutions.com/passive.html` (accessed April 10, 2008)
9. Paarmann, L.: Design and Analysis of Analog Filters. Kluwer Academic Publishers, Dordrecht (2003)
10. Poli, R.: A Simple but Theoretically-Motivated Method to Control Bloat. In: Ryan, C., Soule, T., Keijzer, M., Tsang, E.P.K., Poli, R., Costa, E. (eds.) EuroGP 2003. LNCS, vol. 2610, pp. 204–217. Springer, Heidelberg (2003)
11. Quarles, T.: The SPICE3 Implementation Guide. Memorandum Nro. UCB/ERL M89/44. Electronics Research Laboratory. University of California at Berkeley (1989)
12. Ryan, C., Soule, T., Keijzer, M., Tsang, E.P.K., Poli, R., Costa, E. (eds.): EuroGP 2003. LNCS, vol. 2610. Springer, Heidelberg (2003)
13. Streeter, M.: The Root Causes of Code Growth. In: Ryan, C., Soule, T., Keijzer, M., Tsang, E.P.K., Poli, R., Costa, E. (eds.) EuroGP 2003. LNCS, vol. 2610, pp. 443–454. Springer, Heidelberg (2003)

Thruster Fault-Tolerant for UUVs Based on Quantum-Behaved Particle Swarm Optimization

Jing Liu, Qi Wu, and Daqi Zhu

Abstract. The thruster fault-tolerant approach for Unmanned Underwater Vehicles (UUV) using Quantum-behaved Particle Swarm Optimization (QPSO) is presented in this paper. The QPSO algorithm is a new global convergent stochastic search technique, which is inspired by the fundamental theory of Particle Swarm Optimization (PSO) and quantum mechanics. The corresponding weighting matrix for faulty situations is developed with the faults of the thruster detected, and the QPSO is used to find the solution of the control reallocation problem, which minimizes the control energy cost function. Comparing with the method of the weighted pseudo-inverse, QPSO algorithm does not need truncation or scaling to ensure the feasibility of the solution because its particles search the solution in the feasible space. Both the magnitude error and direction error of the obtained control input vector using QPSO algorithm are equal to zero. The experimental results demonstrate that the proposed scheme based on QPSO algorithm performs an appropriate control reconfiguration.

Keywords: Thruster fault-tolerant Control, Quantum-behaved Particle Swarm Optimization, Control energy cost function, Unmanned Underwater Vehicles.

1 Introduction

In recent years, unmanned underwater vehicles (UUV) have been used extensively due to their low cost and efficiency. And UUVs are liable to faults or failures during underwater missions. Once the fault occurred, UUV would not accomplish its mission, or even lose itself. Thrusters are one of the most common and most important sources of faults. The thruster faults could be caused by the jammed or halted propeller blades, the broken propeller, and the disassembled propeller and so on. Therefore it is important to develop a fault-tolerant control system of the

Jing Liu and Daqi Zhu
Laboratory of Underwater Vehicles, Shanghai Maritime University,
Shanghai, 200135, China

Qi Wu
School of Automation, Southeast University, Jiangsu, 210096, China

B.-C. Chien and T.-P. Hong (Eds.): Opportunities and Challenges, SCI 214, pp. 159–165.
springerlink.com

thruster for the stability of an UUV.The objective of thruster fault-tolerant control is to reallocate control energy (perform an appropriate reconfiguration) among the functioning thrusters based on the fault information provided in such a way that the UUV could still follow the desired task-space trajectories.

The algorithm of weighted pseudo-inverse is a common method to find the solution of the control reallocation problem[1][2]. Pseudo-inverse is a special case of general inverse (GI) because it is simple to compute. However, in many real-life implementations, the thruster constraints must be taken into account. It is difficult for GI approach to handle the constrained controls. The solution obtained by the GI approach is only on a subset of the attainable command set. To handle such cases where attainable control inputs cannot be allocated, the methods of T-approximation (truncation)and S-approximation (scaling) were proposed[1][2]. The solution obtained the above methods is inside the entire attainable command set, but the magnitude error and direction error caused by approximation still exist and UUV would not completely follow the desired trajectories. In this paper, a swarm intelligence (SI) method for control reallocation problem of thruster fault accommodation based on minimum norm criterion is proposed. A control energy cost function is used as the optimization criteria.

This paper is organized as follows. The thruster configuration of UUV and the minimum norm criterion are briefly introduced in section 2. Section 3 presents the Quantum-behaved particle swarm optimization algorithm. Computer simulation and results are discussed in section 4. Finally, the conclusion is given in section 5.

2 Thruster Configuration and Fault-Tolerant Control

In the general case an UUV has p thrusters (^{1}Th, ^{2}Th…^{p}Th). Each thruster exerts the vector of propulsion forces and moments τ . The standard constrained linear control allocation problem can be formulated as follows [3]: for a given τ , find u such that

$$Bu = \tau \tag{1}$$

$$-u_M \le u \le u_M \tag{2}$$

where u is the control vector, $B \in \Re^{6\times p}$ is the thruster control matrix. The UUV (FALCON, 4 thrusters, $\alpha = 45^0$) with the X-shaped thruster configuration is used to demonstrate the performance of the proposed SI approach. To make problem more understandable and easier to visualize and solve, the vectors τ , u and the matrix B are normalized [1]. Then the Eq.(1) and Eq.(2) can be rewritten as:

$$\underline{\tau} = \begin{bmatrix} \frac{\tau_X}{\tau_{XM}} \\ \frac{\tau_Y}{\tau_{YM}} \\ \frac{\tau_N}{\tau_{NM}} \end{bmatrix} = \underline{B} \cdot \begin{bmatrix} \frac{u_1}{u_M} \\ \frac{u_2}{u_M} \\ \frac{u_3}{u_M} \\ \frac{u_4}{u_M} \end{bmatrix} = \begin{bmatrix} 0.25 & 0.25 & 0.25 & 0.25 \\ 0.25 & -0.25 & 0.25 & -0.25 \\ 0.25 & -0.25 & -0.25 & 0.25 \end{bmatrix} \underline{u} \quad \text{and} \quad \begin{bmatrix} -1 \\ -1 \\ -1 \\ -1 \end{bmatrix} \le \underline{u} = \begin{bmatrix} \frac{u_1}{u_M} \\ \frac{u_2}{u_M} \\ \frac{u_3}{u_M} \\ \frac{u_4}{u_M} \end{bmatrix} \le \begin{bmatrix} 1 \\ 1 \\ 1 \\ 1 \end{bmatrix}$$

where $\underline{B}$ is defined as: $\begin{pmatrix} 0.25 & 0.25 & 0.25 & 0.25 \\ 0.25 & -0.25 & 0.25 & -0.25 \\ 0.25 & -0.25 & -0.25 & 0.25 \end{pmatrix}$

And $\underline{\tau} = [\tau_X, \tau_Y, \tau_N]$, $-1 \leq \underline{\tau} \leq 1$ means the controllable DOF surge, sway and yaw. For Eq. (3), there exists infinite number of solutions to thruster control allocation for a given motion. In order to choose a unique best solution from those solutions, the following control energy cost function is introduced [4]:

$$\min_{u \in \psi} \|Wu\|_2 \tag{3}$$

subject to Eq. (1) and Eq.(2). W is the weighting matrix to decide which thruster should be used primarily and usually defined as a diagonal matrix as: $W = \begin{bmatrix} w_1 & 0 & 0 & 0 \\ 0 & w_2 & 0 & 0 \\ 0 & 0 & w_3 & 0 \\ 0 & 0 & 0 & w_4 \end{bmatrix}$ where $w_i > 0$ is the weight of the ith thruster. In the fault-free case, all thrusters have the same priority and W is defined as unit matrix $W = \begin{bmatrix} 1 & 0 & 0 & 0 \\ 0 & 1 & 0 & 0 \\ 0 & 0 & 1 & 0 \\ 0 & 0 & 0 & 1 \end{bmatrix}$. In the fault case, the weighting matrix is increased:

W= $\begin{pmatrix} 1+\Delta w_1 & 0 & 0 & 0 \\ 0 & 1+\Delta w_2 & 0 & 0 \\ 0 & 0 & 1+\Delta w_3 & 0 \\ 0 & 0 & 0 & 1+\Delta w_4 \end{pmatrix}$ where $\Delta w_i = 2\left(\frac{1}{s_i} - 1\right)$, the weight update is introduced to penalize the faulty thruster to compensate restricted usage of the faulty thruster in an optimal way. The types of thruster faults are assumed herein that they could be detected without considering the details of possible nature of thruster faults. $s_i = 1$, which means ith thruster has non-fault; $s_i \to 0$ which means ith thruster has the total fault, and completely ineffective; $0 < s_i \leq 1$, which means that ith thruster has partial fault.

3 Quantum-Behaved Particle Swarm Optimization

To minimize the function of the control energy cost, Quantum-behaved particle swarm optimization is proposed in this work. As demonstrated by F. Van den bergh [5][6], PSO is not a global convergence guaranteed algorithm because the particle is limited in a finite sampling space for each of the iterations. This restriction weakens the global search ability of the algorithm and may lead to premature convergence in many cases. Inspired by PSO and quantum mechanics theories, the Quantum-behaved Particle Swarm Optimization (QPSO) was proposed to increase its global search ability [7].

In the quantum model of a PSO, the state of a particle is depicted by wave function $\Psi(\vec{x},t)$, instead of position $\vec{x}$ and velocity $\vec{v}$. The particles move according to the following iterative Eq.(4)[8]:

$$x(t+1) = p \pm \beta |mbest - x(t)| \ln(1/u) \tag{4}$$

where $mbest = \frac{1}{M}\sum_{i=1}^{M} p_i = \left(\frac{1}{M}\sum_{i=1}^{M} p_{i1}, \frac{1}{M}\sum_{i=1}^{M} p_{i2}, \cdots \frac{1}{M}\sum_{i=1}^{M} p_{id}\right)$, $p = \varphi p_{id} + (1-\varphi)\, p_{gd}, \varphi = rand()$

Mean Best Position (*mbest*) is defined as the mean value of the best position of all particles; p_i is the best position of the *i*th particle; p_g is the position of the best particle among all the particles; *p*, a stochastic point between p_{id} and p_{gd}, is the local attractor on the *d*th dimension of the *i*th particle; φ and u are random numbers distributed uniformly on [0,1]. In this paper, the QPSO is used to minimize the objective function Eq.(3). The particle searches in the range of [-1, +1]. And the parameter β is set from 1 to 0.5.

4 Experiment Results and Discussions

In order to evaluate the results obtained by different methods, two scalar errors are introduced: direction error $\theta = a\cos \tau_d \cdot \tau_d^* / \|\tau_d\|_2 \|\tau_d^*\|_2$ and magnitude error $\|e\|_2 = \|\tau - \tau^*\|_2$. The direction error represents the angle between the desired state τ and the obtained state τ^*, while the magnitude error represents the module of the approximation error vector. The following experiments are simulated for one fault and multi-faults detected in thrusters respectively.

4.1 The Fault-Free Case

Let the given motion $\tau = [0.70\ \ 0.20\ \ 0.25]^T$ for X-shape thruster configuration (FALCON) in the fault-free case. The pseudo-inverse solution $u = [1.15\ \ 0.25\ \ 0.65\ \ 0.75]^T$ is unfeasible because of $u_1 > 1$. The T-approximation is given by $u^* = [1\ \ 0.25\ \ 0.65\ \ 0.75]^T$, and the corresponding $\tau^* = Bu^* = [0.6625\ \ 0.1625\ 0.2125]^T$. The S-approximation is given by $u^* = [1.0000\ \ 0.2174\ \ 0.5652\ \ 0.6522]^T$ and $\tau^* = Bu^* = [0.6087\ \ 0.1739\ \ 0.2174]^T$ [1].While the solution obtained by QPSO algorithm $u^* = [1\ \ 0.1\ \ 0.8\ \ 0.9]$ is feasible without any approximation because *u* is limited in [-1,+1] during the process of searching solution in QPSO algorithm and the corresponding $\tau^* = Bu^* = [0.7\ \ 0.2\ \ 0.25]$.

From the Table 1, it can be seen that the control vector obtained by QPSO algorithm could reallocate the energy function reasonably comparing with the method of pseudo-inverse and preserve the original direction and magnitude.

Table 1 The results of fault-free thruster

	Pseudo-inverse		QPSO
	T-approximation (truncation)	S-approximation (scaling)	
U	[1 0.25 0.65 0.75]	[1 0.2174 0.5652 0.6522]	[1 0.1 0.8 0.9]
$\|e\|_2$	0.065	0.1004	0
θ	2.6363	0	0
$\text{Min}\|Wu\|_2$	1.4309	1.3387	1.5684

4.2 The Single Fault Case

Assumed that the first thruster ([1]Th thruster) is detected faults, and its corresponding value of fault state s1 is 6/7. Let the original state is $\tau = [0.5 \ \ 0 \ \ 0]^T$. Then the weight matrix is $W = \begin{bmatrix} 4/3 & 0 & 0 & 0 \\ 0 & 1 & 0 & 0 \\ 0 & 0 & 1 & 0 \\ 0 & 0 & 0 & 1 \end{bmatrix}$, which penalize the faulty thruster. The solution $u = [0.4615 \ \ 0.4615 \ \ 0.5385 \ \ 0.5385]^T$ obtained by pseudo-inverse is feasible without any approximation. The solution obtained by QPSO algorithm $u^* = [0.4186 \ \ 0.4186 \ \ 0.5814 \ \ 0.5814]$ is also feasible because of its limited search space. The solutions obtained by different methods as shown in Table 2 could perform the appropriate configuration, but the minimum of the control energy cost function of QPSO is better.

Table 2 The results of [1]Th thruster fault

	Pseudo-inverse		QPSO
	T-approximation (truncation)	S-approximation (scaling)	
U	[0.4615 0.4615 0.5385 0.5385]	[0.46150.46150.5385 0.5385]	[0.41860.41860.5814 0.5814]
$\|e\|_2$	0	0	0
θ	0	0	0
$\text{Min}\|Wu\|_2$	1.0824	1.0824	1.0783

4.3 The Multi-fault Case

Let the given motion $\tau = [0.70 \ 0.20 \ 0.25]^T$ for X-shape thruster configuration (FALCON). Let three thrusters ([1]Th, [3]Th and [4]Th) were detected faults, their corresponding s=[0.5 1 0.75 0.25] and $W = \begin{bmatrix} 3 & 0 & 0 & 0 \\ 0 & 1 & 0 & 0 \\ 0 & 0 & 5/3 & 0 \\ 0 & 0 & 0 & 7 \end{bmatrix}$. The experiment results are shown in Table 4. The results show that the solutions obtained by Pseudo-inverse

Table 3 The results of [1]Th, [3]Th and [4]Th thruster faults

	Pseudo-inverse		QPSO
	T-approximation (truncation)	S-approximation (scaling)	
U	[1 0.4579 0.4421 0.5421]	[0.5 0.1686 0.1628 0.1996]	[1 0.1 0.8 0.9]
$\|e\|_2$	0.155	0.4863	0
θ	7.0274	0	0
$\text{Min}\|Wu\|_2$	4.9145	2.0747	7.1048

Table 4 The results of [1]Th , [2]Th, [3]Th and [4]Th thruster faults

	Pseudo-inverse		QPSO
	T-approximation (truncation)	S-approximation (scaling)	
U	[1 0.1145 0.7855 0.8855]	[0.2864 0.0323 0.2218 0.25]	[0.94120.04120.85880.9588]
$\|e\|_2$	0.0063	0.5524	0
θ	0.2377	0	0
$\text{Min}\|Wu\|_2$	8.3108	2.3507	7.4209

still have some error in magnitude or direction, while QPSO algorithm could preserve the original direction and magnitude. Finally let the given motion $\tau = [0.70 \ \ 0.20 \ \ 0.25]^T$ for X-shape thruster configuration (FALCON). Let four thrusters ([1]Th, [2]Th, [3]Th and [4]Th) were all detected faults, their corresponding s=[0.5 0.05 0.75 0.25] and $W = \begin{bmatrix} 3 & 0 & 0 & 0 \\ 0 & 39 & 0 & 0 \\ 0 & 0 & 5/3 & 0 \\ 0 & 0 & 0 & 7 \end{bmatrix}$. From the Table 5, it can be seen that the results obtained by QPSO performs an appropriate control reallocation.

From the above experiments, whether only one thruster was detected fault or several thrusters were detected the different extend faults, QPSO algorithm is demonstrated an efficiency approach to accommodate faults to make UUV follow the desired task-space trajectories.

Acknowledgements. This project is supported by the Shanghai Municipal Education Science Commission(No.2008095) and Shanghai Municipal Chenguang Project (2008CG55).

References

1. Omerdic, E., Roberts, G.N.: Thruster Fault Diagnosis and Accommodation for Open-frame Underwater Vehicles. Control Engineering Practice 12, 1575–1598 (2004)
2. Omerdic, E., Roberts, G.N., Toal, D.: Extension of Feasible Region of Control Allocation for Open-frame Underwater Vehicles. In: IFAC Conference on Control Applications in Marine Systems (2004)
3. Durham, W.C.: Constrained Control Allocation. Journal of Guidance, Control and Dynamics 16(4), 717–725 (1993)

4. Enns, D.: Control allocation approaches. AIAA Guidance, Navigation and Control Conference and Exhibit, 98–108 (1998)
5. Hu, X., Eberhart, R., Shi, Y.: Recent Advances in Particle Swarm. In: IEEE Congress on Evolutionary Computation, pp. 90–97 (2004)
6. Van den Bergh, F.: An Analysis of Particle Swarm Optimizers. Ph.D. thesis, University of Pretoria (2001)
7. Van den Bergh, F.: A New Locally Convergent Particle Swarm Optimizer. IEEE International Conference on Systems, Man and Cybernetics 3, 94–99 (2002)
8. Sun, J., Feng, B., Xu, W.B.: A Global Search Strategy of Quantum-behaved ParticleSwarm Optimization. In: IEEE Conference on Cybernetics and Intelligent Systems, pp. 111–116 (2004)

Intelligent System for Locating, Labeling, and Logging (ISL^3)

Yan Huang, Terry Griffin, and Salomon Lompo

Abstract. As mobile smart devices become ubiquitous in our society, users will be able to receive location based information on the fly. A model that is able to predict a user's next destination will make location based services more effective by providing personalized information to the user. The implementation of such a location prediction model requires a set of correctly labeled destinations collected from users to tune the prediction model to an acceptable level of accuracy. A large collection of data will allow researchers to derive the parameters required to train predication models and also get the trends of user behaviors in general. ISL^3 will allow researchers to do just this by easily allowing them to collect user activity data to create location prediction models.

1 Introduction

As mobile smart devices become ubiquitous in our society, users will be able to receive location based information on the fly. A model that is able to predict a user's next destination will make location based services more effective by providing personalized information to the user. The implementation of such a location prediction model requires a set of correctly labeled

Yan Huang
University of North Texas, USA
e-mail: huangyan@unt.edu

Terry Griffin
Midwestern State University, USA
e-mail: terry.griffin@mwsu.edu

Salomon Lompo
Midwestern State University, USA
e-mail: salomon.lompo@mwsu.edu

B.-C. Chien and T.-P. Hong (Eds.): Opportunities and Challenges, SCI 214, pp. 167–173.
springerlink.com

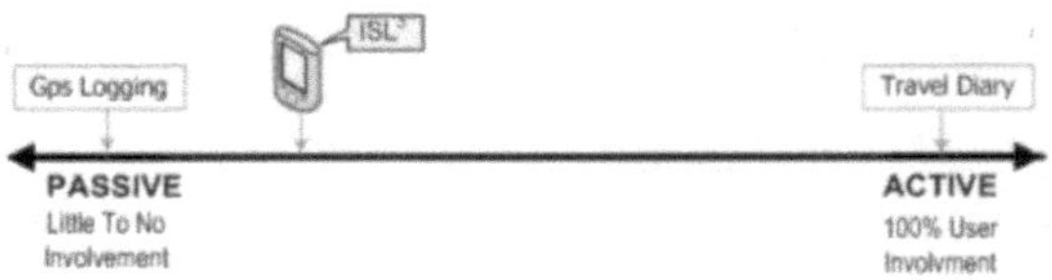

Fig. 1 ISL^3 Interaction Continuum

destinations collected from users to tune the prediction model to an acceptable level of accuracy. Inaccurately labeled or incomplete data sets will adversely affect the performance of a location prediction model. This paper proposes an application software solution for data collection and destination labeling called: ISL^3 - *Intelligent System for Locating, Labeling, and Logging.*

A large collection of data will allow researchers to derive the parameters required to train predication models and also get the trends of user behaviors in general. One key component of any prediction model is a user's history from which the prediction models are built. And the veracity of a collected data set will significantly affect one's ability to ascertain a users history which in turn will impact the overall performance of a prediction model. In addition, researchers tend not to share these types of sensitive data sets, typically due to privacy issues. Therefore, it becomes necessary for researchers to have a convenient and correct tool to ease this burden of data collection. In recent history, many researchers have used GPS devices to collect a user's location history. Location data can be either passively collected without a user's input or actively collected where a user provides labels to their destinations. Passive data collection requires no user attention. However the post labeling (or classification) of destinations was done by hand in a highly interactive manner. The contribution of the tool to be described in this paper is in that it allows researchers to easily collect data that has a correct labeling of a user's destination in a manner that requires little interaction. The relative user involvement can be seen in figure 1.

2 Related Work

In spite of the important part data collection plays in the implementation process of any location prediction model, there have been few papers focusing on data collection and the different methods of labeling / classifying this data. In [3], data was collected for approximately 6 weeks for one user, to be used in a decision tree based prediction algorithm. Ashbrook and Staner [1] collected data from one user for four months and six users for seven months in two different surveys and were able to extract significant locations to aid in the creation of location prediction algorithms. Another technical report [8] from the University of Helsinki gathered a dataset of mobile communication of a small number of people over a long period of time. This data was central to testing an algorithm developed in [7] and [6]. In [2] Hariharam and Toyama provide a data structure for analyzing and generating user histories based on a 2 year collection of personal GPS logs. They also propose an algorithm that will extract interesting information from the raw data collected. In [4] the author was able to compute the likelihood of the user's next destination using the data collected from 200 users in the Seattle Washington area [5]. Collecting data is paramount when creating location prediction algorithms, and ISL^3 was created to ease the burden.

3 Software Development Process Components

3.1 The Graphical User Interface

ISL^3 displays an easy to use and intuitive interface for users. This plays into the overall purpose of the software itself, which is to give the everyday user the ability to easily collect a complete and fully labeled travel diary. To start the program the user clicks on the "play" button. As soon as the GPS enters in contact with the array of GPS satellites, the longitude and latitude of the current location are displayed at the right upper corner (see figure 2). The concept is simple: 1)Add destinations, 2)Record arrivals and departures in respect to those destinations.

To add a destination, the user simply presses the "Add Destination" button (see figure 2-2), showing the keyboard allowing the user to type a short descriptive name for a destination. This adds the destination to a *Destinations* file that contains a unique id for the destination and user, along with the coordinates of that destination. This file populates a drop down list that is accessible on the main screen (see figure 2-4). This drop down list is continuously sorted, where the top of the list is the destination closest to a users current location.

When the user needs to record arrivals and departures, they simply press the *Arrived* or *Leaving* button depending on the circumstance. The action is recorded in the *Actions* file with the specified destination. To specify a destination the user selects it from the drop down list. In addition to a record being appended to a file, visual assurance is given to the user in the form of a log entry on the users screen (see figure 2-5).

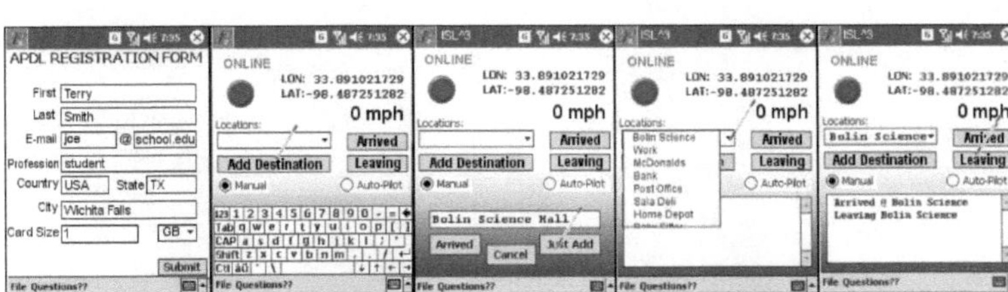

Fig. 2 ISL^3 Graphical User Interface

Initially the interaction by the user is relatively low, but will still diminish as the users typical destination are entered. Typical location entry consists of adding each location as it is encountered (meaning a record would be logged in the *actions* file), but the software does have an option to "Just Add" a location. This allows a user to populate the list with their typical locations before they are actually visited. The one drawback to this method of destination entry is that the destination will have no Lat-Lon pair associated with it until the first time it is physically visited. Consequentially that location would not bubble to the top of the list as the user approaches it (for reasons explained earlier). ISL^3 has two modes of operation: 1)Standard and 2)Autopilot. Standard mode operates as previously discussed. Autopilot mode requires slightly less interaction.

Autopilot mode works much like a GPS logger in the fact that it senses motion. If the "Autopilot" senses that the user is not moving (based on a distance function between contiguously logged points), it starts a timer. When the timer exceeds some predefined threshold, the "Autopilot" will record an arrival at that destination. When the "Autopilot" senses movement, based on the same distance function, it records the user leaving the destination in the Actions file.

3.2 Architecture of the ISL^3 Software

The architecture of the ISL^3 software is shown in figure 3. A smart device communicates with some form of GPS antennae to obtain it location information. Smart devices are not limited to using GPS antennae's to obtain location information, but this software currently depends on one being present. When the device receives a GPS stream, it receives it from a communications port (comm port). Even if a GPS antennae is embedded within the smart device, it still communicates with the operating system using a comm port. Each brand of hardware will determine which comm port is used for GPS antennae communication, and it does vary. To ensure a more robust software package, comm port scanning is necessary. The application scans through the ports of the handheld device to identify the correct port receiving the GPS stream. To do so, the application takes advantage of the .Net framework library. Using the .Net library allows system

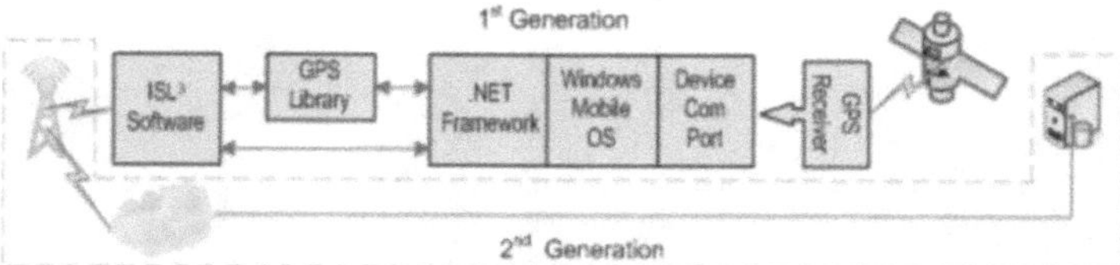

Fig. 3 ISL^3Architecture

to stay in managed mode, which is an option that guarantees a stable system from a windows mobile operating system perspective. After ISL^3 detects the correct comm port, the software sends the GPS stream to a GPS library, where the main objective is to parse the GPS stream and filter out unneeded data.

The main program manages most aspects of the software. Many of the management choices are predetermined by a configuration file. It determines which data are filtered and passed by the GPS library to the main program. The configuration file also determines the frequency and format in which records are logged. Other features that are stored in the configuration file consist of the maximum allocated memory for storage, a radius to help in determining destination arrival, and maximum destination definitions. All of these parameters allow ISL^3 to be highly configurable for the differences in everyday systems.

3.3 Hardware Used

ISL^3 is designed to work on .Net Compact framework 2.0 architecture or above. The application was tested using two type of Pocket PC handheld device. The first handheld was the T610 model from the AIRIS manufacturer. The AIRIS T610 has an integrated GPS chip embedded in the device. The second Pocket PC is a WAYPOINT PDA coupled with an external GPS device via a Bluetooth connection. Both devices were equipped with windows mobile 2005 operating system. The WAYPOINT PDA had shown a better battery life saving and a quick response in getting satellite connection. However after a long period of inactivity, the Bluetooth connection was frequently lost which forced the user to restart the ISL^3 program. The Integrated GPS enabled device seemed more suitable for auto-pilot mode. As long as the device was connected to external power, it remained on and functional. During the testing phase, the AIRIS T610 was set up for a week in a car, on auto-pilot mode. With no interaction, the data collection completed successfully without interruption.

3.4 Managed Files

ISL^3 produces four different files: 1) *User File*-contains information about the user of the software. 2)*Destinations File*-is used to store added destinations by the user. 3)*Actions File*-contains the activities of the user. 4)*Log File*-stores the raw GPS data. The division of stored information was designed in such a way as to ease the extraction of important information from the collected data set, along with reducing storage requirements for the device in use. A relationship between the main data files can be seen in figure 4.

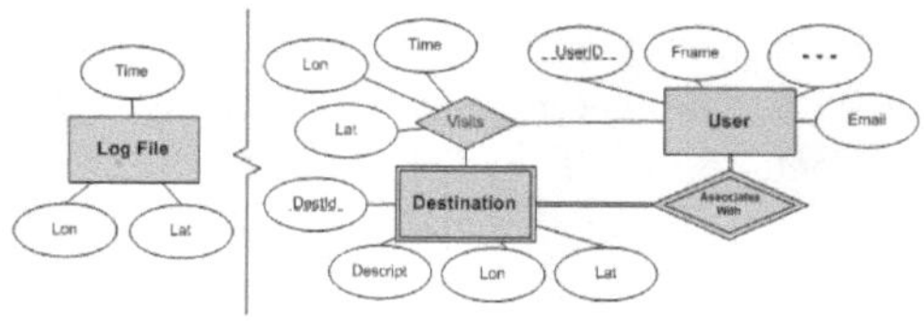

Fig. 4 ISL^3 ER-Diagram

3.5 Data Collected

The 1st generation of the ISL^3 software was tested by four individuals in the North Texas area. The users all recorded data continuously every 5 seconds (approx) as long as the unit was on. The collected data is summarized below:

User	Logged Records	Destinations Defined	Actions Generated
User 1	1855	6	98
User 2	50040	29	256
User 3	14193	22	418
User 4	11153	22	310
Totals	**77241**	**79**	**1082**

4 Future Work

The 2nd Generation of ISL^3 (see figure 3) will be targeting the Smartphone class of handheld devices. This will help satisfy four goals: 1)To reach a much larger audience of potential data collectors. 2)Get easier access to internet based technologies. 3)Implement real time functionality to our software. And most importantly 4)Design and implement location/destination prediction algorithms.

References

1. Ashbrook, D., Staner, T.: Using GPS to Learn Significant Location and Predict Movement Across Multiple Users. Personal and Ubiquitous Computing 7(5), 275–286 (2003)
2. Hariharam, R., Toyama, K.: Project Lachesis: parsing and Modeling Location Histories. In: Egenhofer, M.J., Freksa, C., Miller, H.J. (eds.) GIScience 2004. LNCS, vol. 3234, pp. 106–124. Springer, Heidelberg (2004)
3. Huang, Y., Griffin, T.: A Decision Tree Classification Model to Automate Trip Purpose Derivation. In: Computer Applications in Industry and Engineering (Proceedings), pp. 44–49 (2005)
4. Krumm, J.: Real Time Destination Prediction Based On Efficient Routes. Microsoft Research (2006)
5. Krumm, J., Horvitz, E.: The Microsoft Multiperson Location Survey (MSR - TR-2005-103). Microsoft Research (2005)

6. Laasonen, K.: Clustering and Prediction of Mobile User Routes from Cellular Data. In: Jorge, A.M., Torgo, L., Brazdil, P.B., Camacho, R., Gama, J. (eds.) PKDD 2005. LNCS, vol. 3721, pp. 569–576. Springer, Heidelberg (2005)
7. Laasonen, K., Raenato, M., Toivonen, H.: On-device Adaptive Location Recognition. In: Ferscha, A., Mattern, F. (eds.) PERVASIVE 2004. LNCS, vol. 3001, pp. 287–304. Springer, Heidelberg (2004)
8. Raenato, M.: Mobile Communication and Context Dataset. Technical report. University of Helsinki (2004)
9. Cheng, C., Jain, R., van de Berg, E.: Location Prediction Algorithms for Mobile Wireless Systems. In: Wireless Internet Handbook: Technologies, Standards, and Applications, pp. 245–263. CRC Press, Boca Raton (2003)

Probabilistic Analysis of Information Center Insecurity

Shyue-Liang Wang, Jyun-Da Chen, Tzung-Pei Hong, and Paul A. Stirpe

Abstract. Information security has become a top priority for many organizations due to a growing number of computer threats. Modeling information system security has been studied extensively in recent years and many techniques have been proposed. In this work, we propose a simple model and an algorithm to efficiently calculate the probability of insecurity for each resource in an information center when a single type of threat exists. Numerical simulations showing system insecurity using some common information center topologies are presented. When properly combined with risk management strategy, the proposed technique can effectively calculate the optimal security investment for information centers.

1 Introduction

Information system security has been studied extensively in the past thirty years. There are many approaches proposed to enhance system security. One approach is to build systems that follow certain security policy. Many modeling techniques for this approach have been proposed. For example, access security models such as the Bell and LaPadula model [2], information flow models such as the nondeducibility model by Sutherland [7] and the noninterference model by Goguen and Messeguer [3], etc. The access security model system is secure if its execution satisfies a set of access control conditions such as multilevel security. The information flow models treat system behavior as information flow, and security is mainly about undesired information flows. Another approach is to build systems that reduce the system insecurity or to reduce the system vulnerabilities. For example, the insecurity flow model proposed by Moskowitz and Kang [4] is used to analyze the probability of

Shyue-Liang Wang and Jyun-Da Chen
Department of Information Management

Tzung-Pei Hong
Department of Computer Science and Information Engineering
National University of Kaohsiung , Kaohsiung, Taiwan 81148
e-mail: {slwang,tphong}@nuk.edu.tw

Paul A. Stirpe
Letse, LLC, 45 Oser Avenue, Hauppauge, New York 11788 USA
e-mail: paul.stirpe@letse.com

B.-C. Chien and T.-P. Hong (Eds.): Opportunities and Challenges, SCI 214, pp. 175–180.
springerlink.com

system insecurity, and attack graph models proposed by Sheyner et al. [6], Ammann et. al. [1], Phillips et al. [5] are used to analyze system vulnerability. Once system vulnerabilities are found, they can be used to determine what security measures to deploy to defend the systems.

However, most techniques for analyzing system insecurity and vulnerabilities originate their modeling or analysis from the attacker's perspective or source of the attack [1,5]. This approach generally requires exponential time complexity to produce the graph. In this work, we take a different approach by starting the analysis and calculation of insecurity from the resources to be protected. Based on information system architecture and probabilities of insecurities of system components, we extend the insecurity flow model [4] and propose a naïve algorithm to efficiently calculate the probability of insecurity (or probability of successful attack) for each resource in the system when a single threat exists.

2 Modeling Information Center Security

Given a networked computing environment, organizations can protect their computing assets by establishing protection domains [4], which contain groups of related components. The typical computing assets in an organization may include, for example, resources such as information on web servers, application servers, and database servers. Protection domains can be established by physical and logical components. Each protection domain may possess a security policy and be protected by security mechanisms such as firewalls, access control lists, and intrusion detection systems, etc., to protect their assets.

Based on the concept of computing asset (or resource) and protection mechanism (or filter), a given information center can be heuristically modeled as a graph with nodes representing resources and filters (see Figures 1 and 2). The attack represents the external environment connected to the network or information center.

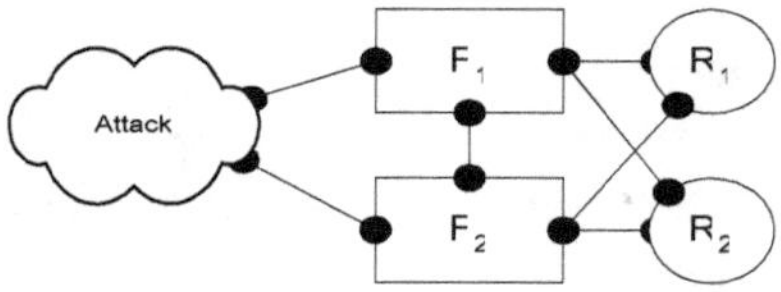

Fig. 1 Common 1-tier Information Center

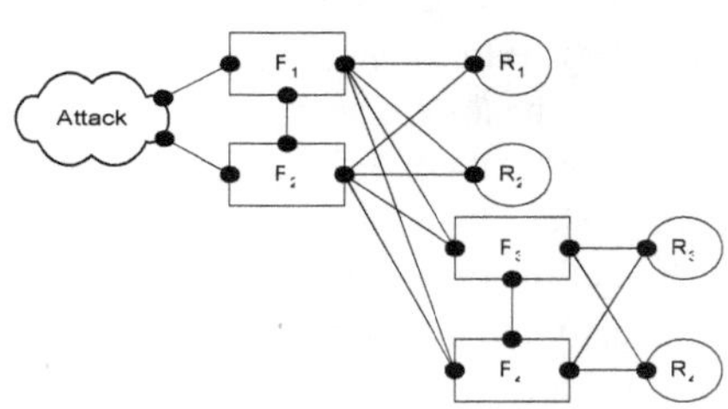

Fig. 2 2-tier Information Center

For a given attack, the probability of insecurity (or successful attack) of a filter is defined as the probability that the given attack successfully passing through the filter. The probability of insecurity (or successful attack) of a resource is defined as the probability of an attack that successfully passing through the resource. For simplicity, in this work we assume that the probabilities are independent from filter to filter and are independent with respect to time. The *accumulative* probability of insecurity (or successful attack) of a resource in an information system can then be modeled by the accumulative probability of insecurity passing through all possible simple paths from the *Attack* to the *Resource*. For example, for resource R_1 in Figure 1, there are six possible simple paths from the *Attack (A)* to R_1, $\{<A, F_1, R_1>, <A, F_2, R_1>, <A, F_1, F_2, R_1>, <A, F_2, F_1, R_1>, <A, F_1, R_2, F_2, R_1>, <A, F_2, R_2, F_1, R_1>\}$. Unlike the previous work in [10] which assumes that once an attacker compromises a resource, the attack process ends. In this work, we allow the attack to be continued after a resource is compromised. This assumption is implemented by assuming the probability of insecurity of a resource to be non-zero.

To calculate the accumulative probability of insecurity of a resource, we consider the following four basic patterns as the fundamental building blocks, namely single filter (1F), multiple serial filters (mSF), multiple parallel filters (mPF), and multiple interface filters (mIF), to model the architecture of an information center.

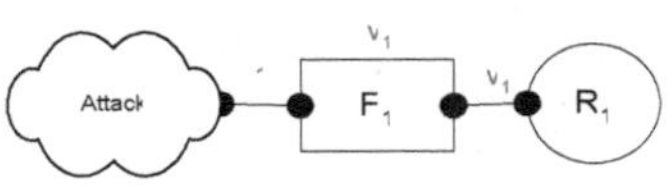

Fig. 3 Single Filter (1F)

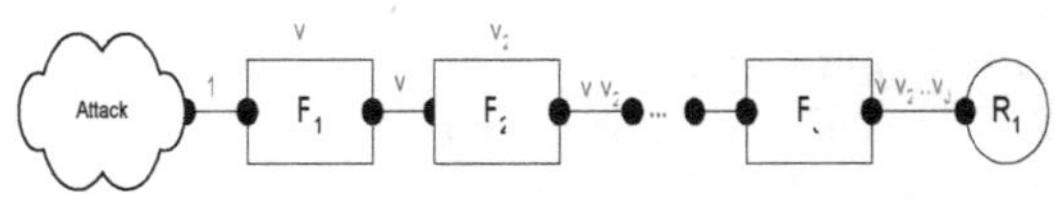

Fig. 4 Multiple Serial Filter (mSF)

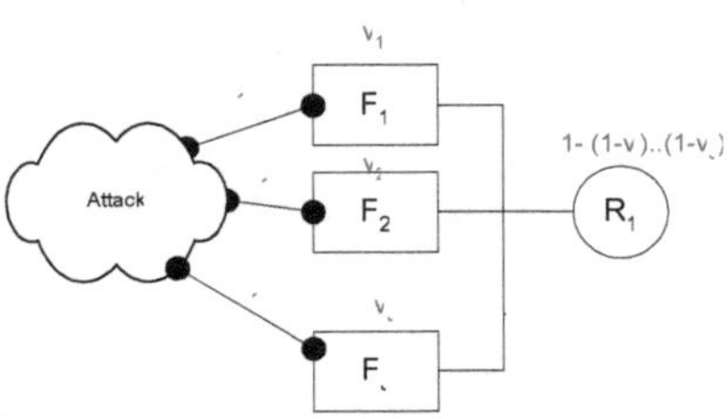

Fig. 5 Multiple Parallel Filter (mPF)

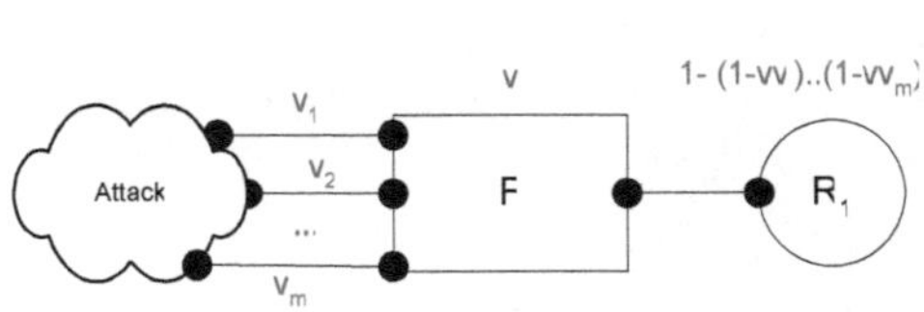

Fig. 6 Multiple Interface Filter (mIF)

3 Proposed Algorithm

To calculate the probability of insecurity for any resource in an information center, we propose an algorithmic approach that generates attack trees that contain all possible simple paths *from* the resource *to* the attack. For a given resource, the algorithm finds simple paths from the resource back to the attack in a top-down manner, based on the system structure, and then adds the path to the tree. A simple path here is defined as a path that does not contain repeated nodes. The leaf node of a path must be the attack. The path from a filter to a non-attack leaf node will be pruned if the filter has an attack as its other child. The algorithm then performs a bottom-up calculation of accumulative probabilities of insecurity passing each filter and resource on the same level of the tree. The probability at the root of the tree will be the accumulative probability of insecurity of the resource. To calculate the probability of insecurity passing through a filter, depending on the number of interfaces, the filter is decomposed into a combination of the basic patterns. The accumulative probability is then calculated according to the combined basic pattern. The proposed algorithm consists of two major functions. The first function builds an attack tree T from a given graph G which is the topology of an information center. It adopts a breadth-first searching approach to construct a tree level-by-level. Once the attack tree is built, a recursive function calculates the accumulative probability of a given resource in the graph. The detail of attack tree construction is given as follow.

Algorithm API (Accumulative Probability of Insecurity)

Input: (1) an information center topology,
(2) probability of insecurity on each filter and resource,
Output: accumulative probability of insecurity for each resource

Function *BFS(G;R;A;T)*
//Build attack tree *T* for attack node *A* to resource node *R* in Graph *G*

```
Initialize a queue BFQ with node R, T=CurrNode;;
While (BFQ is not empty){
    CurrNode = BFQ.dequeue();
NextNodeSet={nodes that can be reached from CurrNode and do not appear
in ancestor nodes};
    If (NextNodeSet is empty) delete CurrNode from the tree;
    For (each node NextNode in NextNodeSet){
        Add an edge from CurrNode to NextNode;
        If (NextNode is not node A) enqueue(BFQ, NextNode);} // end for
};//end while
```

4 Numerical Simulations

This section shows numerical simulations for demonstrating the proposed algorithm in calculating the accumulative probabilities of insecurity for the resources in the common one-tier (Figure 1) and two-tier (Figure 2) information center designs. Figure 7 shows the accumulative probabilities of insecurity on R_1 for the one-tier network when the probabilities of insecurity for the two filters vary from 0 to 1, with $F_1 = F_2$. The higher the probability, the less secure is the resource. It is observed that as the probability of insecurity of filters increases, the accumulative probability of insecurity of resources increases rapidly. In addition, when attacks are allowed to be continued after a resource is compromised, shown as $PI(R) = 1$, it is less secure than attacks that end when a resource is compromised, shown as $PI(R) = 0$. Similar penetration effects on both R_1 and R_3 for the two-tier information center structure are shown in Figures 8 and 9, where $PI(F_3) = PI(F_4) = 0.5$. One possible explanation for such effect could be due to the number of possible attack paths leading to a resource. More attack paths would make the resource less secure. For example, there are 4 attack paths from attack A to R_1 for one-tier information center in Figure 1, where the attack ends once resource is compromised. But there are 6 attack paths if an attack continues after a resource is compromised. However, this is not always true for the multi-home information center structure. Figure 10 shows the cross-over effects on R_1 and R_3. The numbers of possible attack paths are 14 and

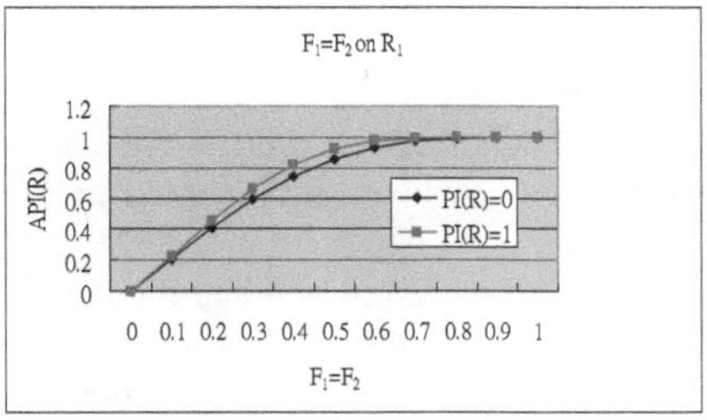

Fig. 7 1-tier Penetration Effects on R_1

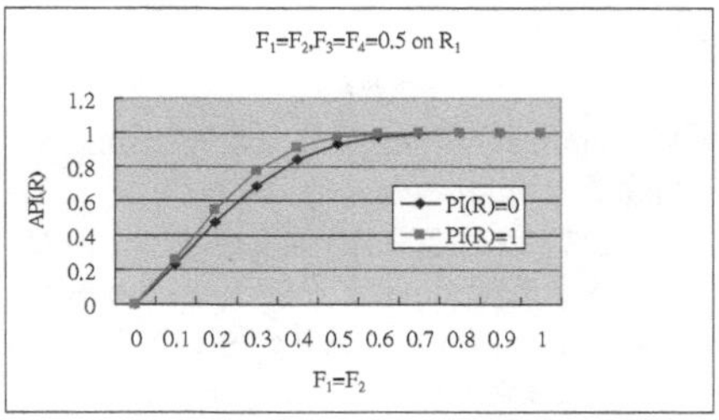

Fig. 8 2-tier Penetration Effects on R_1

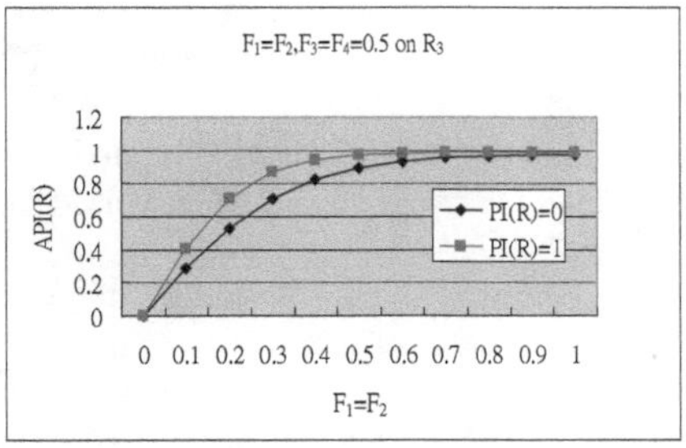

Fig. 9 2-tier Penetration Effects on R_3

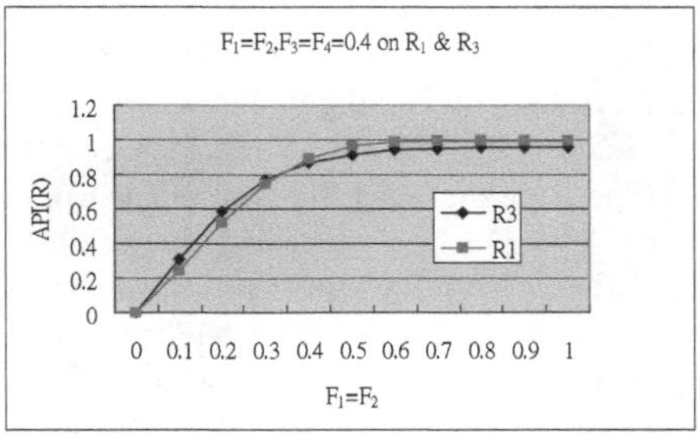

Fig. 10 2-tier Cross-over Effects on R_1 & R_3

46 for resources R_1 and R_3 respectively. et the probabilities of insecurity of F_3 and F_4 be 0.4. The accumulative probability of insecurity on R_1 is larger than R_3 when probabilities of insecurity of filters F_1 and F_2 are larger than approximately 0.34. However, when the probability of insecurity of F_1 and F_2 are less than this cross-over value, resource R_1 is more secure than resource R_3, which is counter intuitive. This may be due to the fact that filters F_3 and F_4 start to take effects on the accumulative probability of insecurity. This is an indication that alternative designs need to be investigated.

5 Conclusion

In this work, we have studied the problem of threats to information centers caused by potential attacks and proposed an algorithm API to calculate the accumulative probability of insecurity for resources in an information center when a single type of attack exists. The proposed API algorithm is based on a threat flow model that models the probabilistic flow of possible attacks on information systems. Numerical simulations on some common information center designs are presented. The calculated results effectively show the accumulative probabilities of insecurity for all the resources and demonstrated the penetration effects and cross-over effects on the resources.

Acknowledgments. This work was supported in part by the National Science Council, Taiwan, under grant NSC-97-2221-E-390-023.

References

1. Ammann, P., Wijesekera, D., Kaushik, S.: Scalable, Graph-Based Network Vulnerability Analysis. In: Proceedings of the 9th ACM Conference of Computer and Communications Security (CCS 2002), pp. 217–224 (2002)
2. Bell, D., LaPadula, L.: Secure Computer Systems: Unified Exposition and multics Interpretation. Technical Report, MTR-2997, MITRE, Bedford, Mass (1975)
3. Goguen, J.A., Meseguer, J.: Security Policies and Security Models. In: Proceedings. of the 1982 IEEE Symposium on Security and Privacy, Oakland, CA, pp. 11–20 (April 1982)
4. Moskowitz, I.S., Kang, M.H.: An Insecurity Flow Model. In: New Security Paradigms Workshop, Langdale, Cumbria, UK (1997)
5. Phillips, C., Swiler, L.P.: A Graph-Based System for Network-Vulnerability Analysis. In: New Security Paradigms Workshop, pp. 71–79 (1998)
6. Sheyner, O., Wing, J.: Tools for Generating and Analyzing Attack Graphs. In: de Boer, F.S., Bonsangue, M.M., Graf, S., de Roever, W.-P. (eds.) FMCO 2003. LNCS, vol. 3188, pp. 344–371. Springer, Heidelberg (2004)
7. Sutherland, D.: A Model of Information. In: Proceedings of the 9th National Computer Security Conference, NSA/NIST, Gaithersburg, MD (September 1986)
8. Wang, S.L., Stirpe, P.A., Hong, T.P.: Modeling Optimal Security Investment of Information Centers. In: DMDRM workshop in PAKDD, Osaka, Japan, pp. 293–304 (May 2008)

Application of Combined Forecasting Models to Intelligent Transportation Systems

Yang Zhang and Yuncai Liu

Abstract. Several combined forecasting techniques are investigated in this paper. Six baseline individual predictors are selected as basic combination components. Experimental results demonstrate that the combined predictors can significantly reduce error rates and provide a large improvement in stability and robustness. It reveals that the techniques are practically promising in the traffic domain.

1 Introduction

The prediction of traffic states is fundamental to the success of intelligent transportation systems (ITS). Some papers appearing as extensive review arouse significant scientific interest in more flexible methodological approaches. The empirical approaches can be approximately divided into two types: parametric and nonparametric techniques. Both techniques have shown their advantages on different occasions in recent years [1].

The paper aims to generate and compare various hybrid forecasting models by integrating different predictors. The basic idea of the combined method is to apply each model's unique feature to capture different patterns in the data [2, 3]. The complement in capturing patterns of data sets is essential for more accurate prediction. Due to its practicability, it seems more important to traffic engineering.

Applying six individual predictors, we are motivated to propose the combined methods to predict the obtained traffic data. Our paper focuses on two-model based linear combining forecasts that belong to the traditional univariate approach. The linear least squares regression (LLSR), autoregressive moving average (ARMA), historical-mean (HM), radial basis function neural network (RBF-NN), support vector regression (SVR) and least squares support vector machines (LS-SVMs) are selected as individual predictors. Based on their forecasts, four combined predictors including equal weights (EW), optimal weights (OW),

Yang Zhang and Yuncai Liu
Research Center of Intelligent Transportation Systems, Shanghai Jiao Tong University, 200240 Shanghai, P.R. China
e-mail: {zhang-yang, whomliu}@sjtu.edu.cn

B.-C. Chien and T.-P. Hong (Eds.): Opportunities and Challenges, SCI 214, pp. 181–186.
springerlink.com

minimum error (ME) and minimum variance (MV) methods are detailedly investigated. The forecast performance is measured by different indices of forecast accuracy.

2 Several Combined Forecasting Methods

Suppose there are N forecasts such as $\hat{V}_{P1}(t)$, $\hat{V}_{P2}(t)$, …, $\hat{V}_{PN}(t)$, where $\hat{V}_{Pi}(t)$ represents the forecasting result obtained from the ith model during the time interval t. The combination of the different forecasts into a single forecast $\hat{V}_P(t)$ is

$$\hat{V}_P(t) = \sum_{i=1}^{N} w_i \hat{V}_{Pi}(t) \tag{1}$$

where w_i denotes the assigned weight of $\hat{V}_{Pi}(t)$. Commonly, the sum of the weights is equal to one, i.e., $\sum_i w_i=1$.

2.1 Equal Weights (EW) and Optimal Weights (OW) Methods

Applying a simple arithmetic average of the individual forecasts, the EW method is a relatively robust one with low computational efforts. Each w_i is equal to 1/N (i=1, 2, …, N), where N is the number of forecasts. The beauty of using the simple average is that it is easy to understand and implement, not requiring any estimation of weights or other parameters.

For the OW method, using a MV criterion can determine the weights to adequately apply the additional information hidden in the discarded forecast(s) [3]. Assuming that the individual forecast errors are unbiased, we can calculate the vector of weights to minimize the error variance of the combination according to

$$\boldsymbol{w} = \boldsymbol{M}_V^{-1}\mathbf{I}_n(\mathbf{I}_n'\boldsymbol{M}_V^{-1}\mathbf{I}_n)^{-1} \tag{2}$$

where $\mathbf{I}_n$ is the n×1 matrix with all elements unity (i.e. n×1 unit vector) and $\boldsymbol{M}_V$ is the covariance matrix of forecast errors.

2.2 Minimum Error (ME) and Minimum Variance (MV) Methods

A solution for the ME method applies linear programming (LP) whose principle and computational process are described as follows [2]. Set the sum of absolute forecasting error (i.e., $\sum_i E_i(t)$ during the time interval t) as

$$F_{LP} = \sum_{i=1}^{N} |E_i(t)| = \sum_{i=1}^{N} \left| w_i(t)\left(\hat{V}_{Pi}(t) - V_O(t)\right)\right|, \quad t = 1, 2, \cdots, \mathrm{T} \tag{3}$$

where F_{LP} is the objective function of LP; $V_O(t)$ denotes the observed value during the time interval t and T the number of forecasting periods. To eliminate the absolute sign of the objective function, assume that

$$u_i(t)=\frac{|E_i(t)|+E_i(t)}{2}=\begin{cases}E_i(t), & E_i(t)\geq 0\\ 0, & E_i(t)<0\end{cases},\quad v_i(t)=\frac{|E_i(t)|-E_i(t)}{2}=\begin{cases}0, & E_i(t)\geq 0\\ -E_i(t), & E_i(t)<0\end{cases}. \tag{4}$$

The introduction of $u_i(t)$ and $v_i(t)$ aims to transform the absolute sign of the objective function so as to be consistent with the standard form of LP. Obviously, $|e_i(t)|=u_i(t)+v_i(t)$, $e_i(t)=u_i(t)-v_i(t)$. Then, the LP model can be constructed as follows:

$$\begin{cases}Min\ O=\sum_{i=1}^{\mathrm{N}}\left(u_i(t)+v_i(t)\right),\\ \sum_{i=1}^{\mathrm{N}}w_i(t)\left(\hat{V}_{Pi}(t)-V_O(t)\right)-u_i(t)+v_i(t)=0,\\ \sum_{i=1}^{\mathrm{N}}w_i(t)=1,\\ w_i(t)\geq 0,\ u_i(t)\geq 0,\ v_i(t)\geq 0,\ i=1,2,\cdots,\mathrm{N},\ t=1,2,\cdots,\mathrm{T}\end{cases} \tag{5}$$

where i denotes the number of individual forecasts, and t represents the forecasting periods. In the equation group, assuming $w_i \geq 0$ aims to make every forecast method contribute to the combined forecasting results.

For the MV method [4], the main ideas can be described as

$$\begin{cases}Min\ (w_i \boldsymbol{M}_V w_i^{\mathrm{T}}),\\ \sum_{i=1}^{\mathrm{N}}w_i=1,\quad i=1,2,\cdots,\mathrm{N}\\ w_i\geq 0,\end{cases} \tag{6}$$

where $\boldsymbol{M}_V$ is the matrix of error variance. By solving the quadratic programming (QP) problems, an optimal weight set can be obtained for the combining forecasts.

3 Research Approach and Experiments

Data for this study come from the Performance Measurement System (PeMS), which can be accessed through the Internet [5]. The travel time index (TTI) expresses the average amount of extra time it takes to travel in the peak relative to free-flow travel. It can present congestion levels in a format that is easy to understand and communicate to the general public.

The traffic data of 24 weeks from May 1 to Oct. 15, 2006 are used. The data for a particular day start every 1 hour between 00:00 am and 23:00 pm. For simplicity, six individual predictors including LLSR (1), ARMA (1, 1), HM (K=3), RBF-NN (d=1), SVR (d=3) and LS-SVMs (d=4) are selected as the preparation of the research. The performance of combined methods is compared with that of the conventional individual ones. We focus on predicting the 24th week (168 time points), from Oct. 9 to Oct. 15, on the basis of the information from the former 23 weeks. Combining each two individual forecasts, six combined models (EW, OW, ME I, ME II, MV I and MV II) are then investigated. Different from the ME and MV models that directly determines $w_i(t)$ (weights during time interval t, i=1, 2) by solving the LP and QP problems, ME II and MV II models apply the average of $w_i(t)$, $w_i(t$-1) and $w_i(t$-2) as the weight $\widehat{w}_i(t)$ to calculate the combined forecasts. Thus, the weights at the first time point are chosen as 0.5 for OW, ME I and ME II models, and the weights at the first 2 time points are set to 0.5 for ME II and MV II models. Moreover, in the four ME and MV models we set the same threshold $w_i(t) \in [0.15, 0.85]$ to ensure the adequate effectiveness of both individual models in the combining process. In the study, the mean absolute percentage error (MAPE), variance of absolute percentage error (VAPE) and percentage error (PE) are used as measures of forecast accuracy to assess the quality of the forecasts.

Table 1 Performance comparison in MAPE & VAPE using EW & OW method (%)

EW / OW	MAPE					
	LLSR	ARMA	HM	RBF-NN	SVR	LS-SVMs
LLSR	**0.9729**	1.0366	0.9486	*0.9730*	0.8450	*0.8808*
ARMA	0.9995	**1.2267**	0.9527	*1.0797*	0.9097	0.9508
HM	0.9188	1.0123	**1.1589**	0.9532	0.9297	0.9280
RBF-NN	0.9741	1.0926	0.9603	**1.0629**	0.8564	0.9047
SVR	*0.8380*	*0.8997*	0.9510	0.8711	**1.0042**	0.7978
LS-SVMs	0.8853	0.9184	0.8884	0.8965	0.7937	**0.9315**
EW / OW	**VAPE**					
	LLSR	ARMA	HM	RBF-NN	SVR	LS-SVMs
LLSR	**0.8704**	0.9545	1.1038	*0.9178*	0.8145	*0.8173*
ARMA	0.9037	**1.3050**	1.1062	*1.0901*	0.8854	0.8676
HM	0.8494	1.0668	**1.6777**	1.0370	1.1162	1.0615
RBF-NN	0.9105	1.1349	1.0331	**1.2363**	0.8269	0.9024
SVR	*0.8081*	*0.8950*	0.9912	0.8745	**1.0768**	0.8088
LS-SVMs	0.8281	0.8373	0.9036	0.9014	0.8157	**0.9773**

Table 2 Performance comparison in MAPE & VAPE using ME I & ME II method (%)

ME I / ME II	MAPE					
	LLSR	ARMA	HM	RBF-NN	SVR	LS-SVMs
LLSR	**0.9729**	1.1303	*0.8656*	0.9858	0.8879	0.9335
ARMA	1.0592	**1.2267**	0.9434	1.1234	0.9337	0.9394
HM	0.8786	*0.9222*	**1.1589**	0.9394	0.8464	*0.8247*
RBF-NN	0.9936	1.0992	*0.9310*	**1.0629**	0.8917	0.9605
SVR	0.8437	0.9422	*0.8420*	*0.8473*	**1.0042**	0.8179
LS-SVMs	0.9003	0.9336	0.8483	0.9048	*0.7843*	**0.9315**
ME I / ME II	**VAPE**					
	LLSR	ARMA	HM	RBF-NN	SVR	LS-SVMs
LLSR	**0.8704**	1.1270	*0.8610*	0.9622	0.8236	0.9379
ARMA	1.0013	**1.3050**	0.9910	1.2517	0.9463	0.9094
HM	0.8764	*0.9938*	**1.6777**	1.0475	0.8538	*0.7667*
RBF-NN	0.9775	1.2076	*0.9574*	**1.2363**	1.0527	1.0354
SVR	0.8398	0.9798	*0.8710*	*0.8717*	**1.0768**	0.8519
LS-SVMs	0.8551	0.8510	0.8181	0.9203	*0.7987*	**0.9773**

Table 3 Performance comparison in MAPE & VAPE using MV I & MV II method (%)

MV I / MV II	MAPE					
	LLSR	ARMA	HM	RBF-NN	SVR	LS-SVMs
LLSR	**0.9729**	0.9763	0.9393	0.9932	0.8438	0.8981
ARMA	*0.9760*	**1.2267**	1.0317	1.0893	0.9062	*0.9123*
HM	0.9540	1.0572	**1.1589**	0.9748	0.9589	0.8999
RBF-NN	1.0000	1.0843	0.9872	**1.0629**	0.8762	*0.8955*
SVR	0.8474	0.9083	0.9902	0.8852	**1.0042**	0.7980
LS-SVMs	0.9021	0.9182	0.9261	0.8956	0.7983	**0.9315**
MV I / MV II	VAPE					
	LLSR	ARMA	HM	RBF-NN	SVR	LS-SVMs
LLSR	**0.8704**	0.8596	0.8907	0.9616	0.8143	0.8508
ARMA	*0.8598*	**1.3050**	1.0795	1.1676	0.8912	*0.8671*
HM	0.9695	1.1862	**1.6777**	1.0133	1.0098	0.9353
RBF-NN	0.9766	1.1557	1.0603	**1.2363**	0.9540	*0.9027*
SVR	0.8163	0.9107	1.1629	0.9726	**1.0768**	0.8257
LS-SVMs	0.8536	0.8745	1.0404	0.9035	0.8247	**0.9773**

The prediction performance of different approaches is listed in Table 1, 2 and 3 respectively. Each table presents the error rates obtained from six individual models and two combined models. Indicated in bold, the six pairs of results from the component predictors are shown along the diagonal lines that divides the upper or lower half of each table into two parts. Based on the permutations and combinations of the six individual predictors in the two-model combining process, 15 categories may exist for each combined model. Its results are listed above or below the diagonal line. For instance, the combined result of SVR and LLSR from EW method is shown above the diagonal of the upper half of Table 1 with MAPE=0.8450. Its counterpart measured in VAPE is displayed above the diagonal of the lower half of the table with VAPE=0.8145. Measuring the performance by either error rate, we can notice that over 80% combined forecasts are better than that directly obtained from the individual predictors. Selecting the best results of the combined forecasts (ME II, LS-SVMs~SVR), we can indicate that the predictor can reduce at least 15.80% and 18.27% in MAPE and VAPE respectively compared with its individual counterpart with the best performance (LS-SVMs).

For comparison, we pick out the best two-model combined forecasts measured by MAPE in each category. As shown in Table 1 to 3, these error rates are indicated in italic. The EW, OW, ME I, MV I and MV II models provide 3, 2, 2, 2 and 1 best results respectively. The ME II model produces 5 such results, which adequately demonstrates its effectiveness. We also divide the upper or lower half of each table into 4 parts using cross lines. The lines classify these combined models based on different combination elements. For any upper or lower part of each table, the results in the left upper part of it are calculated from the combination of two parametric techniques (Situation I). The combined forecasts from two non-parametric techniques are presented in its lower right part (Situation II). And the rest two parts contain the error rates from the combination of one parametric technique and one nonparametric technique (Situation III). Obviously, the ME and MV methods can produce relatively stable results in all situations.

We choose to analyze three representative models: ARMA~HM (Situation I), RBF-NN~SVR (Situation II) and LLSR~LS-SVMs (Situation III). After calculating the PEs for each model, we analyze the numbers of the forecasts lying in different ranges of |PEs| (the absolute value of PEs). The range boundaries are set as 1%, 2%, and 4%. Fig. 1 shows the comparisons in three classes. The performance of combined models can be greatly improved when nonparametric technique is chosen as one component. All combined models in Situation II can produce over 120 points with |PEs|∈ [0, 1%]. The ME and MV methods perform well in Situation I and II respectively; the EW and OW models perform well in Situation III.

Generally, the combination of forecasts outperforms individual forecasts and shows its superiority in our experiments. The ME and MV methods generally perform better than the others in different situations. Its extraordinary ability determines the effectiveness and robustness of combined forecasts, which has been proved in the empirical studies.

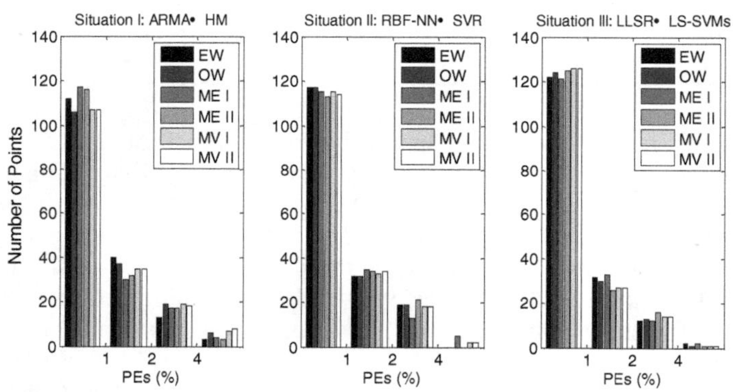

Fig. 1 The numbers of predicted time points lying in different ranges of |PEs|

Acknowledgements. This research effort was supported by the National Science & Technology Supporting Program of the 11th Five-year Plan (No. 2006BAJ18B02), P. R. China.

References

1. Lam, W.H.K., Chan, K.S., Tam, M.L., Shi, J.W.Z.: Short-term Travel Time Forecasts for Transport Information System in Hong Kong. J. Adv. Transp. 39(3), 289–305 (2005)
2. Yu, L., Wang, S., Lai, K.K.: A Novel Nonlinear Ensemble Forecasting Model Incorporating GLAR and ANN for Foreign Exchange Rates. Comp. & Oper. Res. 32(10), 2523–2541 (2005)
3. Bates, J.M., Granger, C.W.J.: The Combination of Forecasts. Oper. Res. Quart. 20(4), 451–468 (1969)
4. Yu, L., Wang, S., Lai, K.K., Nakamori, Y.: Time Series Forecasting with Multiple Candidate Models: Selecting or Combining? J. Syst. Sci. & Complexity 18(1), 1–18 (2005)
5. Freeway Performance Measurement System (PeMS), `http://pems.eecs.berkeley.edu`

Fuzzy Performance Analysis Model Based on Grid Environment

Huey-Ming Lee, Chia-Hsien Chung, Tsang-Yean Lee, and Jin-Shieh Su

Abstract. In grid computing environment, job requirements are so large scale and complex that we need the allocating mechanism to manage the resources and schedule the job. So that, a well-allocated mechanism is needed to enhance the grid resources be more useful and scalable. In this paper, we propose a resource performance analysis model for grid resources under the grid computing environment. By this model, we can analyze the information about CPU usage, memory usage by fuzzy inferences, and number of running jobs of each grid resource node to achieve load-balancing and make the plans and allocations of the resources of collaborated nodes optimize. There are three modules in the proposed model, namely, resource detecting module, resource estimator module, and resource assignment module. According to the result of experiment, the mechanism can achieve the best resources allocation, and enhance the overall grid computing performance.

1 Introduction

The term "Grid" was coined in the mid 1990s to denote a proposed distributed computing infrastructure for advanced science and engineering [4]. In grid environment, user may access the computational resources at many sites [8]. The functions of information systems based on grid computing architectures are resources (e.g., CPUs, memory, storages, etc.) sharing, collaborative processing, reliable and secure connection, etc. However, each resource of coordinate nodes in the grid environment, (e.g., CPU loading, memory rate of utilization, etc.) changes dynamically, Therefore, how to optimize these resources usages is an important issue.

Miller et al. [12] proposed Paradyn parallel performance measurement tools which can identify the heaviest loading process by heuristic method and find out the bottleneck point. Lee et al. [9] proposed a dynamic supervising model that can

Huey-Ming Lee, Chia-Hsien Chung, Tsang-Yean Lee, and Jin-Shieh Su
Department of Information Management, Chinese Culture University
55, Hwa-Kung Road, Yang-Ming-San, Taipei (11114), Taiwan
e-mail: hmlee@faculty.pccu.edu.tw, jonguser@gmail.com,
tylee@faculty.pccu.edu.tw, sjs@faculty.pccu.edu.tw

B.-C. Chien and T.-P. Hong (Eds.): Opportunities and Challenges, SCI 214, pp. 187–192.
springerlink.com

utilize the gird resources more flexible and optimal. Lee et al. [10] presented an optimal analyzing resources model that can receive the information of grid nodes to achieve load-balancing and make the plans and allocations of the resources of collaborated nodes optimize.

Moreover, some researches also presented the computational economy for the grid to solve the resource allocation problem. Abramson *et al.* [1, 2, 3] proposed the Nimrod-G Resource Broker which is a grid-enabled resource management and scheduling system based on the concept of computational economy. Nimrod-G uses the Monitoring and Discovery System (MDS) services for resource discovery and GRAM API dispatches jobs over grid resources. The users can specify deadline by which the results of their experiments are needed. Nimrod-G broker tries to find the cheapest resources available that can do the job and meet the deadline.

To achieve load-balancing and make the plans and allocations of the resources of collaborated nodes optimize, we have to collect the related information of nodes and analyze these information by fuzzy inferences to calculate the best allocation of jobs.

In this paper, we propose a resource performance analysis model for grid resource under the grid computing environment. By this model, we can evaluate the information about CPU usage, memory usage, and number of running jobs of each grid resource node to achieve load-balancing and make the plans and allocations of the resources of collaborated nodes optimize. According to the result of experiment, the proposed mechanism can achieve the best resources allocation, and enhance the overall grid computing performance.

2 Framework of the Proposed Model

We built grid node capacity information locally. It contains node name, CPU speed, CPU usage, memory size, memory usage and disk size, etc. This information can be used to select a correct grid node to process new job. Then, we presented a resources performance analysis model (RPAM) based on grid computing architecture. It was built on the grid node. There are three modules in this model, namely, resource detecting module (RDM), resource estimator module (REM), and resource assignment module (RAM), as shown in Fig. 1.

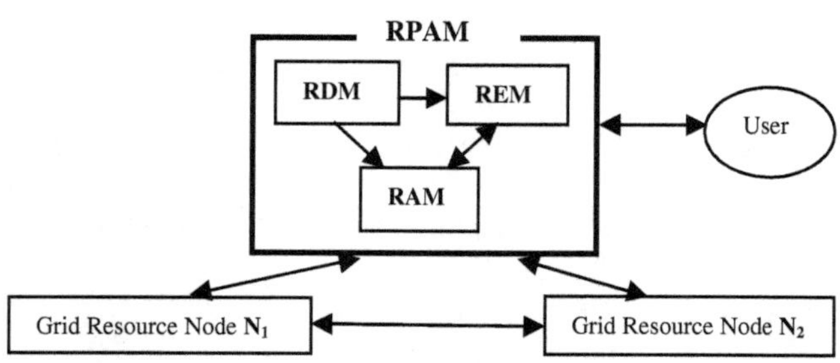

Fig. 1 Framework of the proposed model (RPAM)

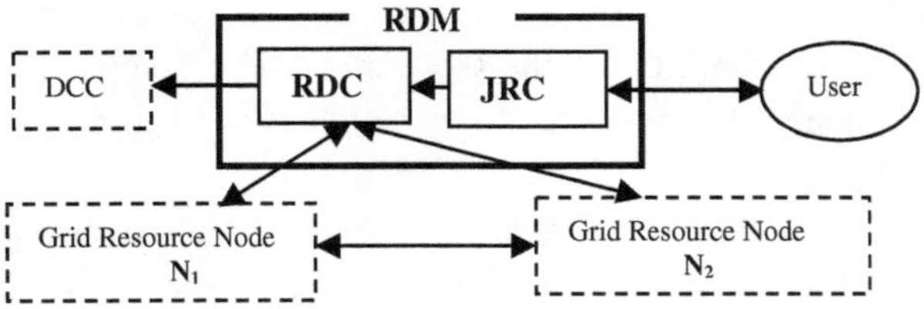

Fig. 2 Framework of the RDM

2.1 *Resource Detecting Module (RDM)*

The RDM comprises two components which are job receiving component (JRC) and resource detecting component (RDC), as shown in Fig. 2. The JRC deals with the user's job request, and sends it to RDC. The RDC will search and detect the available nodes and then collect these related information of nodes (e.g., CPU speed, CPU usage, memory size, memory usage, disk size, etc.).

2.2 *Resource Estimator Module (REM)*

The REM comprises four components, which are data collection component (DCC), trust degree component (TDC), resource price component (RPC), computing performance component (CPC), and trust degree data base (TDDB), as shown in Fig. 3. The DCC integrates and collects the related information of grid nodes from RDC and then sends the integrated information to TDC, RPC, and CPC, as the evaluation data. In addition, DCC also receives the information of evaluation from TDC, RPC, and CPC, and then sends it to the Resource Assignment Module (RAM), as the basis for the allocation of plan. The TDC can adjust trust degrees of other collaborators by the workload, and provide these trust degrees for RAM to plan the works processes. The RPC evaluates the price of grid node by its computer hardware information (e.g., CPU, memory, storages, etc.).The CPC can evaluate computing performance of others collaborators by their CPU of utilization and memory of utilization. We use assessment criteria which were divided into five levels by fuzzy inferences.

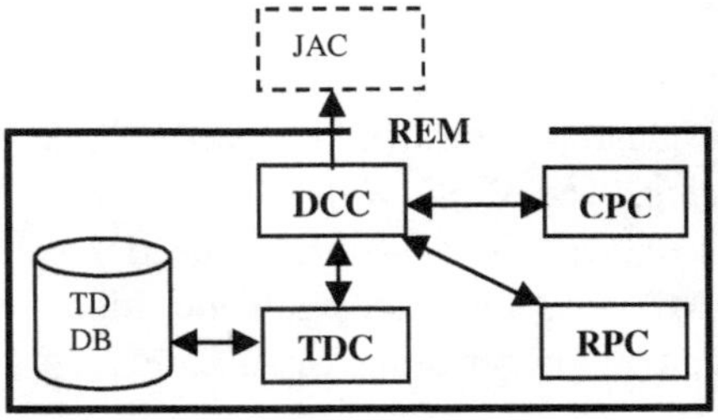

Fig. 3 Framework of the REM

The criteria ratings of CPU usage are linguistic variables with linguistic values C_1, C_2, C_3, C_4, C_5, where C_1= very light, C_2= light, C_3= middle, C_4= heavy, C_5= very heavy. These linguistic values are treated as fuzzy numbers with trapezoid membership functions.

The criteria ratings of Memory usage are linguistic variables with linguistic values M_1, M_2, M_3, M_4, M_5, where M_1= very light, M_2= light, M_3= middle, M_4= heavy, M_5= very heavy. These linguistic values are treated as fuzzy numbers with trapezoid membership functions.

The criteria ratings of computing performance are linguistic variables with linguistic values P_1, P_2, P_3, P_4, P_5, where P_1= very low, P_2= low, P_3= middle, P_4= high, P_5= very high. These linguistic values are treated as fuzzy numbers with triangular membership functions.

(I) Evaluating computing performance

We set zero as an initial value of computing performance of the collaborated nodes. When has a new job, the value of computing performance will be re-evaluated. There are twenty-five rules in our fuzzy inference rule base, as shown in Table 1.

Table 1 fuzzy inference rule base

Memory \ CPU	very light	light	middle	heavy	very heavy
very light	very high	very high	high	middle	middle
light	very high	high	middle	middle	low
middle	high	high	middle	low	low
heavy	middle	middle	low	low	very low
very heavy	low	low	very low	very low	very low

(II) Computing Performance Application

We can plan and allocate the priority of demanders depending on computing performance.

2.3 Resource Assignment Module (RAM)

The RAM comprises two components: job arranging component (JAC), and plan mode component (PMC), as shown in Fig. 4. The JAC receives the grid node evaluation information from DCC, and sends it to PMC for the allocation of planning. The PMC will allocate the job to proper nodes according to the user demands. There are three allocation methods which we can choose in modes bases (MB), saying, average allocation, optimize cost allocation, and optimize computing performance allocation.

Fig. 4 Framework of the RAM

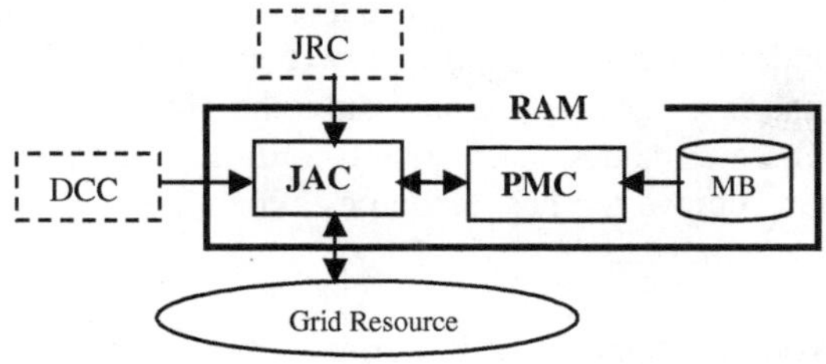

3 Model Implementation

For easy implementation, we build a very small scenario. It is the simplest grid environment in local area network (LAN) which intended to illustrate the concepts and components behind the Grid. And an Ethernet LAN and three Intel® CPU machines also were used.

After detecting available resources of node on grid environment, the RPAM receives related resource information of the node. Then we can choose three allocation methods in modes bases (MB) which are average allocation, optimize cost allocation, and optimize computing performance allocation. The RPAM is based on the user's demand and then the work will be allocated to the appropriate node. Under the same type of work, these three allocation methods will come up with three computed results, as shown in Fig 5.

According to the results of the experiment, we can reduce the overall computing time of the grid computing environment, and improve the overall computing performance effectively.

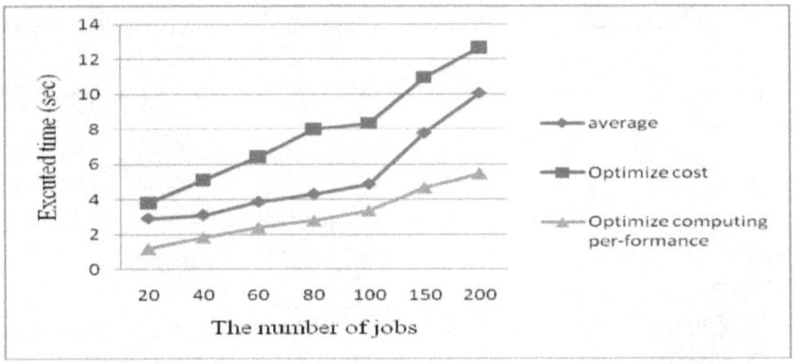

Fig. 5 The computed results by the three allocation methods

4 Conclusion

At present, in grid computing environment, job requirements are so large scale and complex that we need the allocating mechanism to manage the resources and schedule the jobs. Therefore, well-allocated mechanism is needed to enhance the grid resource be more useful and scalable.

In this paper, we propose a resource performance analysis model for grid nodes under the grid computing environment. By this model, we can analyze the information of grid nodes and make the plans and allocations of the resources of collaborated nodes optimize. According to the result of experiment, the mechanism can achieve the best resources allocation, and enhance the overall grid computing performance.

Acknowledgments. This work was supported in part by the National Science Council, Taiwan, under grant NSC-96-2745-M-034-002–URD.

References

1. Abramson, D., Buyya, R., Giddy, J.: A Computational Economy for Grid Computing and its Implementation in the Nimrod-G Resource Broker. Future Generation Computer Systems Journal 18(8), 1061–1074 (2002)
2. Buyya, R., Abramson, D., Giddy, J.: Nimrod/G: An Architecture for a Resource Management and Scheduling System in a Global Computational Grid. In: Proceedings of the 4th International Conference and Exhibition on High Performance Computing in Asia-Pacific Region, Beijing, China, vol. 1, pp. 283–289 (2000)
3. Buyya, R., Abramson, D., Giddy, J., Stockinger, H.: Economic Models for Resource Management and Scheduling in Grid Computing. The Journal of Concurrency and Computation: Practice and Experience (CCPE) 14, 1507–1542 (2002)
4. Foster, I., Kesselman, C.: The Grid 2: Blueprint for a new computing infrastructure. Morgan Kaufmann, San Francisco (2004)
5. Foster, I., Kesselman, C., Tuecke, S.: The anatomy of the grid: Enabling scalable virtual organizations. Int. J. High Perform. Comput. Appl. 15(3), 200–222 (2001)
6. Foster, I., Kesselman, C.: Gloubs: A Metacomputing Infrastructure Toolkit. International Journal of Supercomputer Application 11(2), 115–128 (1997)
7. Krauter, K., Buyya, R., Maheswaran, M.: A Taxonomy and Survey of Grid Resource Management Systems for Distributed Computing. International Journal of Software: Practice and Experience 32(2), 135–164 (2002)
8. Kandagatla, C.: Survey and Taxonomy of Grid Resource Management Systems, University of Texas, Austin (2003), `http://www.cs.utexas.edu/~browne/cs395f2003/projects/KandagatlaReport.pdf` (accessed June 25, 2008)
9. Lee, H.-M., Hsu, C.-C., Hsu, M.-H.: A Dynamic Supervising Model Based on Grid Environment. In: Khosla, R., Howlett, R.J., Jain, L.C. (eds.) KES 2005. LNCS, vol. 3682, pp. 1258–1264. Springer, Heidelberg (2005)
10. Lee, H.-M., Lee, T.-Y., Yang, C.-H., Hsu, M.-H.: An Optimal Analyzing Resources Model Based on Grid Environment. WSEAS Transactions on Information Science and Applications 5(3), 960–964 (2006)
11. Lee, H.-M., Lee, T.-Y., Hsu, M.-H.: A Process Schedule Analyzing Model Based on Grid Environment. In: Gabrys, B., Howlett, R.J., Jain, L.C. (eds.) KES 2006. LNCS, vol. 4253, pp. 938–947. Springer, Heidelberg (2006)
12. Miller Barton, P., Callaghan Mark, D., Cargille Jonathan, M., Hollingsworth Jeffrey, K., Bruce, I.R., Karavanic, K.L., Kunchithapadam, K., Tia, N.: The Paradyn Parallel Performance Measurement Tool. IEEE Computer 28(11), 37–46 (1995)

Emotional Processes in Computer Games

Khalil Shihab and Nida Chalabi

Abstract. Expressing emotions in computer games has become a popular focus for game research and development. Many research projects and papers emerged in the fields of game studies, psychology and HCI and others attempting to understand games and emotions. In this work, we present an emotional model that explains the emotional decision-making processes. The model is designed to explore people's behavior in certain circumstances, while under specified emotional states. Special attention was given to the thought process and actions displayed in the hypothetical scenarios. Also, we developed an experimental game program for the evaluation of our emotional decision making model.

1 Introduction

The concepts of an emotion and an emotional game may refer to very different things depending on one's viewpoint. Emotions may be states into which players indulge without consideration or they may have a functional role in carefully calculated processes. Some are interested in certain particularly emotional moments that stand out from the playing experiences while some see emotions as means of making sense of the experience. It is not either impossible to imagine someone to whom emotions, regardless of why they emerge, are stepping stones in making more compelling games or selling more games to a wider audience [1, 2].

We, however, considered emotion as a factor in the decision-making process and actions taken by an agent. Human emotions play a large part in how an individual thinks and acts. For example, decisions made in anger can often be different from those made otherwise. Likewise, trying to perform an action like throwing a ball can also be affected by the mood an individual is in, which is governed by emotions. Emotions can be a driving force behind the types of decisions

Khalil Shihab
School of Computing and Design, Swinburne University of Technology
e-mail: kshihab@swinburne.edu.my

Nida Chalabi
Department of Computer Science, SQU, Box 36, Al-Shod, 123, Oman
e-mail: nida@squ.edu.om

B.-C. Chien and T.-P. Hong (Eds.): Opportunities and Challenges, SCI 214, pp. 193–198.
springerlink.com

and actions and individual makes [3]. Depending on ones emotional state, the individual can make better or worst decisions and perform action more or less effectively [3, 4, and 5]. Therefore to bring artificial intelligence to the next level, that is closer to human, emotions need to be incorporated in the decision-making process and actions of agents. If agents can be made to behave with emotion then they will appear more human, which is exactly what is wanted (computer controlled agents simulate a human opponent).

Adopting this emotion approach to agent, artificial intelligence may not always result in an optimal decision or action [4, and 5]. Rather it will result in the best possible decision or action given the agents emotional state. Human players get angry, nervous and frustrated and this affects the way they play.

2 Research in Computer Games

Artificial intelligence (AI) has been growing and maturing in the passing years and the domain of video games has become an increasingly popular platform for artificial intelligence research [6]. As games become more complex and realistic, so too does the AI that drives these games. Games may be simplified when compared to the real world but none the less they provide complex, dynamic environments and situations which even human players find challenging. Although AI in videogames has been constantly improving, it is still at a stage where inflexible and predictable behavior is exhibited.

Gordon and Logan [5, and 6] have proposed GRUE, which uses teleo-reactive programs (TPRs) that basically consist of a series of rules, with each rule containing some number of conditions and actions.

In recent years, game theory and decision theory have had a profound impact of artificial intelligence in video games [6]. Traditionally, multi-agent systems using game-theoretic analysis for decision making use a normative approach [2]. It is here that decisions are derived rationally from the game description. However, this approach is believed to be insufficient and it does not capture the decision making process of real life agents. Real life agents (real people) may be partially irrational or may use models other than the real world (the game model) to make decisions.

3 The Emotional Decision Making Model

As shown in Figure 1, there are seven key stages in the 'emotional decision making model'.

We begin at point (1), the game agent. The game agent represents any computer-controlled entity. Moving on to point (2), referred to as 'emotion'. It is here that the agent's current emotional state is stored, which will continue to change as the game progresses and the game agent makes decisions and performs actions. Next we have point (3), referred to as 'decision'. Here with game agent will store all possible decision available while under a particular emotional state. The decision with the highest percentage value will always take precedence over decisions

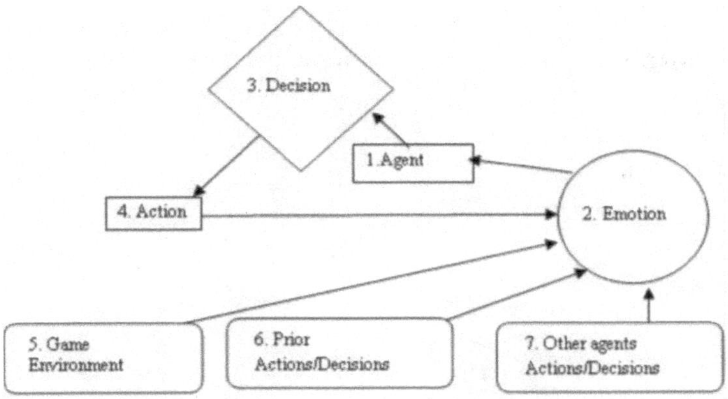

Fig. 1 The emotional decision making model

with lower percentage values, which will be executed. If there are two or more decisions with equal percentage values, the first decision in the list out of the possible decisions will be selected and executed. Decisions are stored as a list and are traversed until a suitable decision is found.

4 Model Implementation

In this experiment we used two agents that simulate the human reasoning process. When people reason about the behavior of others they often express their emotion (i.e., feeling sorry for someone, feeling happy for them, resenting their good fortune, or gloating over their bad fortune). To do this, agents maintain a list of cases establishing points of view of other agents and use these cases to take future actions.

The agents are able to participate in a multi-stage game in which one intelligent agent (1) observes and interacts with a naïve agent (2) express feelings about other agent's actions. The naïve agent uses those emotions to take the right action.

Our naïve agent can learn through the feedback from the intelligent agent. The agent can pass one room to another but has no knowledge of the environment. It does not know which sequence of doors the agent must pass to go outside the building.

The game environment is a simple evacuation of an agent from any room in the building, see Figure 2. At the start of the game, the agent in allocated in Room C and we want the agent to learn to reach outside the house (F).

We consider each room (including outside the building) as a state. Agent's movement from one room to another room is called action, see Figure 3.

Our intelligent agent will learn through experience by applying Q-learning technique. Figure 2 and Table 1 show the state diagram and the instant reward values respectively. The minus sign in the table says that the row state has no action to go to column state.

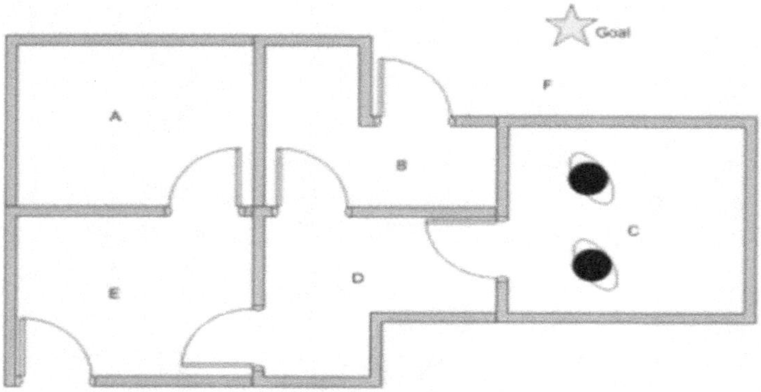

Fig. 2 A simple house evacuation

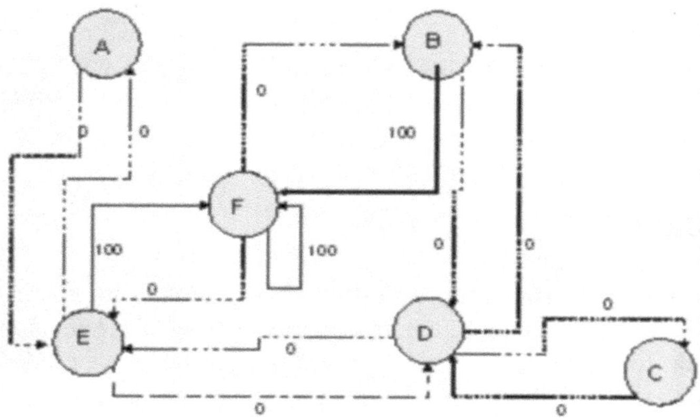

Fig. 3 The state diagram

The transition rule of this Q learning is given by the following formula
Q(state, action)=R(state, action)+α. Max[Q(next state, all actions)] (1)

Q Learning

Given : State diagram with a goal state (represented by matrix **R**)

Find : Minimum path from any initial state to the goal state (represented by matrix **Q**)

Q Learning Algorithm goes as follow

1. Set parameter **α**, and environment reward matrix **R**
2. Initialise matrix **Q** as zero matrix
3. For each episode:
   - Select random initial state
   - Do while not reach goal state
     - Select one among all possible actions for the current state
     - Using this possible action, *consider* to go to the next state

- Get maximum Q value of this next state based on all possible actions
- Compute the formula (1)
- Set the next state as the current state

End Do

Table 1 State reward values

	Action to go to state					
	A	B	C	D	E	F
A	-	-	-	-	0	-
B	-	-	-	0	-	100
C	-	-	-	0	-	-
D	-	0	0	-	0	-
E	0	-	-	0	-	100
F	-	0	-	-	0	100

The above algorithm is used by our intelligent agent to learn from experience or training. In each training session, the agent explores the environment (represented by Matrix **R**), get the reward (or none) until it reach the goal state. The purpose of the training is to enhance the 'brain' of our agent that represented by **Q** matrix. More training will give better **Q** matrix that can be used by the agent to move in *optimal* way.

Algorithm used by the naïve agent

Input: list of **emotions** (during the first play, the list is empty)

1. Set current state = initial state.
2. From current state, move to the next state.
3. Add the emotional expression of the intelligent agent to the list.
4. Set current state = next state

Go to 2 until current state = goal state

The algorithm above will return a list of sequence of states and their associated emotional expressions from initial state until goal state. For the next game sessions, the naïve will use the list of emotions that is produced by the first play to make its future moves.

5 Conclusion

This paper addresses the possibility of incorporating emotion in game agents. It begins by proposing a model, the emotional decision making model, and then applying emotional data to drive our emotional decision making model.

An experimental study is also presented showing the possibility of implementing our emotional model in a computer game.

References

1. Bererton, C.: State estimation for game AI using particle filters. In: Fu, D., Henke, S., Orkin, J. (eds.) Challenges in Game Artificial Intelligence: Papers from the 2004 AAAI Workshop, pp. 36–40. AAAI Press, Menlo Park (2004); Technical Report WS-04-04
2. Fielding, D., et al.: Using embodied agents to report on online computer games. In: Jennings, N.R., Sierra, C., Sonenberg, L., Tambe, M. (eds.) Proceedings of the Third International Joint Conference on Autonomous Agents and Multi-Agent Systems (AAMAS 2004), vol. 3, pp. 1530–1531. IEEE, New York (2004)
3. Orkin, J.: Symbolic representation of game world state: Toward real-time planning in games. In: Fu, D., Henke, S., Orkin, J. (eds.) Challenges in Game Artificial Intelligence, pp. 26–30. AAAI Press, Menlo Park (2004); Technical Report WS-04-04
4. Leesa, M., Theodoropoulos, G.K.: Agents, games and HLA. Simulation Modeling Practice and Theory 14(6), 752–767 (2006)
5. Gordon, E., Logan, B.: Managing goals and real world objects in dynamic environments. In: Davis, D. (ed.) Visions of Mind: Architectures for Cognition and Affect (2004)
6. Freeman, D.: Creating Emotion in Games. New Riders Publisher (2004) ISBN: 1592730078

Probability Reasoning Internet Assessment System with Rule Usage Structures and Similarity Comparisons

Yuan-Horng Lin

Abstract. The purpose of this study is to implement an Internet assessment on probability reasoning which provides graphs of rule usage and similarity coefficient in comparison with expert. Ordering theory (OT) combined with interpretive structural modeling (ISM) is the method to construct structural graphs of rule usage on probability reasoning. Set operation is adopted to calculate the similarity coefficient for graphs of rule usage. An empirical study for pupils shows that rule usage varies with the total score and there is significant difference on similarity coefficient based on age, gender and interaction. Finally, some recommendations and suggestions for future research are discussed.

Keywords: Internet assessment system, ordering theory, probability reasoning, rule assessment.

1 Introduction

Probability reasoning is one of the logic thinking which will influence cognitive development [1, 2, 12]. There are defective rules and correct rule as to probability reasoning test [3, 11, 13]. Total score from paper-pencil test provides limited information on cognition diagnosis and knowledge structures of probability reasoning [10]. Therefore, development of Internet assessment system for probability reasoning with diagnostic information of rule usage should be a prospective research.

Internet assessment system of probability reasoning test which is extended from research of Siegler will be implemented in this study. Ordering theory (OT) combined with interpretive structural modeling (ISM) is used to calculate the subordinate relationship among rules. Similarity coefficient for graphs of rule usage between examinee and expert will also be developed. An empirical study for pupils will be investigated and discussed.

Yuan-Horng Lin
Department of Mathematics Education, National Taichung University
140 Min-Sheng Rd., Taichung City 403, Taiwan
e-mail: lyh@mail.ntcu.edu.tw

B.-C. Chien and T.-P. Hong (Eds.): Opportunities and Challenges, SCI 214, pp. 199–205.
springerlink.com

2 Literature Review

Rules of problem-solving for probability reasoning marble test will be discussed. OT and ISM are introduced in view of their algorithm and application.

2.1 Probability Reasoning and Rule Usage of Marble Test

One well-known probability reasoning is called marble test [3]. Fig. 1 is an example and it is represented as (3, 4) vs. (5, 6). It means that there are 3 black marbles and 4 white marbles in set A and 5 black marbles and 6 white marbles in set B.

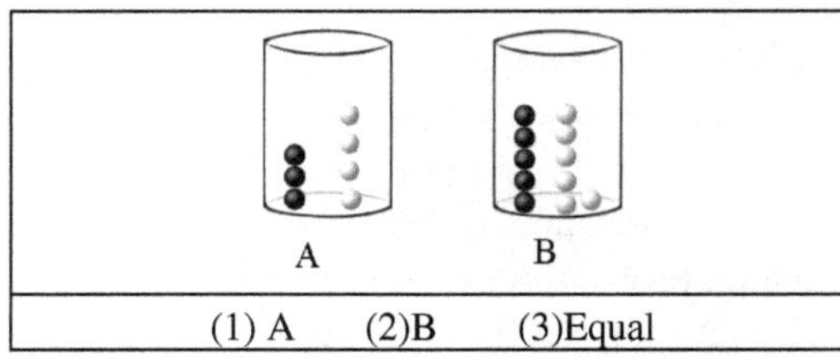

Fig. 1 Example of marble test item

Students are asked to imagine picking one marble randomly from the two sets and must decide which set provides greater chance of picking a black marble, set A or set B or "equal." Siegler indicates that there are four rules when students respond to marble test items [3]. Rule 1 to Rule 3 are defective rules and Rule 4 is correct rule.

2.2 Ordering Theory

OT is mainly to determinate the hierarchical relationship among items [6]. Suppose there be n examinee responding two dichotomous item i and item j. n_{01} is the frequency of examinee who respond item i incorrectly but respond item j correctly. $r_{ij} = n_{01}/n$ is ordering coefficient and smaller value means more possibility that item i is the precondition of item j [14]. Tolerance level ε $(0 < \varepsilon < 1)$ is to decide whether the binary precondition relationship between two items exists or not. It is

$$r_{ij}^{*} = \begin{cases} 1 \ , & r_{ij} < \varepsilon \\ 0 \ , & r_{ij} \geq \varepsilon \end{cases} \tag{1}$$

$r_{ij}^{*} = 1$ means item i is the precondition of item j with linkage from item i to item j. On the other hand, there is no linkage from item i to item j [7, 8, 14]. OT will be used to calculate the ordering coefficient for subordinate relationship of rule usage in this study.

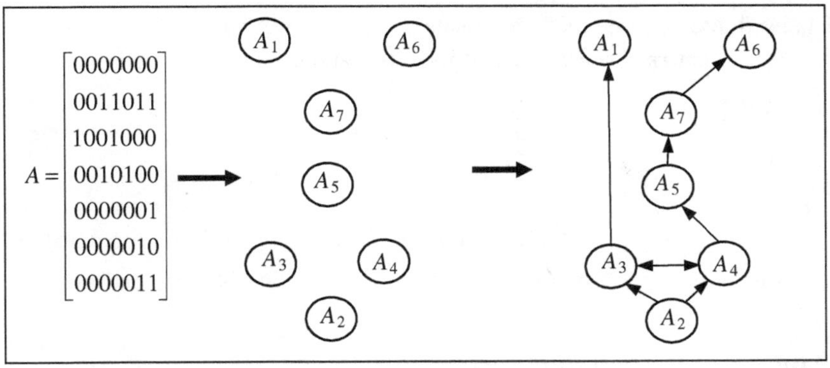

Fig. 2 Construction of hierarchical graph by ISM

2.3 Interpretive Structural Modeling

ISM aims to arrange elements by hierarchical graph [5, 16]. Binary relationship among K elements is denoted by binary relation matrix $A=(a_{ij})_{K\times K}$. $a_{ij}=1$ means A_i is the precondition of A_j. Otherwise, means $a_{ij}=0$ presents A_i is not the precondition of A_j. An example of the construction of ISM graph is depicted in Fig.2.

3 Research Design and Data Resource

Calculation of items is discussed and Internet assessment system design with sample will be discussed.

3.1 Ordering Calculation for Rule Usage

If the response of student n on item m conforms with rule r , it is denoted by $s_{mr}=1$; otherwise it is $s_{mr}=0$. For rule r and rule r', contingency table for student n responding f items is depicted in Table 1.

Table 1 Contingency table of frequency for two rules

		Rule r'		Sum
		1	0	
Rule r	1	f_{11}	f_{10}	$f_{1\bullet}$
	0	f_{01}	f_{00}	$f_{0\bullet}$
Sum		$f_{\bullet 1}$	$f_{\bullet 0}$	f

OT is used to determine the subordinate relation between rule r and rule r'. The ordering coefficient is $f_{rr'} = f_{01}/f$ and its binary subordinate relationship is

$$f_{rr'}^{*} = \begin{cases} 1 \quad , \quad f_{rr'} < \varepsilon \\ 0 \quad , \quad f_{rr'} \geq \varepsilon \end{cases} \tag{2}$$

$F_n^{*} = (f_{rr'}^{*})_{R\times R}$ is the binary relation matrix of student n on R rules to construct graphs of rule usage and the algorithm of graph construction is based on ISM.

3.2 Similarity Calculation for Graph of Rule Usage

Similarity coefficient for graphs of rule usage between student n and expert is

$$s_{n(expert)} = (\frac{1}{R})\sum_{r=1}^{R} \frac{\#\left(G_n(v_r) \cap G_{expert}(v_r)\right)}{\#\left(G_n(v_r) \cup G_{expert}(v_r)\right)} \tag{3}$$

It is $G_n(v_r) = \{v_r \mid f_{rr'}^{*} = 1\}$ and $\#(G_n(v_r) \cap G_{expert}(v_r))$ is the number of rules belonging to $(G_n(v_r) \cap G_{expert}(v_r))$; while $\#(G_n(v_r) \cup G_{expert}(v_r))$ is the number of rules belonging to $(G_n(v_r) \cup G_{expert}(v_r))$. Larger similarity coefficient $s_{n(expert)}$ means more similar in graphs of these two rule usage.

3.3 Probability Reasoning Items and Internet Assessment System Design

Probability reasoning test includes 20 marble items. Operation procedure of the system and subjects are depicted in Fig. 3 and Table 2.

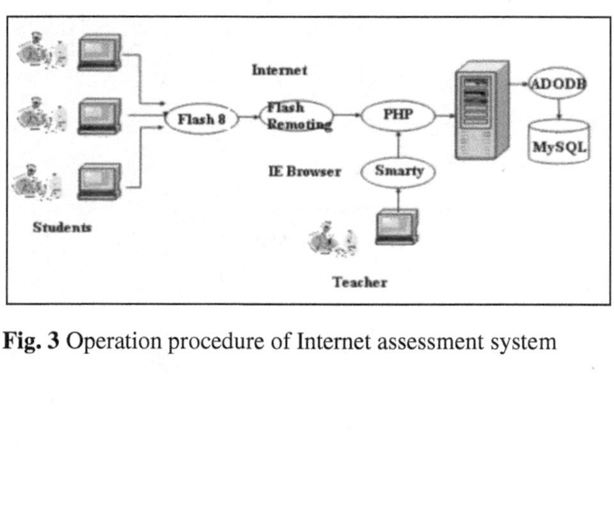

Fig. 3 Operation procedure of Internet assessment system

Table 2 Subjects with grade and gender

Grade	Gender		Total
	Male	Female	
5	438	395	833
6	371	339	710
7	543	535	1078
8	373	345	718
Total	1725	1614	3339

4 Results

Graphs of rule usage according to different total score show the characteristics of rule usage. Two way analysis of variance (two way ANOVA) based on grade and gender display whether there are significant differences respective to similarity coefficient.

4.1 Rule Usage of Different Total Score

Subjects within the highest 27% total score belong to high score group and subjects within the lowest 27% total score belong to low score group. The others belong to middle total score group. Three students are randomly selected from the above three groups and tolerance level ε is decided $\varepsilon = 0.1$. Fig. 4 shows graphs of rule usage for expert and three students. Student A, B, C, won distinct graph of rule usage. Their similarity coefficients compared with expert are 1, .71 and .33 respectively.

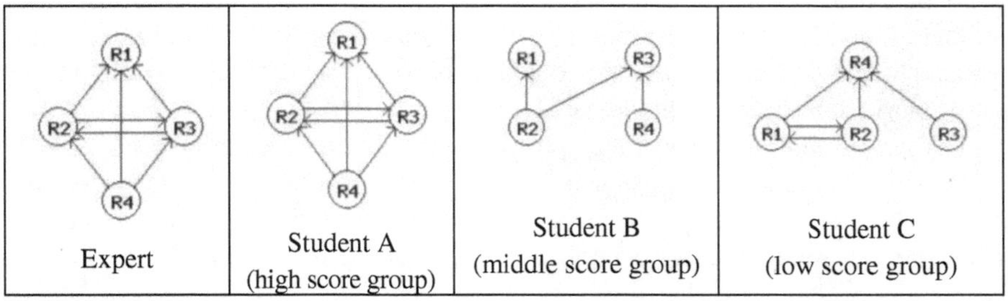

Expert | Student A (high score group) | Student B (middle score group) | Student C (low score group)

Fig. 4 Graphs of rule usage for expert and three students of different groups

4.2 Comparisons on Similarity Coefficient

Age and gender are considered as to variance of rule usage. As shown in Table 3, there are significant differences on grade and gender and there also exists interaction between grade and gender. Table 4 displays post hoc comparison for simple main effect of grade × gender.

Table 3 Two way ANOVA on similarity coefficient for grade and gender

Source of variance	SS	df	MS	F	Post hoc comparison
Grade	16.70	3	5.57	139.25***	7 > 5, 8>5, 7>6, 8>6, 8>7
Gender	.20	1	.20	5.00*	female > male
Grade × Gender	.41	3	.14	3.50*	
Error	138.13	3331	.04		
Total	155.44	3338			

*p<.05 ***p<.001

Table 4 Post hoc comparison for simple main effect of grade × gender

Source of variance		SS	df	MS	F	Post hoc comparison
Grade	male	9.81	3	3.27	78.07***	8 > 7 > 6 > 5
	female	7.39	3	2.46	60.00***	7 > 5, 7 > 6, 8 > 5, 8 > 6
Gender	5	.01	1	.01	.25	
	6	.48	1	.48	11.67**	female > male
	7	.11	1	.11	2.36	
	8	.01	1	.01	.03	

**p<.01 ***p<.001

5 Conclusions

One major result is that the Internet assessment system for graph of rule usage are established. Rule usage also provides information of hierarchies and relationship. Secondly, formula of similarity coefficient for graphs of rule usage is developed so that comparisons among graphs of rule usage are feasible. Thirdly, two variables, age and gender, are considered and it shows there are significant differences on similarity coefficient as grade, gender and interaction.

The assessment system with graph of rule usage and similarity coefficient is beyond the limitation of paper-pencil test. Further research could consider rule assessment methodology. Other cognitive issue on rule usage like proportion reasoning and relation reasoning could be a prospective research [4, 9, 15].

References

1. Piaget, J., Inhelder, B.: The Origin of the Idea of Chance in Children. Routledge and Kegan Paul (1975)
2. Konold, C.: Informal concepts of probability. Cognition and Instruction 6, 59–98 (1989)
3. Siegler, R.S.: Developmental of sequences within and between concepts. Society for Research in Child Development Monographs 46, Whole No. 189 (1981)

4. Siegler, R.S.: The rule-assessment approach and education. Contemporary Educational Psychology 7, 272–288 (1982)
5. Warfield, J.N.: Interpretive structural modeling (ISM). In: Olsen, S.A.W.E., Walter, W.V., Lehner, W. (eds.) Group Planning & Problem Solving Methods in Engineering, pp. 115–201. Wiley, New York (1982)
6. Bart, W.M., Krus, D.J.: An ordering-theoretic method to determine hierarchies among items. Educational and Psychological Measurement 33, 291–300 (1973)
7. Bart, W.M., Williams-Morris, R.: A refined item diagraph analysis of proportional reasoning test. Applied Measurement in Education 3, 143–165 (1990)
8. Bart, W.M., Post, T., Behr, M., Lesh, R.: A diagnostic analysis of a proportional reasoning test item: An introduction to the properties of a semi-dense item. Focus on Learning Problems in Mathematics 16, 1–11 (1994)
9. Jansen, B.R.J., Han van der Maas, M.L.J.: Statistical test of the rule assessment methodology. Developmental Review 17, 321–357 (1997)
10. Lin, Y.H., Hung, W.L.: Robust clustering on rule usage of probability reasoning with raw rule score. In: The 4th International Conference on Fuzzy Systems and Knowledge Discovery, Haikou, China, pp. 251–255 (2007)
11. Cosmides, L., Tooby, J.: Are humans good intuitive statisticians after all? Rethinking some conclusions from the literature on judgment under uncertainty. Cognition 58, 1–73 (1996)
12. Fischbein, E., Gazit, A.: Does the teaching of probability improve probabilistic intuitions? Educational Studies in Mathematics 15, 1–24 (1984)
13. Tarr, J.E., Jones, G.A.: A framework for assessing middle school students' thinking in conditional probability and independence. Mathematics Education Research Journal 9, 39–59 (1997)
14. Bart, W.M., Rothen, W., Read, S.: An ordering-analytic approach to the study of group differences in intelligence. Educational and Psychological Measurement 46, 799–812 (1986)
15. Goldsmith, T.E., Johnson, P.J., Acton, W.H.: Assessing structural knowledge. Journal of Educational Psychology 83, 88–96 (1991)
16. Warfield, J.N.: Crossing theory and hierarchy mapping. IEEE Transactions on System, Man, and Cybernetics 7, 505–523 (1977)

Combining Genetic Algorithm and Simulated Annealing to Design H_2/H_∞ Deconvolution Filter with Missing Observations

Jui-Chung Hung* and Wei-Chi Lee

Abstract. In this paper, we combine GA (genetic algorithm) and SA (simulated annealing) to approach H_2/H_∞ deconvolution filter with missing observations. The missing observations model is based on a probabilistic structure. The probability of occurrence of missing data is unknown prior. The aim of H_2/H_∞ criterion is to achieve the H_2 optimal reconstruction and subject to the H_∞ norm constraint on the transfer function from the channel input to the filter error. In this situation, the design deconvolution filter becomes a complicated nonlinear estimation problem. In this paper, we combine the selected features form GA and SA to achieve weak dependence on initial parameters and fast convergence to treat the signal reconstruction problem with missing observations. Finally, a numerical example is presented to illustrate the design procedure and confirm the robustness performance of the proposed method.

1 Introduction

Deconvolution filters have widespread applications in the field of engineering literature, especially in signal processing applications. This problem is widely ranged of application in the field of engineering such as equalization, seismology, and image restoration [1-3].

The deconvolution problems are generally mainly two frameworks, one is H_2, and the other is H_∞ Most of the deconvolution problems are generally solved via H_2 optimal method in the time domain [4], and frequency [1]. When the channel are perturbed the inadequacy of H_2 optimal filter. A robust deconvolution filter design based on H_∞ theory has received great attention for its robustness property against the system uncertainties [2]. However, the H_∞ robustness design can not achieve the optimal performance. The proposed H_2/H_∞ optimal deconvolution

Jui-Chung Hung and Wei-Chi Lee
Department of Information Technology, Ling Tung University, Taiwan
e-mail: juichung@seed.net.tw

* Corresponding author.

B.-C. Chien and T.-P. Hong (Eds.): Opportunities and Challenges, SCI 214, pp. 207–212.
springerlink.com

filter design is to minimize an H_2 reconstruction performance subject to a robustness requirement based on the H_∞ norm to attenuate the performance degradation due to the system's uncertainty [5]. This problem can be interpreted as a problem of optimal reconstruction filter design subject to a robustness constraint against the deterioration due to parameter variation of channel.

The GA has been introduced for optimization searching [6-8]. It is a parallel global search technique that emulates natural genetic operators such as reproduction, crossover, and mutation. In particular, The GA need not assume the search space being differentiable or continuous, and can also iterate several times on each data received.

The SA algorithm [9] is a numerical simulation method based on the dynamics of crystallization. Consider heating a solid until its constituents can move freely and it melts. Then, the melt is allowed to cool very slowly until it solidifies in a certain arrangement. SA's strategy is essentially serial. Geman [9] proved that if the cooling schedule is slow enough, the desired global extreme can always be found. Comparing GA and SA, we can find that GA exhibits fast initial convergence, but its performance deteriorates as it approaches the desired global extreme. Interestingly, SA shows a complementary convergence pattern, in addition to high accuracy. We combines the selected features from GA and SA to achieve weak dependence on initial parameters, parallel search strategy, fast convergence and high accuracy.

2 H_2/H_∞ Deconvolution Filter under Missing Observations

Consider a discrete deconvolution system with missing observations as shown in Fig. 1. The received signal $y_m(k)$ is as

$$\begin{aligned} y_m &= g(k)y(k)+v(k) = g(k)\left[x(k)+r(k)\right]+v(k) \\ &= g(k)\left[H(z^{-1})C\left(z^{-1}\right)\omega(k)+N(z^{-1})n(k)\right]+v(k) \end{aligned} \tag{1}$$

where z^{-1} is the inverse of z (i.e., unit delay), $v(k)$, $\omega(k)$, and $n(k)$ are assumed toe be zero-mean white Gaussian noises. The $g(k)$ is the model of missing observations, such as

$$g(k) = \begin{cases} 0 & \text{if } y(k) \text{ is missing} \\ 1 & \text{otherwise} \end{cases} \tag{2}$$

Thus y_m can be regarded as the measurements of $y(k)$ with missing observations. The sequence $g(k)$ is assumed to be asymptotically stationary and independent of $y(k)$. Furthermore, they are mutually independent. The probability of missing measurement is as

$$E\left[g(k)\right] = \Pr\left[g(k)=1\right] = p \tag{3}$$

where p is a fixed probability, independent of time. Let $\hat{u}(k)$ denote the estimate of $u(k)$. From Fig. 1, the power spectral density of $e(k)$ is given by

$$\begin{aligned}\Phi(z^{-1}) = p\left[1 - D(z^{-1})H(z^{-1})\right] \; C(z^{-1})\sigma_{\omega}^{2}C^{*}(z^{-1})\times\left[1 - D(z^{-1})H(z^{-1})\right]^{*} + \\ p\cdot D(z^{-1})N(z^{-1})\sigma_{n}^{2}N^{*}(z^{-1})\times D^{*}(z^{-1}) + D(z^{-1})\sigma_{v}^{2}D^{*}(z^{-1})\end{aligned} \tag{4}$$

where superscript $*$ denotes the complex conjugate and $\Phi(z^{-1})$ is the power spectral density of $e(k)$.

In the fixed-order H_2 optimal deconvolution filter design case, a stable filter $D(z^{-1})$ will be specified to minimize mean square error (MMSE) [4]. By the Residue theorem, the H_2 optimal deconvolution problem in (5) becomes the following minimization problem

$$\min_{D(z^{-1}),\, p} \; I_2 = \min \sum_{j=1}^{n} \operatorname{Re}s_{zj}\left(\frac{\Phi(z^{-1})}{z}\right) \tag{5}$$

In practical deconvolution systems, the noise variances σ_v^2, σ_ω^2, and σ_n^2 may vary and the observations may missing. How to eliminate the performance degradation due to noise uncertainties and channel perturbation to guarantee the reconstruction performance is an important topic in practical signal deconvolution problem. The H_∞ robustness design can guarantee the worst-case effect of these noise uncertainties or channel perturbation on the reconstruction performance to be less than a prescribed level [5], i.e., if the H_∞ norm of error spectrum $\Phi(z^{-1})$ is less than ε, the sensitivity of reconstruction error $e(k)$ to the noise uncertainties or channel perturbation must be less than ε from the energy point of view. In this study, in order to take advantage of both H_2 optimal reconstruction and less sensitivity to noise uncertainties and channel perturbation, a fixed-order deconvolution filter $D(z^{-1})$ is specified to achieve the H_2 minimization of reconstruction error in (6) and at the same time to satisfy the following H_∞ robustness requirement,

$$I_\infty = \| \Phi(z^{-1}) \|_\infty =: \sup_{w\in[0,\, \pi]} | \Phi(e^{-jw}) | < \varepsilon \tag{6}$$

where ε is a positive value, i.e., the optimal reconstruction in (6) and H_∞ robustness in (7) should be satisfied simultaneously.

How to search the coefficients of $D(z^{-1})$ via genetic algorithm to solve the mixed H_2 / H_∞ deconvolution filter design problem is the key point in our design. This nonlinear complicated design problem will be treated by GA/SA in the next section.

3 Deconvolution Filter Design

The genetic algorithm is composed of three operations: (1) reproduction, (2) crossover, and (3) mutation [6-8]. These operations are implemented by performing the basic tasks of copying strings, exchanging portions of strings, and changing the state of bits from 1's to 0's or vice-versa. These operations ensure that the "fittest" members of the population survive and their information is preserved and combined to generate still better offspring. The result is an improvement in the next generation's performance. Therefore, they are more suitable than most searching algorithms for H_2/H_∞ optimal deconvolution filter. A simple GA for the H_2/H_∞ optimal deconvolution filter design can be found in [7].

The SA's consists of three components: (1) generation of neighbor candidate solution by perturbation of current solution (2) acceptance test of solution by Boltzmann probability (3) iterative procedure.
The design procedure of GA/SA parameter estimation with noises and missing observations is divided into the following steps.

1. Given the received data $y_m(k)$, the order n of the deconvolution filter $D(z^{-1})$, the robustness constraint ε, and generate random population $(a_1,\cdots, a_n, b_0,\cdots, b_n, p)$ of Γ chromosomes.
2. Check the H_∞ robustness constraint in (7). If the robustness constraint is not satisfied, then renew the chromosomes.
3. Compute the H_2 performance in (6).
4. Compute the corresponding fitness value.
5. Use the GA operators (reproduction, crossover, mutation) to produce chromosomes of next generation.
6. Repeat the produce from step 2 to step 5 until the fittest individual remains the same for L_g generations.
7. Given the best chromosome, C_d, C_i, C_t, k, T, B and δ_a^2 of SA initial values.
8. Generate the new-state by

$$(a_1,\cdots, a_n, b_0,\cdots, b_n, p)' = (a_1,\cdots, a_n, b_0,\cdots, b_q, p) + Norm(0, \delta_a^2)$$

9. Check the H_∞ robustness constraint $I_\infty(a_1\cdots a_l, b_0\cdots b_n, p)<\varepsilon$. If the robustness constraint is not satisfied, then renew the new-state.
10. Compute the new-state energy K from(8).
11. Let $p_s(K') = \frac{1}{1+\exp(-K'/BT)}$; if $p_s(K')$ >random [0,1) or new-state energy (K') < original-state energy (K) then the current-state = new-state.
12. If the temperature T is the same in k iterations, then decrease the temperature as $T=C_t T$.

Then, repeat the procedure from step 8 to step 12 until system has been frozen.

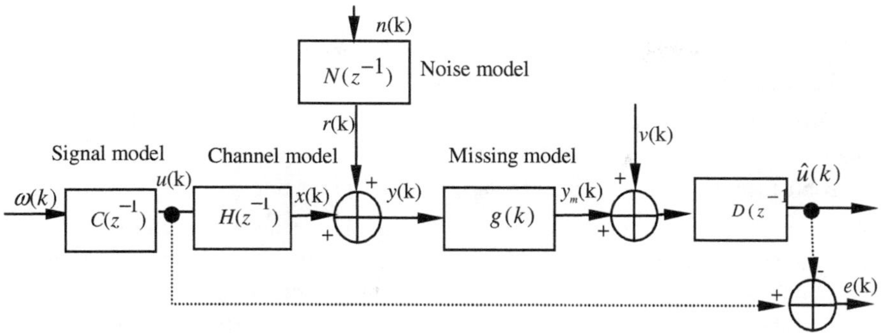

Fig. 1 Deconvolution filter system model with irregular missing observations

4 Design Example

Consider a non-minimum phase deconvolution system in Fig.1, the missing sequence $g(k)$ is a random process using Bornoulli modulations given in (3). The robustness constraint, signal, channel, missing model and noise model are described as follows:

$$C(z^{-1})=\frac{1+0.85z^{-1}}{1+0.5z^{-1}},\ H(z^{-1})=\frac{(1+1.45z^{-1}+0.5z^{-2})}{(1-0.1z^{-1})(1+z^{-1})}\ N(z^{-1})=\frac{(1+1.65z^{-1}+0.72z^{-2})}{(1-0.1z^{-2})}$$

The driving signal $\{\omega(k)\}$ and disturbance noise $\{n(k)\}$, and measurement noise $v(k)$ are also assumed to be independent, stationary and white with zero mean and variances as given by $\sigma_\omega^2=1$, $\sigma_n^2=0.1$, $\sigma_v^2=0.1$, respectively. The convergence of cost function K of the H_2/H_∞ optimal filter via GA/SA is shown in Fig. 2 with respect to $p=0.9$. Note that the cost functions of the GA-based estimation method are with exponential and rapid convergence at the beginning of generation, but its performance is improved very slowly as it approaches the desired global extreme. The proposed GA/SA based estimation algorithm always starts the search procedure as a pure-GA parameter estimation and ends as a pure-SA parameter estimation.

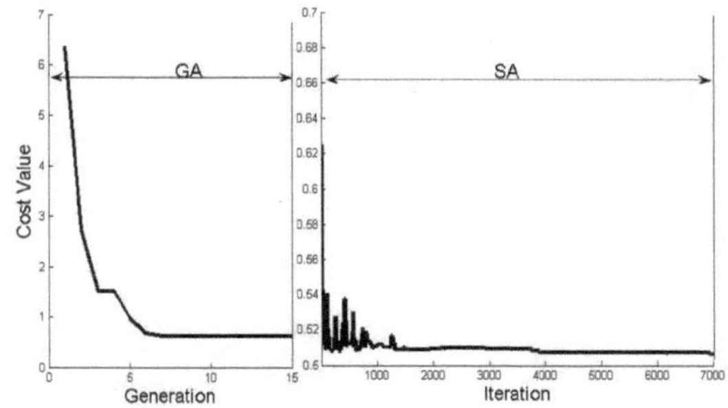

Fig. 2 The cost function by proposed GA/SA estimation algorithm with $p=0.9$

5 Conclusion

In this paper, design methods of mixed H_2/H_∞ optimal deconvolution filters with missing data have been introduced via GA/SA. This deconvolution filter design method takes advantage of H_2 optimal reconstruction performance and H_∞ robustness against the channel variation and noise uncertainties. It is obvious that the reconstruction performance is improved significantly if the missing probability is considered in the proposed deconvolution filter design procedure in the case of data missing. The GA/SA design algorithm rapidly converges and the reconstruction performance is acceptable even if the order of the deconvolution filter is lower.

References

1. Silvia, M.T., Robinson, E.A.: Deconvolution of geophysical time series in the exploration for oil and natural gas. Elsevier, New York (1979)
2. Vural, C., Sethares, W.A.: Blind image deconvolution via dispersion minimization. Digital Signal Processing 16, 137–148 (2006)
3. Chen, B.S., Peng, S.C.: Optimal deconvolution filter design based on orthogonal principle. Signal Process 25, 361–372 (1991)
4. Gelfand, S.B., Krogmeier, J.V., Wei, Y.: Uniform observability and exponential convergence rate of the Kalman filter for the FIR deconvolution problem. Signal Process 81, 593–607 (2001)
5. Wang, S., Xie, L., Zhang, C.: Mixed H_2/H_∞ deconvolution of uncertain periodic FIR channels. Signal Process 81, 2089–2103 (2001)
6. Holland, J.H.: Outline for a logical theory of adaptive systems. J. ACM 3, 297–314 (1962)
7. Hung, J.C.: A genetic algorithm approach to the spectral estimation of time series with noise and missed observations. Information Sciences 178, 4632–4643 (2008)
8. Metropolis, N.A., Rosenbluth, M., Rosenbluth, A., Teller, A.: Equation of State Calculations by Fast Computing Machines. Journal of Chemical Physics 21j, 1087–1092 (1953)
9. Geman, S., Geman, D.: Stochastic relaxation, Gibbs distributions and the Bayesian restrotation of images. IEEE Trans. Pattern Anal. Mach. Intell., 721–741 (1984)

The Hybrid Model Development of Clustering and Back Propagation Network in Printed Circuit Board Sales Forecasting

Yen-Wen Wang, Chen-Hao Liu, and Chin-Yuan Fan

Abstract. Reliable prediction of sales can improve the quality of business strategy. This research develops a hybrid model by integrating K-mean cluster and Back Propagation Network (KBPN) to forecast the future sales of a printed circuit board factory. Base on the K-mean clustering technique, the history data can be classified into different clusters, thus the noise of the original data can be reduced and a more accurate prediction model can be established. Numerical data of various affecting factors and actual demand of the past 5 years of the printed circuit board (PCB) factory are collected and input into the hybrid model for future monthly sales forecasting. Experimental results show the effectiveness of the hybrid model when comparing it with other approaches.

1 Introduction

Printed circuit board (PCB) industry plays an important role in Taiwan economy, but severe inventory stacking and material lacking problems still exist. How to predict PCB customer's demand and prepare material flows in advance to reduce the cycle time has become a pressing issue to be dealt with. Therefore, it becomes indispensable to build a forecasting model to predict the monthly sales in PCB industry through an efficient and effective manner.

Sales forecasting is a very general topic of research. When dealing with the problems of sales forecasting, many researchers have used hybrid artificial intelligent

Yen-Wen Wang
Department of Industrial Engineering and Management, Chin Yun Tech. University, 229 Chien-Hsin Rd., Taoyuan 320, Taiwan, R.O.C.
e-mail: ywwang@cyu.edu.tw

Chen-Hao Liu
Department of Information Management, Kai-Nan University,Taoyuan, Taiwan, R.O.C.

Chin-Yuan Fan
Department of Industrial Engineering and Management,
Yuan-Ze University, Taoyuan, Taiwan, R.O.C.

B.-C. Chien and T.-P. Hong (Eds.): Opportunities and Challenges, SCI 214, pp. 213–218.
springerlink.com

algorithms to forecast, and the most rewarding method is the application integrating artificial neural networks (ANNs). This method is applied by incorporating the experience-based principal and the capacity of memory and error-allowance of ANNs, as well as self learning by numeral data. This research focuses on the sales forecasting of PCB and modifies the back-propagation network system (BPN) with the purpose of improving the forecasting accuracy and using this information to help managers make decisions.

2 Literature Review

Although the traditional sales forecasting methods have been proved effective, they still have certain shortcomings. As ref. [6], the new developed Artificial Intelligent (AI) models have more flexibility and can be used to estimate the non-linear relationship, without the limits of traditional Time Series models. Therefore, more and more researchers tend to use AI forecasting models to deal with problem.

ANN appear to be particularly suited for financial time series forecasting, as they can learn highly non-linear models, have effective learning algorithms, can handle noisy data, and can use inputs of different kinds (see ref.[3]). Furthermore, complex non-linear models based on exponential GARCH processes [1] show similar results (in terms of out-of-sample prediction performance) to those obtained by much simpler ANN based on multi-layer perception (MLP) architectures [2].

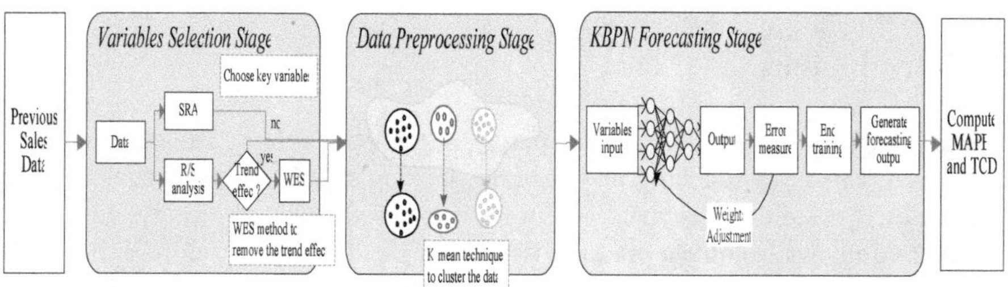

Fig. 1 Architecture of This Research

3 Methodology

There are three main stages in this research (as shown in fig.1) and the first stage is the variables selection stage. This stage is to select many possible variables, which may influence PCB product sales amount. In order to eliminate the unrelated variables, Stepwise Regression Analysis (SRA) was used to choose the key variables to be considered in the forecasting model. The second stage is the data preprocessing stage, K-mean cluster technique would be adopted. The parameter of cluster number k need to be determined first, after experimental design, the best cluster number will be adopted to measure the testing data. The last stage is the K-mean BPN (KBPN) forecasting stage, which was developed to forecast the demand of PCB sales amount in this research and will be described in details in the following section. After being compared with other three forecasting models, the superior model will be recommended to the decision makers. The details of each stage will be described as follows:

3.1 Variable Selection Stage

In this stage, fewer factors were considered in order to increase the efficiency of network learning. Many researchers have used several methods to select key factors in their forecast system (Ref. [5]). In this research, the SRA method was used to determine the main factors that would influence the PCB sales amount.

3.1.1 Stepwise Regression Analysis (SRA)

Stepwise regression procedure determines the set of independent variables that most closely determine the dependent variable. This is accomplished by the repetition of a variable selection. At each of these steps, a single variable is either entered or removed from the model. Each of these regressions is subjected to an 'F-test'. This general procedure is easily applied to polynomials by using powers of the independent variable as pseudo-independent variables.

3.1.2 Winter's Exponential Smoothing (WES)

In order to take the effects of seasonality and trend into consideration, Winter's Exponential Smoothing (WES) is used to preliminarily forecast the quantity of PCB production. According to this method, three components to the model are assumed: a permanent component, a trend, and a seasonal component. Each component is continuously updated using a smoothing constant applied to the most recent observation and the last estimate. In this research we assume $\alpha = 0.1$, $\beta = 0.1$ and $\gamma = 0.9$.

3.2 Data Preprocessing Stage

The K-means clustering method starts with k initial seeds of clustering, one for each cluster. All the n objects are then compared with each seed by means of the Euclidean distance and assigned to the closest cluster seed. The accuracy of the K-means procedure is very dependent upon the choice of the initial seeds. To obtain better performance the initial seeds should be very different among themselves. One efficient strategy to improve the K-means performance is to use, for example, the Ward's procedure first to divide the n objects into k groups and then use the average vector of each of the k groups as the initial seeds to start the K-means. As all the agglomerative clustering procedures, this method is available in a majority of statistical software.

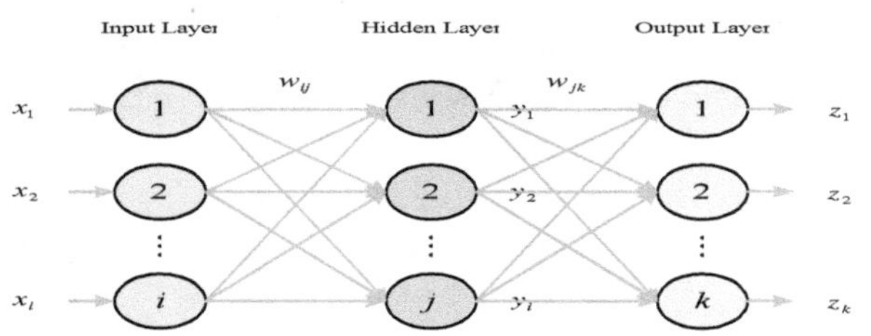

Fig. 2 The Structure of Back-Propagation Neural Network

3.3 Back Propagation Network Forecasting Stage

An Artificial Neural Network (ANN) is a simplified simulation of biological neural networks in human brains. The back-propagation network (BPN) is an ANN using back-propagation algorithm and is one of the popular ANNs, which has been widely applied to many scientific and commercial fields for non-linear analysis and prediction. The structure of BPN contains three layers: input, hidden, and output layers as shown in fig. 2. Each layer contains I , J , and K nodes denoted respectively by circles. The node is also called neuron or unit. The circles are connected by links, denoted by arrows in fig. 2, each of which represents a numerical weight. The w_{ij} is denoted as numerical weights between input and hidden layers and so is w_{jk} between hidden and output layers as also shown in fig. 2. The processing or the computation is performed in each node in the hidden and output layers.

3.4 Evaluating Performance Index

In order to evaluate the accuracy and performance of different forecasting models, this research adopts three evaluating indexes: Mean Absolute Percentage Error and Total Cost Deviation. The calculating formula are as follows:

1. MAPE (Mean Absolute Percentage Error)

$$MAPE = \frac{1}{n}\sum_{t=1}^{n}\frac{|F_t - A_t|}{A_t}$$

2. TCD (Total Cost Deviation)

$$TCD = \frac{1}{n}\sum_{t=1}^{n}\left(|F_t - A_t| \times \frac{S_t}{A_t}\right)$$

Where, F_t is the expected value for period t, A_t is the actual value for period t, S_t the actual sales amount of PCB for period t, S_t/A_t is the unit cost of each item, n is the number of periods.

The smaller the values of the above three indexes are, the better the forecasting models will be; smaller values means that the calculating results are closer to the historic data.

4 Comparison and Experimental Results

The data in this research are from an electronic company in Taiwan within 5 years. Monthly sales amount is considered as an objective of the forecasting model. The variations of the historical monthly sales data from the subject PCB Company are shown in Fig.3. This research develops a clustering BPN for sales forecasting in PCB industries and we will compare this method with other traditional methods such as Winter's Exponential Smoothing (WES), Multiple Regression Analysis (MRA), Back-propagation network (BPN) and Genetic Algorithm with Neural Network (GANN), which method has been applied in our previous research (ref. [4]).

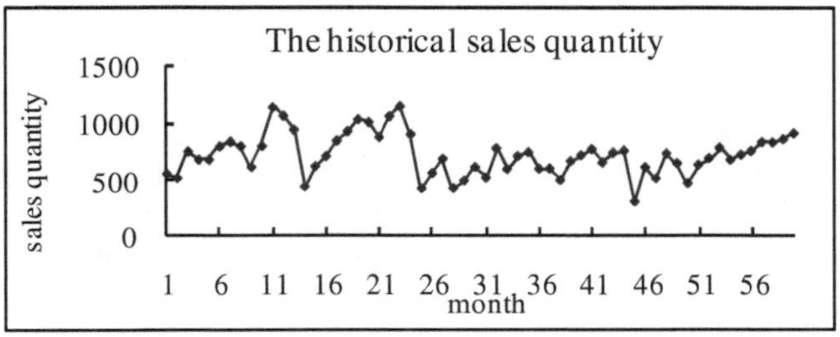

Fig. 3 Variations of the Historical Monthly Sales in Taiwan PCB Company

Table 1 MAPE of Different Forecasting Models

Method	MRA	WES	BPN	GANN[4]
Average	7.14%	8.77%	6.57%	3.06%
Std	4.78%	10.84%	5.15%	2.61%
Max	16.85%	29.23%	15.36%	8.40%
Min	1.11%	0.10%	0.19%	0.01%

Table 2 MAPE and TCD of GANN and KBPN with different clusters

Method	GANN	KBPN (K=2)	KBPN (K=3)	KBPN (K=4)
MAPE Average	3.06%	2.99%	2.44%	2.16%
MAPE Std	2.61%	2.21%	1.69%	1.92%
TCD (million)	18.9	18.8	14.3	12.2

MAPE and TCD were applied as a standard performance measure for all different models in this research. After the intensive experimental test (table 1), GANN is found to be quite accurate. These results are found to be superior to other three forecasting models. By the advanced non-linear problem forecasting ability of neural network and the powerful searching capability of genetic algorithm, GANN model achieves very good accuracy for forecasting PCB production quantity. Since GANN performs better than other models, following we compare GANN with KBPN and the results can be found in table 2.

5 Conclusions

The experimental results in section 4 demonstrated the effectiveness of the KBPN that is superior to other traditional approaches. The KBPN approach also provides another informing tool to the decision maker in PCB industries. In summary, this research has the following important contribution in the sales forecasting area and these contributions might be interested to other academic researchers and industrial practitioners:

This research applies two different performance measures, MAPE and TCD (Total Cost Deviation) of forecasting to compare the KBPN with other methods, i.e., WES, MRA, BPN and GANN. The intensive experimental results show the following: 1. In encompassing test, KBPN, GANN and BPN models are superior to WES and MRA. 2. As for MAPE and TCD, KBPN (K=4) is the most accurate model with the smallest MAPE and it can also save a great total cost (amount of 12.2 million per month) in the application. Therefore, KBPN model by combining K-mean cluster and BPN model is a very powerful and effective forecasting tool.

References

1. Bollerslev, T.: Generalized Autoregressive Conditional Heteroskedaticicty. Journal of Econometrics 52, 307–327 (1986)
2. Campbell, J.Y., Lo, A.W., MacKinlay, A.C.: The Econometrics of Financial Markets. Princeton University Press, Princeton (1997)
3. Chang, P.C., Wang, Y.W.: Fuzzy Delphi and Back-Propagation Model for Sales Forecasting in PCB Industry. Expert Systems with Applications 30(4), 715–726 (2006)
4. Chang, P.C., Fan, C.-Y., Liu, J.Y.-C., Huang, W.-H.: Sales forecasting for thin film transistor liquid crystal display products with data clustering and an evolving neural network model. Proceedings of the Institution of Mechanical Engineers, Part B, Journal of Engineering Manufacture 222(5), 625–635 (2008)
5. Hsu, C.C., Chen, C.Y.: Applications of Improved Grey Prediction Model for Power Demand Forecasting. Energy Conversion and Management 44, 2241–2249 (2003)
6. Kuo, R.J., Xue, K.C.: A Decision Support System for Sales Forecasting through Fuzzy Neural Networks with Asymmetric Fuzzy Weights. Decisions Support Systems 24, 105–126 (1998)

Toward the Knowledge Circulation between Institutions and Public Audiences with Virtual Agents

Hung-Hsuan Huang, Hidekazu Kubota, and Toyoaki Nishida

Abstract. This paper proposes the CINOVA framework that supports the knowledge circulation between the institutions which possess large amount of knowledge and would like to disseminate it to public audiences. This framework proposes the use of visualized knowledge management systems and virtual agents for knowledge presentations. Two VKMSs and two virtual agent based presentation systems cooperate on the framework with a common contents presentation media, knowledge cards. The prototype system is still being implemented but the two virtual agent based presentation systems are already deployed to our client, NFRI in real-world exhibitions and Web based on-line services.

1 Introduction

Institutions demand an effective way to disseminate their knowledge and information to public audiences. The Ministry of Health wants people to notice the spreading infectious diseases and know how to prevent it, the Meteorological Agency wants people to pay attention to a coming typhoon or understand the mechanism of earthquakes, a science museum wants its visitors to understand and experience the principles of mechanics, research institutes want to introduce their results and make difficult theories easily understandable to the public. At the same time, institutes want to get the feedbacks from the public, what people want to know and what was not clearly conveyed. In a large institution, usually there are many experts who

Hung-Hsuan Huang and Toyoaki Nishida
Graduate School of Informatics, Kyoto University, Japan
e-mail: huang@ii.ist.i.kyoto-u.ac.jp, nishida@i.kyoto-u.ac.jp

Hidekazu Kubota
National Institute of Advanced Industrial Science and Technology, Japan
e-mail: h.kubota@aist.go.jp

B.-C. Chien and T.-P. Hong (Eds.): Opportunities and Challenges, SCI 214, pp. 219–224.
springerlink.com

possess specific aspects of knowledge but do not know the others well. The scattered institution knowledge has to be stored, well managed and organized to be useful and can be reused to create new values.

Two essential issues emerged in the knowledge circulation, the first one is how to efficiently store, organize and reuse large amount of knowledge that is scattered among many experts, the second one is how to efficiently disseminate information to and get feedback from public audiences. This paper presents the Circulating Knowledge with Virtual Agents (CINOVA) Framework that proposes the integration of visualized knowledge management systems (VKMS) and life-like virtual agents for these two issues. The status of the deployment of the prototype systems to our client institute, NFRI (National Food Research Institute) is also presented.

2 The CINOVA Framework

The basic requirements of a knowledge circulation framework is the storage and a common presentation of knowledge. The knowledge representation should be able to describe various principles of knowledge and can be easily accessed by many experts who work on different computer systems and have different preferences on user interfaces. The core of CINOVA framework is a back-end knowledge repository of the whole institution and is shared by all of the experts (Fig. 1). The basic unit of the common knowledge representation stored in the knowledge base and exchanged among the subsystems is so called *knowledge cards*. As proposed in [4], describing pieces of knowledge into card media is an efficient way for one or a group to organize known information and to create new thoughts. A knowledge card in CINOVA is a metaphor of such a card that represents a piece of knowledge and is composed with a fragment of XML text and one image. It is simple but is a general representation of knowledge in any principle and can be processed by various applications on various operating systems. Multiple relevant cards can be further linked sequentially to be a *story* to form a presentation of specific topic.

When the amount of the knowledge contents gets large, they become difficult to be handled and be thoroughly understood. Therefore, information visualization techniques are applied to provide efficient interfaces for the operations like uploading, organizing and authoring of the knowledge repository that may contain many thousands of knowledge cards. Several such visualized knowledge management systems (VKMS) can be connected to the same shared repository and provide different abstract views for the experts' convenience.

These knowledge contents are then presented by life-like virtual agents as the interface toward end public audiences. Life agents are considered particularly effective and intuitive for non-expert public users because no extra training is required and allow people to use daily-life communication skills to interact with them. Two ways of presentations are anticipated, the presentation on the Web which is more limited in functionalities but has broader audiences, on-site presentation in exhibitions which is more interactive and allows the visitors to directly try and

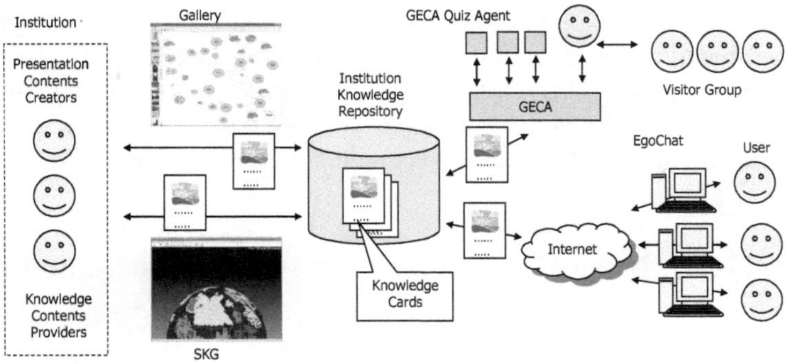

Fig. 1 The concept diagram of the CINOVA Framework

experience so that deeper understanding can be expected. There are four user classes in CINOVA framework.

Knowledge contents providers. They are the experts in the institution who possess specific knowledge in their minds and are willing to contribute it to the others in the institute or to disseminate it to the public. For example, in the case of NFRI, they are the researchers of food science. One provider may describe a piece of knowledge as a knowledge card and upload it to the shared knowledge base by using one of the VKMSs.

Presentation contents creators. They are the people who belong to the institution and create agent presentation contents (stories) by authoring the knowledge cards stored in the shared knowledge base by using one of the VKMSs. Depending on the target presentation agent system, the knowledge of how to compose expressive and natural non-verbal behaviors of the agent is required, they may be or may not be the knowledge contents providers.

Grouped exhibition visitors. They are the users who actually visited the exhibitions of the institution or the museum. From our observations in NFRI, the visitors go to exhibitions are usually in groups like students in the same class, friends, couples or families. In the CINOVA framework, we meant to provide these visitors immersive and multi-modal interactions with the knowledge presenting virtual agents. The setting of sensor devices, microphones or cameras that capture the activities of the visitors and 3D graphics that required high-end machine are possible.

Individual Web visitors. They are the people who access the Web site of the institution remotely. In the Web environment, the setting of sensor devices and the timing control of the agent's behaviors are not practical and thus the agent's functionalities are more suppressed.

These users exchange, share and acquire knowledge via knowledge card media through the CINOVA framework. The experts provide their knowledge to the knowledge base, the creators author the cards to presentation contents (stories), the knowledge is then presented by virtual agent systems instead of the staff of the institution. The knowledge consumers (visitors on-site or from remote) acquire their demand knowledge via the interactions with the virtual agents who are never tired and can serve queries in all aspects as long as the answers can be found in the knowledge repository rather than a human exhibitor who is usually only an expert of certain area. This forms a circulation of knowledge and is considered to be able to facilitate the communication between the institution and the public audience. The knowledge is made actionable and can also facilitate the institution to create new knowledge.

3 Visualized Knowledge Management and Agent Presentation Subsystems

There are two knowledge card based and visualized knowledge management subsystems implemented in CINOVA framework up to now, a zoomable 2D implementation, Gallery [3] and a 3D implementation Sustainable Knowledge Globe (SKG) [6]. Both of the two system share the basic ideas of utilizing zooming user interface for browsing large image collections and humans' spatial memory, that is, the human ability to remember the location of a stored item.

Gallery presents its contents on a smoothly zoomable 2D surface forming a memory space. It has virtually unlimited size and comprises concept nodes that represent the user's thoughts while knowledge cards are displayed as image thumbnails within a node. New concept node are created by dragging the mouse cursor out from an existing node and entering a filter string. Gallery uses this string to match keywords, annotations, the file path, and date of the items in the parent node. Items coinciding with the filtering string will then become the contents of the newly created node. The user retrieves information from the memory space and places newly generated nodes repeatedly and then a logically organized tree structure of the user's view over the whole repository will finally be constructed.

SKG is a 3D implementation of CINOVA compatible VKMSs and features the metaphor of a planet. Knowledge cards are represented as image thumbnails on the surface of the planet and are organized to tree structures by the user's direct manipulation. Virtual landscape with mountains and islands is used as the abstract representation of knowledge cards clusters. Because the surface of a sphere is limited, the sphere will expand to make space for accommodating new cards when there is no space (sea) left.

Gallery is 2D and is implemented in OS independent Java language while SKG is 3D and is implemented in Microsoft Windows' native API. This caused major difference of the possibilities of the two systems. Despite the argument on infinite 2D surface or 3D is better which is out of the scope of this paper, Gallery can be run virtually anywhere including the browsers but SKG is bounded to Microsoft Windows. Gallery is also light-weighted and can hold larger scale of contents. On the other hand, SKG's interface is much more fancy and rich. It can also deal with OS dependent contents like Word, PowerPoint files or multimedia clips while Gallery only accepts most simple knowledge cards with text fragments and images. They provide the varieties of choices and allow the users to pick their favorite one depending on their environments and needs.

Two virtual agent presentation subsystems are the front-end of the knowledge repository of the institution toward public visitors either from the Internet or in an exhibition.

EgoChat [5] is a Web based avatar presentation system. One or two avatars stand for the contents creator and present the stories composed with knowledge cards. A Q&A module is also attached and allow the contents user to feedback a question when there is something that they do not understand. An answer will be selected from a Q&A stories related to the channel of that story. If there is no appropriate answer found, an answer request will be sent to the contents creator via an e-mail.

Life-like virtual agents who can do real-time and multi-modal face-to-face interactions with human users involve many research disciplines and are very difficult to develop. Our previously proposed Generic Embodied Conversational Agent (GECA) framework [1] introduced the concept to distribute agent functionalities like sensor data processing, deliberation or character animator to simple, general purpose and reusable standalone modules that are connected with a common platform. In the reference implementation, the possible user-agent interactions are defined as stimulus-reaction pair based script language. One knowledge card is converted to one scene of a GSML script and the image of the knowledge card is associated to the background image of that scene.

EgoChat agents and GECA based agents have obviously different usages. Limited to the Web environment and remote access nature, EgoChat agents have suppressed GUI but it can have very broad audiences. As a contrast, the GECA agents are displayed on-site and user activity information from sensor devices are possible. They thus can aim high-degree multi-modal interactions with user groups.

4 Deployment Status

The National Food Research Institute (NFRI) is one of the typical institutions seeking for the solutions of the knowledge disseminating channel with public audiences. It is executing research programs that contribute to secure supply of safe food, and technical innovation in agriculture and food industries. It stands for the Japanese government and bears the responsibility to be the source of dispatching food related information and arouse the public's awareness on food safety.

Among the four introduced subsystems, the two virtual agent presentation systems are already deployed to NFRI in real uses. EgoChat is being used as a poster presenting agent since an exhibition in November 2006. It is reported by the staff of NFRI that the presentation done by the avatar agent is more understandable and more efficient to convey information to the audiences in short time period than presenting by themselves. By using EgoChat agents, they also don't need to worry about failures. It is also used for on-line quiz and in the homepage of certain working groups of the institute. The average page views per month in last year was 1,543 with a peak in May (3,943) This probably came from the effect of the open lab event held at the end of April.

This preliminary version of the GECA based quiz agent kiosk is equipped with a touch panel user interface and equips an emotion dynamics module changing the background melodies. This kiosk was displayed in four NFRI open lab exhibitions from April 2007. Each time there are totally around 2,000 people visited these events and averagely 250 (78 groups) of them played with the quiz kiosk. The typical visitors of NFRI exhibitions were the people who live in the neighborhood or teenage students come from nearby high schools. Almost during the whole day, there were dozens of visitors waiting for playing the game. Therefore, we considered that the basic idea was very successful in attracting the visitors. Besides, from questionnaire investigation, most of the visitors reported that they enjoyed the game and felt the knowledge explained by the agent is more trustable.

5 Conclusions

The four subsystems were developed with keeping knowledge card / story concepts in mind, but due to the fairly heterogeneous natures of them. The needs of knowledge representation are different one to one. For example, in the highly interactive GECA tour guide agent, the user can point to a specific area and ask the agent to introduce that place, the coordinates of areas on the background image is essential to the GECA agent but is meaningless to the present only EgoChat agent. Therefore, a clear definition of ontology of knowledge card mediated systems is required. The GECA based quiz agent is the current research focus in our group. We are trying to improve it to be aware of the status of user groups via video and audio information. The rough idea and preliminary results have been presented in [2].

References

1. Huang, H.H., Cerekovic, A., Nakano, Y., Pandzic, I.S., Nishida, T.: The design of a generic framework for integrating eca components. In: Padgham, L., Parkes, D., Muller, J.P. (eds.) The 7th International Conference of Autonomous Agents and Multiagent Systems (AAMAS 2008), Inesc-Id, pp. 128–135 (2008)
2. Huang, H.H., Furukawa, T., Ohashi, H., Ohmoto, Y., Nishida, T.: Toward a virtual quiz agent who interacts with user groups. In: The 7th International Workshop on Social Intelligence Design (SID 2008), Puerto Rico (2008)
3. Huang, H.H., Sumi, Y., Nishida, T.: Personal image repositories as externalized memory spaces. International Journal of Knowledge-based and Intelligent Engineering Systems 10(2), 169–180 (2006)
4. Kawakita, J.: The KJ Method. A Scientific Approach to Problem Solving. Kawakita Research Institute (1975)
5. Kubota, H., Kurohashi, S., Nishida, T.: Virtualized egos using knowledge cards. Electronics and Communications in Japan 88(1), 32–39 (2004)
6. Kubota, H., Nomura, S., Sumi, Y., Nishida, T.: Sustainable memory system using global and conical spaces. Journal of Universal Computer Science 13(2), 135–148 (2007)

Muon Tomography Algorithms for Nuclear Threat Detection

Richard Hoch, Debasis Mitra, Kondo Gnanvo, and Marcus Hohlmann

Abstract. In this article on *Muon Tomography* we report our work on the development of an intelligent pattern detection system for materials with high atomic numbers (Z) for Homeland Security application. Muons are naturally produced in the upper atmosphere by primary cosmic rays and are used as passive probes of a cargo volume. By sensing the incoming and outgoing tracks and measuring the momentum of each muon for a probed volume one may derive the scattering parameters. A statistical algorithm is being used to estimate scattering densities of the material in each unit volume (voxel) of the probed target. The article describes the algorithm and some results from our simulation experiments.

1 Introduction

Nuclear materials that pose a homeland security threat typically have high atomic numbers ($Z > 82$). It is of vital importance to develop smart, efficient, and inexpensive systems to detect such high-Z materials without opening a container.

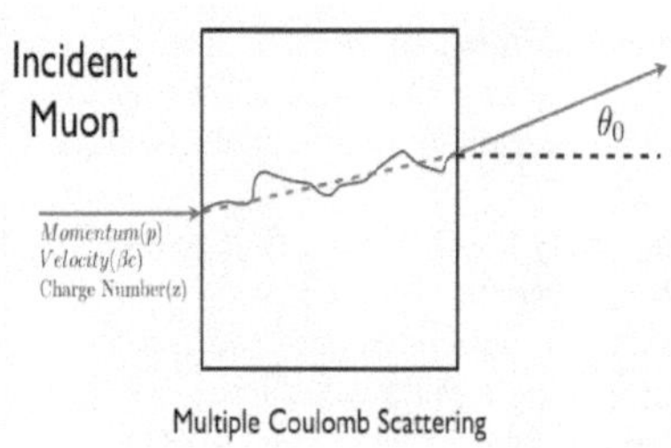

Fig. 1 Scattering of a particle

Muons, a type of elementary particle, produced by primary cosmic rays at the upper atmosphere provide an excellent source of passive probes for discriminating materials with different Z, without extra radiation or incurring any extra cost for the probe generation. It is highly penetrating compared to many other type of rays. A muon track may suffer from multiple scatterings by Coulomb interaction with the nuclei of atoms on its path. The amount of scattering depends on the charge Z of the corresponding nucleus [3] (Fig. 1). The incoming and outgoing tracks for each muon via the probed volume may be detected by appropriate sensors.

Richard Hoch and Debasis Mitra
Department of Computer Science
Florida Institute of Technology, Melbourne, Florida, USA
e-mail: rhoch@fit.edu, dmitra@cs.fit.edu

Kondo Gnanvo and Marcus Hohlmann
Department of Physics and Space Sciences
Florida Institute of Technology, Melbourne, Florida, USA
e-mail: {hohlmann,kgnanvo}@fit.edu

B.-C. Chien and T.-P. Hong (Eds.): Opportunities and Challenges, SCI 214, pp. 225–231.
springerlink.com

For the purpose of our simulations, the z-axis is the vertical direction, the x-axis is the axial direction for any mobile object's (vehicle) direction of movement, and the y-axis is perpendicular to the xz-plane. A typical geometry of a probed volume contains sensor planes (typically three) above and below and parallel to the xy-plane.

2 Reconstruction Algorithms

Our first algorithm for reconstruction of scattering points makes a naïve assumption: each scattering is a single event, or only one atomic nucleus (a point) is involved in scattering. This *Point-of-closest-approach* is called the POCA point [6]. We assign the scattering angle to that point instead of distributing it to multiple points on a muon track as would be the case in multiple scattering. This is a purely geometric algorithm that ignores any underlying physics of scattering. The corresponding *POCA-algorithm* is shown in Fig. 2.

First, the lines corresponding to incoming and outgoing tracks are computed from the corresponding three sensor points where the muons are detected above and below the probed volume, respectively (three sensor-array planes above and three below). We presume that the sensor-electronics will be able to associate the muon detection points to a single muon path by using the timing information of muon detections on the sensor arrays.

Algorithm POCA
Input: A list of {for each muon i, three incoming sensor points (a_i, b_i, c_i) where the muon is detected, and three corresponding exiting sensor points (d_i, e_i, f_i)}
Output: Corresponding list of {for each muon i, point of closest approach P_i between each incoming and respective exiting tracks, and the scattering angle θ_i at that point}

(1)For each muon i=1 to M in the list
(2) create incoming track I_i, and exiting track E_i by least-square-fitting the respective three points each
(3) using analytical formula, find two closest pts s_i and t_i, respectively, on I_i & E_i
(4) compute middle pt P_i between s_i & t_i
(5) compute angle θ_i between lines I_i & E_i
(6)return the list of $\{(P_i, \theta_i) \mid 1 \le i \le M\}$

Fig. 2 POCA Algorithm

In 3D, the incoming and the outgoing tracks are not necessarily coplanar due to scattering and measurement errors, and they are unlikely to meet at a single point. Consequently, for each line (incoming or exiting) the point closest to the other line is computed using a linear algebraic formulation. The mid-point to these two points is the POCA-point corresponding to each muon (Fig. 3). Also the scattering angle for each muon is computed in line 5. POCA-point and scattering angle pairs are returned for all muons where the angle is not very close to zero (POCA point does not exist for parallel lines or where a muon has traversed without any scattering). Complexity of POCA is $O(M)$ for M tracks.

The POCA algorithm is a simple algorithm with a very strong assumption of single-point scattering. A better algorithm, originally proposed by Verdi et al. [7], and subsequently adapted by Schultz et al. [5] utilizes both the scattering angle and the measured linear displacement of a muon-track over the xy-plane [Fig. 3]. Actually, the scattering angle has a near normal distribution that depends on the

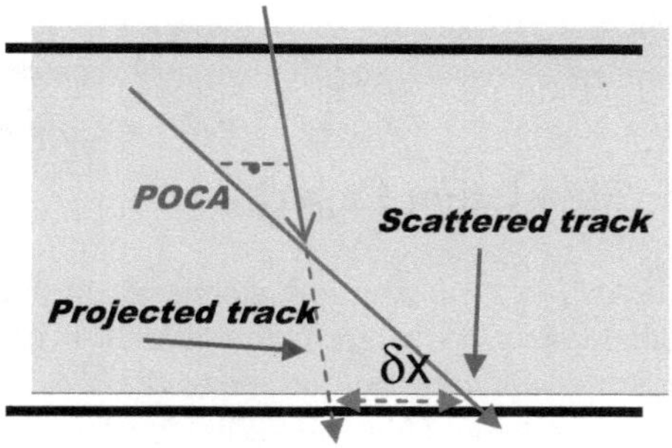

Fig. 3 Linear deviation of a track

material and distance of traversal within the material [Eq. 1]. Our next algorithm, Expectation Maximization *(EM)-reconstruction,* uses both the information – scattering angles and linear deviations as input. Here, the scattering angle θ_i is measured between the incoming and outgoing track-vectors of a muon. The linear deviation δ_i is measured between the point *E* representing the actual emergent track at the topmost bottom detector plane and the point *F* where the projected-incoming track hits the same horizontal plane of *E* (Fig. 3). We use the two x and y components for each of the two parameters (θ_i, δ_i) that improves the chance of determining scattering location by adding extra information. The *EM-reconstruction* algorithm (Fig. 4) attempts to distribute the scattering location along the POCA-track instead of assigning the scattering event to a single point as the *POCA* algorithm does. The track of a muon connects a representative entering point to the POCA point and then the POCA point to a representative exiting point (typically the detection point on the respective nearest sensor- array plane to the volume).

This algorithm views a discretized volume for the interrogated space. Each unit of volume is called a voxel and its dimension is predetermined. Scattering is presumed to have happened over some voxels along the POCA track of the muon. The conditional probability (likelihood) of the observed data $D_i \equiv (\theta_i, \delta_i)$ for a muon *i,* given the scattering density distribution λ (a vector) over the voxels (*j*) is given by equation (1).

$$P(D_i \mid \lambda) = \frac{1}{2\pi|\Sigma_i|^{1/2}} \exp\left(-\frac{1}{2} D_i^T \Sigma_i^{-1} D_i\right) \quad with \quad \Sigma_i = p_{r_i}^2 \sum_{j=1}^{n} \lambda_j W_{ij} \tag{1}$$

where the sum is taken over all *n* voxels along the *i*-th muon track, p_{r_i} is the momentum ratio inversely proportional to momentum p_i, λ_j is the scattering density of the *j*-th voxel, and *W* is the symmetric 2x2 covariance matrix between scattering angle θ and linear deviation δ. Elements of *W* depend on the path length of muon *i* through voxel *j,* and the vertical height of the voxel *j* from the bottom plane [5].

Maximizing the total likelihood of observation *D*, (by equating a partial derivative of the total log-likelihood with respect to λ to zero, under an assumption of independence between voxels), we get the update equation for the scattering density

$$\lambda_j^{k+1} = \lambda_j^k + \left(\lambda_j^k\right)^2 median\left[(C_{ij}^k = D_i^T \Sigma_i^{-1} W_{ij} \Sigma_i^{-1} D_i - Trace(\Sigma_i^{-1} W_{ij}))\right], \tag{2}$$

where the median is taken over all tracks *i* that go through voxel *j*, and *k* indicates the iteration index. Asymptotic complexity of the *EM-reconstruction* is *O*(*IMN*), where I is the number of iterations, M is the number of muons, and N is the number of voxels, and the memory requirement is $O(M + N)$.

3 Simulation Experiments

GEANT4 [1], a common stochastic physics toolkit for simulating the passage of subatomic particles through material, is used for our experimental set up. For generating cosmic ray muons we have used a package called CRY, developed at Lawrence Livermore National Lab [8, 10].

The geometry of our standard simple scenario that is used for testing has a probed area with the dimensions 4mX4m in X and Y and 3m in Z. The rectangular targets of five different materials are centered as following: Aluminum at (-1000mm, -1000mm, 0mm), Iron at (1000mm, -1000mm, 0mm), Lead at (0mm, 0mm, 0mm), Tungsten at (-1000mm, 1000mm, 0mm), and Uranium at (1000mm, 1000mm, 0mm). Each of these boxes is of size 40cm×40cm×20cm.

Algorithm EM-reconstruction
Input: A list {for each muon i, (D_i, p_{ri}), where $D_i = (\theta_i, \delta_i)$} & θ_i is the scattering angle, δ_i is the displacement of the track, and p_{ri} is the muon momentum parameter;
Initial λ_j-value for each voxel;
Maximum number of iterations I;
Output: Final λ_j-value for each voxel

(1) set initial vector λ^{new}
(2) for each iteration $k = 1$ to I do
(3) set vector $\lambda^{old} = \lambda^{new}$
(4) for each muon-track $i = 1$ to M do
(5) compute C_{ij}, using eq. (2)
(6) for each voxel $j = 1$ to N do
(7) find median of correction term $\Delta\lambda_j$
(8) $\lambda_j^{new} = \lambda_j^{old} + \Delta\lambda_j$, using eq. 3
(9) return vector λ

Fig. 4 The expectation maximization algorithm

4 Implementation and Results

For reconstruction, we have run the *POCA* algorithm first, which returns a set of POCA points and scattering angles at each point. In Fig. 5, the color of a point may indicate the value of the angle assigned to the scattering event at that point.

We run the *EM-reconstruction* algorithm on the same simulation data after appropriate pre-processing of the input. One of the major challenges in implementing the algorithm is the median calculation for the correction factor of λ_j for each voxel in each iteration. Typically, each such computation requires $O(H)$ amount of memory (and steps),

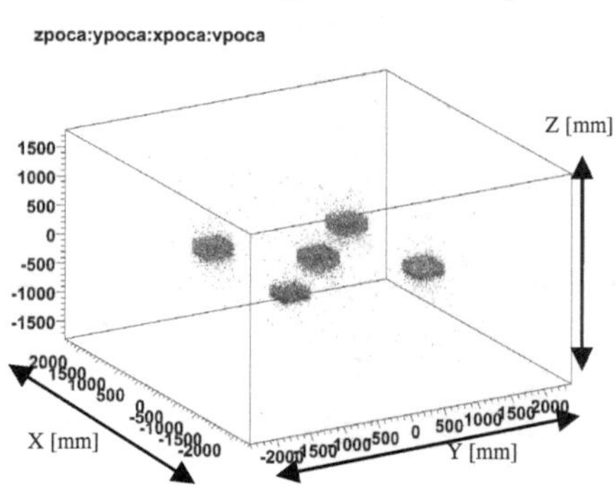

Fig. 5 POCA reconstruction

where H is the number of tracks through the voxel. This blows up the resource requirement to an impractical level. In order to avoid this we have developed an approximate technique for the median calculation.

The total range of the λ-correction factor is divided into bins of fixed sizes d. For each bin, the number of data points in the bin and the mean value over the bin is stored which reduces the requirement of storing all data points. Subsequently, the frequency parameters on the binned data points are used to find the median bin and the corresponding mean of that bin is used as the median of the whole data set. The complexity for this computation (for each voxel in each iteration) is $O(K)$, where K is the number of bins, which is far smaller than H. The error incurred in this approximation is less than the bin size d. The smaller d the better the accuracy is, but a smaller value of d will increase the value of K and will consequently decrease computation time and increase memory consumption. To the best of our knowledge this approximate median calculation is new.

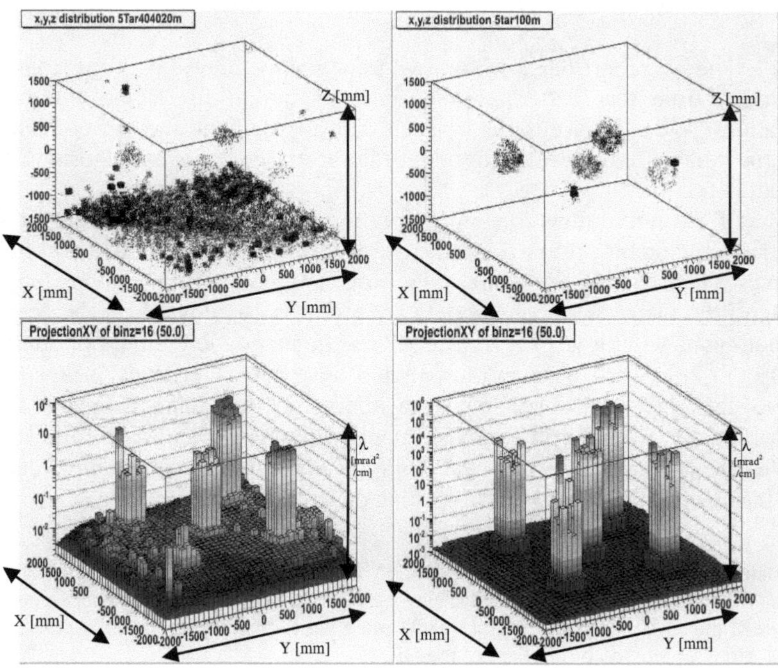

Fig. 6 Reconstruction from Expectation Maximization algorithm (average method left; approximate median method right). The top figures are a 3D representation of the lambda values of each voxel. The bottom figures represent lambda values in the plane z=50mm

The result from the *EM-reconstruction* algorithm for the experimental setup discussed in the previous section is shown in Fig. 6. Note that here the output is a voxel-wise λ value, an important difference from Fig. 5. The top plots in Fig. 6 are 3D representations of the λ value for each voxel. Statistically more accurate method is to use average in eq. 2 but median provides better result. The results from the average method are shown on the left. The targets in this plot are barely visible and there is much noise at the bottom of the volume. With the median method shown on the right the reconstruction is much clearer. The targets are clearly visible (though the box shape is distorted) and there is no noise at the bottom. However, the plots at the bottom of Fig. 6, which show the λ values of the voxels in the plane z=50mm, indicate that there are still improvements to be made. Both methods discriminate between the targets and the surrounding vacuum, yet for the average method the absolute λ values appear too low and for the median method the values appear too high. Both methods also reconstruct the outer target voxels with higher λ values compared to the interior voxels. The median method does appear to reconstruct the scenario better than the average method, but there is work left to do to improve the overall discriminatory power and to reproduce actual physical λ values.

5 Discussion, Future Direction, and Summary

In this article we report our work on two tomography algorithms for reconstructing images from the scattering of cosmic ray muons for Homeland security applications. We also discuss an efficient but approximate median computation technique that we have developed to make the *EM-reconstruction* algorithm using the median feasible.

Some of the future directions of this work are to develop an integrated algorithm that will combine the two reconstruction algorithms described here for better efficiency and accuracy. We will also develop an online *anytime-good* reconstruction algorithm that will run as the data collection is ongoing where the accuracy will continuously improve over time to the maximum possible limit. For homeland security purposes such an algorithm should have a high practical value, where resource consumption is very important and cargo interrogation time is of the essence. Finally, we will run our experiments with more complex real life scenarios. We are also waiting for the availability of actual noisy experimental data [4] rather than using somewhat pure simulation data, to test the algorithms.

Acknowledgements. We acknowledge support from the U.S. Department of Homeland Security (DHS) and some students. The views and conclusions contained in this document are those of the authors and should not be interpreted as necessarily representing the official policies, either expressed or implied, of the DHS.

References

1. Agostinelli, S., et al.: GEANT4 – a simulation toolkit. Nucl. Instrum. Meth. A 506, 250–303 (2003)
2. Allison, J., Amako, K., Apostolakis, J., Araujo, H., Dubois, P., Asai, M., et al.: GEANT4 developments and applications. IEEE Trans. Nuclear Sc. 53(1), 270–278 (2006)
3. Bethe, H.: Moliere's theory of multiple scattering. Physical Review 89(6), 1256 (1953)

4. Hohlmann, M., Ford, P., Gnanvo, K., Helsby, J., Pena, D., Hoch, R., Mitra, D.: GEANT4 Simulation of a Cosmic Ray Muon Tomography System with Micro-Pattern GasDetectors for the Detection of High-Z Materials. In: Proc. SORMA WEST 2008 Conf., Berkeley, CA, Trans. Nuclear Science (article in press) (2008)
5. Schultz, L.J., Blanpeid, G.S., Borozdin, N., Fraser, A.M., Hengartner, N.W., Klimenko, A.V., Morris, C.L., Orum, J.C., Sossong, M.J.: Statistical reconstruction for cosmic ray muon tomography. IEEE Trans. Image Processing 16(8), 1985–1993 (2007)
6. Sunday, D.: Distance between Lines and Segments with Their Closest Point of Approach (2006), `http://geometryalgorithms.com/Archive/algorithm0106/algorithm0106.htm`
7. Verdi, Y., Shepp, L.A., Kaufman, L.: A statistical model for positron emission tomography. l. Of American Statistical Association 80(389), 8–20 (1985)
8. Wright, D., others from the Cosmic-ray Physics Team at the Lawrence Livermore National Laboratory: Monte Carlo Simulation of Proton-induced (2006)
9. Cosmic-ray Cascades in the Atmosphere. Lawrence Livermore National Lab., CA, Tech. Rep. LA-UR-06-8497
10. Hagmann, C., Lange, D., Wright, D.: Cosmic-ray shower generator (CRY) for Monte Carlo transport codes. In: 2007 Proc. IEEE Nucl. Sci. Symp., Honolulu, HI, vol. 2, pp. 1143–1146 (2007)

A Model-Based Software Reasoning Approach to Software Debugging*

Rui Abreu, Peter Zoeteweij, and Arjan J.C. van Gemund

Abstract. Current model-based approaches to software debugging use static program analysis to derive a model of the program. In contrast, in the software engineering domain diagnosis approaches are based on analyzing dynamic execution behavior. We present a model-based approach where the program model is derived from dynamic execution behavior, and evaluate its diagnostic performance on the Siemens software benchmark, extended by us to accommodate multiple faults. We show that our approach outperforms other model-based software debugging techniques, which is partly due to the use of De Kleer's intermittency model to account for the variability of software component behavior.

1 Introduction

Two major approaches to software fault localization can be distinguished, (1) the spectrum-based fault localization (SFL) approach, a statistical approach that correlates dynamic software component activity (i.e., execution traces) with program failures [1,8,9,13], and (2) the model-based diagnosis or debugging (MBD) approach, which deduces component failure through logic reasoning over a static model of the program [4,5,6,7,10,12,14].

Because of its low computational complexity and absence of modeling requirements, SFL has gained large popularity in the software engineering community. Although inherently not restricted to single faults, in most cases these statistical techniques are applied and evaluated in a single-fault context, such as the Siemens benchmark set, which is seeded with only 1 fault per program (version). In practice, however, the defect density of even small programs typically amounts to multiple faults. Although the root cause of a particular program failure need not constitute multiple faults that are acting *simultaneously*, many failures will be caused by *different* faults. Hence, the problem of multiple-fault localization (diagnosis) deserves detailed study.

Rui Abreu, Peter Zoeteweij, and Arjan J.C. van Gemund
Embedded Software Lab, Delft University of Technology, The Netherlands
e-mail: {r.f.abreu,p.zoeteweij,a.j.c.vangemund}@tudelft.nl

* This work has been carried out as part of the TRADER project under the responsibility of the Embedded Systems Institute. This project is partially supported by the Netherlands Ministry of Economic Affairs under the BSIK03021 grant.

B.-C. Chien and T.-P. Hong (Eds.): Opportunities and Challenges, SCI 214, pp. 233–239.
springerlink.com

Unlike SFL, MBD inherently considers multiple faults. However, the logic models of software systems that are used in the diagnostic inference are typically based on static program analysis. Consequently, they do not consider dynamic execution behavior, such as (data-dependent) conditional control flow, which, in contrast, forms the essence of the SFL approach. Aimed to combine the best of both worlds, in this paper we present an approach that exploits the dynamic, execution trace-based observation approach from SFL, to derive models and observations as input to MBD to produce multiple-fault diagnoses.

This paper makes the following four contributions. (1) We present our multiple-fault diagnosis method which combines a dynamic modeling and observation approach known from SFL with a diagnostic reasoning approach from MBD. (2) We evaluate our approach using the Siemens set benchmark, extended by us to accommodate multiple faults. (3) We evaluate the merit of two specific strategies for updating the probabilities of diagnosis candidates, based on De Kleer's intermittent fault model [4], to account for the fact that faulty (software) components very often exhibit nominal behavior. (4) We compare our approach to related reasoning approaches (AIM [11], Δ-slicing [7], and `explain` [7]) for the Siemens set program `tcas` (their common benchmark).

Our experiments show that strategies that exploit (intermittency) information to exonerate components involved in passed runs outperform those that do not include such information. Furthermore, experiments using the `tcas` program show that our approach finds more bugs with less effort required.

2 Observation-Based Modeling

Model-based diagnosis approaches are dependent on the existence of a model of the program, which would have to be derived from the system specifications. Even if a model were available for each component (statement), only for the simplest of programs a program model could be extracted based on static dependence analysis. In this section we present our dynamic, observation-based diagnosis approach.

2.1 Observations

Observations are collected as abstractions of execution traces, also called program spectra. This data typically consists of a number of counters or flags for the different components of a program. In the context of this paper we use hit spectra, which indicate whether a component (statement) was involved in a run. Let M be the number of components, and N the number of execution runs. Let O denote the Nx$(M+1)$ *observation matrix*. For $j \leq M$, the element o_{ij} is equal to 1 (*true*) if component j was observed to be involved in the execution of run i, and 0 (*false*) otherwise. The element $o_{i,M+1}$ is equal to 1 (*true*) if run i failed, and 0 (*false*) otherwise. The rightmost column of O is also denoted as e (the error vector).

2.2 Computing Diagnoses

Unlike many MBD approaches, which statically deduce information from the program source, O is the *only*, dynamic source of information, from which *both* a model, and the input-output observations are derived. Apart from exploiting dynamic information, this approach only requires a generic component model, avoiding the need for detailed functional modeling or relying, e.g., on invariants or pragmas for model information. Note, however, that this default model can easily be extended when more detailed information is available.

Abstracting from particular component behavior, each component c_j is modeled by the weak model $h_j \Rightarrow (x_j \Rightarrow y_j)$, where h_j models the health state of c_j and x_j, y_j model its input and output variable value *correctness* (i.e., abstracting from actual variable *value*). This weak model implies that a healthy component c_j translates a correct input x_j to a correct output y_j. However, a faulty component or input *may* lead to an erroneous output.

As each row in O specifies which components were involved, we interpret a row as a "run-time" model of the program as far as it was considered in that particular run. Consequently, O is interpreted as a sequence of typically different models of the program, each with its particular observation of input/output correctness. The overall diagnosis can be viewed as a sequential diagnosis approach that incrementally takes into account new structural program (and pass/fail) evidence with increasing N. Every failed run $O_{n,*}$ yields a conflict and, as in the former MBD approach, the conjunction of conflicts are then subject to a hitting set algorithm that generates the diagnostic report $D=\{d_1,\ldots,d_k,\ldots,d_K\}$ with K diagnosis candidates (refer to [2,3] for details). From a passing computation nothing can be inferred (apart from the exoneration when it comes to probabilistically ranking the diagnosis candidates as explained in next section). Note that, e.g., unlike constraint-based models [11], we do not exploit actual data dependencies between components but execution patterns.

2.3 Ranking Diagnoses

Similar to the incremental compilation of conflicts per run we compute the posterior probability for each candidate d_k based on the pass/fail observation *obs* for each sequential run using Bayes' rule [5,6]

$$\Pr(d_k|obs) = \frac{\Pr(obs|d_k)}{\Pr(obs)} \cdot \Pr(d_k)$$

where $Pr(\{j\})=p$ (p arbitrarily set to 0.01) denote the *a priori* probability that a component c_j is at fault. $Pr(obs|\, d_k)$ is defined as follows

$$\Pr(obs|d_k) = \begin{cases} 0 & \text{if } d_k \text{ and } obs \text{ are inconsistent} \\ 1 & \text{if } d_k \text{ implies } obs \\ \varepsilon & \text{if neither holds} \end{cases}$$

Many policies ε exist [2,4]. In the following, we distinguish between three ε policies. The first policy, denoted $\varepsilon^{(0)}$ (classical MBD policy) and is defined as follows

$$\varepsilon^{(0)} = \begin{cases} \frac{2^M}{2^M+(2^l-1)\cdot 2^{M-l}} & \text{if run passed} \\ \frac{(2^l-1)\cdot 2^{M-l}}{2^M+(2^l-1)\cdot 2^{M-l}} & \text{if run failed} \end{cases}$$

where $l = |d_k|$ is the number of faulty components in the diagnosis. A disadvantage of this classical policy is that passed runs, apart from making single faults more probable than multiple faults, do not help much in pinpointing the fault location. This has to do with the fact that all diagnoses are possible when a run passes due to the weak fault model (the 2^M term in the equation above). In addition, there is no way to distinguish between diagnoses with the same cardinality, because the terms are merely a function of the cardinality of the diagnosis candidate.

An approach to account for the fact that, similar to statistical approaches for fault localization, components involved in passed computations should to some extent be exonerated is by extending the component model with an intermittent failure model, as introduced by De Kleer [4]. As in software components it is quite usual that a faulty component exhibits correct behavior, we include statistical information on the probability that a faulty component c exhibits correct behavior. Let $g(d_k)$ denote the aforementioned ("goodness") probability that faulty components in d_k are exhibiting good behavior. We distinguish between two different policies, which we refer to as $\varepsilon^{(1)}$, and $\varepsilon^{(2)}$, which are defined as follows

$$\varepsilon^{(1)} = \begin{cases} g(d_k) & \text{if run passed} \\ 1 - g(d_k) & \text{if run failed} \end{cases} \qquad \varepsilon^{(2)} = \begin{cases} g(d_k)^t & \text{if run passed} \\ 1 - g(d_k)^t & \text{if run failed} \end{cases}$$

where $t = \prod_{j \in dk} [oij = 1]$ is the number of faulty components according to d_k involved in the run i. We propose policy $\varepsilon^{(2)}$ as a variant of $\varepsilon^{(1)}$, which is due to De Kleer [4]. It approximates the probability $\sum_{j \in dk} gj$ that the components in d_k all exhibit good behavior by $g(d_k)^t$, assuming that all components of d_k have equal goodness probabilities. In both strategies we use

$$g(d_k) = \frac{\sum_{i=1..N} [(\bigvee_{j \in d_k} o_{ij} = 1) \wedge e_i = 0]}{\sum_{i=1..N} [\bigvee_{j \in d_k} o_{ij} = 1]}$$

where [.] is Iverson's operator ([*true*]=1], [*false*]=0).

3 Experimental Evaluation

In this section we assess the diagnostic capabilities of the dynamic modeling approach for the well-known Siemens set[1]. It contains 132 faulty versions of 7 C programs (LOC varies between 174 to 539) with extensive test suites (1052-5542 runs). For our experiments, we extended the Siemens set with program versions in which we can activate arbitrary combinations of faults. For this purpose, we limit

[1] http://sir.unl.edu

ourselves to a selection of 102 out of the 132 faults, based on criteria such as faults being attributable to a single line of code, to enable unambiguous evaluation. The observation matrices are obtained using the GNU `gcov`[2] profiling tool.

Using this extended Siemens set, we evaluate our dynamic modeling approach in two ways: first, in Section 3.2, we measure its diagnostic performance on single and multiple-fault programs for the three ε strategies outlined in Section 2.3. Next, in Section 3.3 we compare this performance against other diagnosis techniques. Here we use single-fault versions of the `tcas` program, which is the common program used in literature to evaluate these other techniques.

3.1 Performance Metric

We quantify the diagnostic quality as the amount of code a developer would have to inspect before (but not including) finding the fault cause, *wasted effort* W. It is defined as the number of inspected components divided by the total number of components (M). For example, suppose a triple-fault program ($M=6$, and c_1, c_2, and c_3 faulty) for which the following diagnosis $D = \{\{1,2,6\},\{3,4,5\}\}$ is obtained. This induces a wasted effort of $W=33\%$ as c_6 in the first candidate is inspected in vain, as well as, on average two out of three inspections in the second candidate.

3.2 Results

The wasted effort W incurred by the dynamic modeling approach and strategies $\varepsilon^{(0)}$, $\varepsilon^{(1)}$, and $\varepsilon^{(2)}$ for debugging single, double, and multiple-fault programs have been evaluated. We aimed at $C=5$ for the multiple fault-cases, but for `print_tokens` insufficient faults are available, and for `print_tokens2` and `replace` our current implementation of the hitting set algorithm practically prevents analyzing combinations of more than four and three faults, respectively. The hitting set computation is aborted after all diagnosis candidates with cardinality C' have been generated. To simulate a more or less realistic debugging scenario, where the actual number of faults is unknown, we set $C'=max(C,3)$. All measurements are averages over 100 versions of randomly combined C faults, or over the maximum number of combination available, where we verified that all faults are active in at least one failed run. The improvement of $\varepsilon^{(2)}$ over $\varepsilon^{(1)}$ is marginal at best. We expect that this can be explained by using $g(d_k)^t$ to approximate the product of the goodness parameters of the individual components in a diagnosis d_k, as explained in Section 2.3. In the context of strategy $\varepsilon^{(2)}$, this entails using different goodness parameters for the same component as it occurs in different diagnoses, converging to the fraction of all runs that have passed as the diagnosis cardinality increases. For interested readers, refer to [2] for a more detailed analysis.

---

[2] http://gcc.gnu.org/onlinedocs/gcc/Gcov.html

3.3 Comparison

In the following we compare the diagnostic performance of our approach with AIM [11], nearest neighbor [13] (NN), `explain` [7] and Δ-slicing [7] techniques. For compatibility with results reported for those techniques, we will use the effort, or `score` metric [1,13] instead of wasted effort W which amounts to the percentage of lines of code that need *not* be examined when the diagnosis results are used to guide the search for the fault. Our current implementation of the dynamic modeling approach only supports C programs, while the AIM technique has mainly been evaluated for Java programs. The only C program that has been taken into account is `tcas`, which happens to be a common benchmark among the other techniques as well, so for this comparison we limit ourselves to that program. Furthermore, the other techniques have only been evaluated for single faults, so we set $C'=C=1$, and therefore $\varepsilon^{(1)} = \varepsilon^{(2)}$.

Similar to the results in [10], we compare our approach with AIM and NN on `tcas`. As expected, $\varepsilon^{(0)}$ is the worst performing technique. AIM consistently outperforms NN. For an effort of less than 1%, $\varepsilon^{(1,2)}$ outperform AIM, which yields the best results if 10% of the code is inspected. Both techniques find all faults by inspecting less than 20% of the code. For a detailed analysis, refer to [2,3].

4 Conclusions and Future Work

In this paper we present a dynamic modeling approach to software fault localization based on abstraction of program traces. The model, along with the set of traces for pass/fail executions is used to reason about observed failures.

We have evaluated the diagnostic performance of three Bayesian probability update policies, including De Kleer's intermittency model and an extension proposed by us. Empirical results obtained from the widely-used Siemens set of programs, extended by us to accommodate multiple fault programs, show that policies that are able to exonerate components that are involved in passing runs clearly outperform the probability update scheme that is traditionally used in model-based diagnosis.

For future work, we plan to study whether the multiple-fault diagnosis candidates' information can be used to efficiently engage several developers to repair the defect(s) in parallel.

References

1. Abreu, R., Zoeteweij, P., van Gemund, A.J.C.: On the accuracy of spectrum-based fault localization. In: Proc. TAIC PART 2007 (2007)
2. Abreu, R., Zoeteweij, P., van Gemund, A.J.C.: A Dynamic Modeling Approach to Software Multiple-fault Localization. In: Proc. DX 2008 (2008)
3. Abreu, R., Zoeteweij, P., van Gemund, A.J.C.: An observation-based model for fault localization. In: Proc. WODA 2008 (2008)
4. De Kleer, J.: Diagnosing intermittent faults. In: Proc. DX 2007 (2007)

5. De Kleer, J., Mackworth, A.K., Reiter, R.: Characterizing diagnoses and systems. Artif. Intell. 56, 197–222 (1992)
6. De Kleer, J., Williams, B.C.: Diagnosing multiple faults. Artif. Intell. 32(1), 97–130 (1987)
7. Groce, A.: Error explanation with distance metrics. In: Jensen, K., Podelski, A. (eds.) TACAS 2004. LNCS, vol. 2988, pp. 108–122. Springer, Heidelberg (2004)
8. Jones, J.A., Harrold, M.J.: Empirical evaluation of the tarantula automatic fault-localization technique. In: Proc. ASE 2005 (2005)
9. Liu, C., Yan, X., Fei, L., Han, J., Midkiff, S.P.: Sober: Statistical model-based bug localization. In: Proc. ESEC/FSE-13
10. Mayer, W., Stumptner, M.: Evaluating models for model-based debugging. In: Proc. ASE 2008 (2008)
11. Pucel, X., Bocconi, S., Picardi, C., Dupre, D., Massuyes, L.: Diagnosability analysis for web services with constraint-based models. In: Proc. DX 2007 (2007)
12. Reiter, R.: A theory of diagnosis from first principles. Artif. Intell. 32(1), 57–95 (1987)
13. Renieris, M., Reiss, S.P.: Fault localization with nearest neighbor queries. In: Proc. ASE 2003 (2003)
14. Wotawa, F., Stumptner, M., Mayer, W.: Model-based debugging or how to diagnose programs automatically. In: Hendtlass, T., Ali, M. (eds.) IEA/AIE 2002. LNCS, vol. 2358, p. 746. Springer, Heidelberg (2002)

Learning Rules from Multiple Criteria Decision Tables

Chien-Chung Chan

Abstract. This paper introduces algorithms for learning decision rules from Multiple Criteria Decision Tables (MCDT) in the context of dominance-based rough sets introduced by Greco et al. Our method enables the use of existing rule-learning algorithms by transforming a MCDT into a family of discretized decision tables. The transformation is based on the indexed block representation of dominance-based approximation spaces. When the family of indexed blocks forms a partition on the set of objects in a MCDT, we have certain rules; otherwise, the set of rules learned may include uncertain rules. The method is demonstrated by using the ROSE2 and BLEM2 learning tools.

1 Introduction

In rough sets theory [9, 10, 11], information of objects in a domain is represented by an information system $IS = (U, A, V, f)$, where U is a finite set of objects, A is a finite set of attributes, $V = \cup_{q\in A} V_q$ and V_q is the domain of attribute q, and $f : U \times A \to V$ is a total information function such that $f(x,q) \in V_q$ for every $q \in A$ and $x \in U$. In many applications, data sets are represented by a special case of information systems called *decision tables*. In a decision table $(U, C \cup D = \{d\})$, there is a designated attribute $\{d\}$ called *decision attribute*, and attributes in C are called *condition attributes*. Each attribute q in $C \cup D$ is associated with an equivalence relation R_q on the set of objects of U such that for each x and $y \in U$, $xR_q y$ means $f(x,q) = f(y,q)$. For each x and $y \in U$, x and y are *indiscernible* on attributes $P \subseteq C$ if and only if $xR_q y$ for all $q \in P$.

Dominance-based Rough Set Approach (DRSA) is the application of rough set theory to multiple criteria decision analysis. It was introduced by Greco, Matarazzo and Slowinski [2, 3, 4]. In DRSA, attributes with totally ordered domains are called *criteria*. Each criterion q in C is associated with an outranking relation [16] S_q on the set of objects of U such that for each x and $y \in U$, $xS_q y$ means

Chien-Chung Chan
Department of Computer Science
University of Akron, Akron, OH, 44325-4003, USA
e-mail: chan@uakron.edu

B.-C. Chien and T.-P. Hong (Eds.): Opportunities and Challenges, SCI 214, pp. 241–246.
springerlink.com

$f(x,q) \geq f(y,q)$. For each x and $y \in U$, x *dominates* y on criteria $P \subseteq C$ if and only if $xS_q y$ for all $q \in P$. Dominance relations are totally pre-ordered, i.e., strongly complete and transitive binary relations [6]. A consistent preference model is taken to be one that obeys the dominance principle when assigning actions (objects) to the preference ordered decision classes. Action x is said to dominate action y if x is at least as good as y under all considered criteria. The monotonicity of dominance principle requires that if action x dominates action y, and then x should be assigned to a class not worse than y.

The DRSA has been shown to be an effective tool for MCDA and has been applied to solve multi-criteria sorting problems [5, 6]. Algorithms for inducing decision rules consistent with dominance principle were introduced in [7, 8]. However, it is not clear how to apply existing rule learning algorithms to dominance-based decision tables. In this paper, we show that by applying proper transformation or preprocessing, existing learning tools can be used to generate rules from dominance-based decision tables. For a given multiple criteria decision table with K decision values, the basic idea is to compute a family of indexed blocks [17] to represent the table, then indexed blocks with consistent decision values are used to transform each criterion into K criteria. If there are M criteria in the original table, we will have $M * K$ criteria in the transformed table. For criteria with continuous values, the transformation serves as a discretization process. For ordinal criteria, the transformation is to arrange their values into groups of values. Once we have a transformed table, the learning of rules can proceed using existing rough set tools to find reducts of criteria and to generate rules. In our experiments, the ROSE2 tool [14, 15] was used to reduce the number of criteria from a MCDT and the BLEM2 tool [1] was used to generate rules.

The remainder of this paper is organized as follows. The concept of indexed blocks is reviewed in Section 2. In Section 3, we present the algorithms for transforming a multiple criteria decision table into a family of discretized decision tables. In Section 4, an algorithm for learning rules from multiple criteria tables is presented. Finally conclusions are given in Section 5.

2 Indexed Blocks

The inconsistencies in a MCDT caused by violating of dominance principle can be represented by the concept of indexed blocks [17]. *Indexed blocks* are sets of objects indexed by pairs of decision values. Blocks with indices (*i, i*) are consistent, for any decision value *i*; other blocks are inconsistent. Detailed computing of indexed blocks were given in [17, 19].

Let $(U, C \cup D = \{d\})$ be a multiple criteria decision table where condition attributes in C are criteria and decision attribute d is associated with a total preference ordering. For each condition criterion q and a decision value d_i of d, let $\min_q(d_i)$ denote the minimum value of q among objects with decision value d_i, and $\max_q(d_i)$ denote the maximum value.

For each condition criterion q, the mapping $I_q(i, j): D\times D \to \wp(V_q)$ is defined as

$I_q(i, j) = \{f(x,q) = v \mid v \geq \min_q(d_j)$ and $v \leq \max_q(d_i)$, for $i < j; i, j = 1,\ldots,V_D; x \in U\}$,

$I_q(i, j) = I_q(j,i)$ if $i > j$, and $I_q(i,i) = \{f(x,q) \mid f(x,d) = i \wedge$

$f(x,q) \notin \cup_{i<j} I_q(i, j)\}$, where $\wp(V_q)$ denotes the power set of V_q. For simplicity, the set $I_q(i, j)$ of values is denoted as $[\min_q(j), \max_q(i)]$ or simply as $[\min_j, \max_i]$ for a decision value pair i and j with $i < j$.

For each $I_q(i, j)$ and $i \neq j$, the corresponding set of *ordered pairs* $[I_q(i, j)]$: $D\times D \to \wp(U\times U)$ is defined as

$[I_q(i, j)] = \{(x, y) \in U\times U \mid f(x,d) = i, f(y,d) = j$ such that $f(x,q) \geq f(y,q)$ for $f(x,q)$, $f(y,q) \in I_q(i, j)\}$.

For each set $[I_q(i, j)]$ of ordered pairs, the restrictions of $[I_q(i, j)]$ to i and j are defined as:

$[I_q(i, j)]_i = \{x \in U \mid$ there exists $y \in U$ such that $(x, y) \in [I_q(i, j)]\}$ and

$[I_q(i, j)]_j = \{y \in U \mid$ there exists $x \in U$ such that $(x, y) \in [I_q(i, j)]\}$.

The corresponding indexed block $B_q(i, j) \subseteq U$ of $[I_q(i, j)]$ is defined as

$$B_q(i, j) = [I_q(i, j)]_i \cup [I_q(i, j)]_j.$$

3 Algorithms for Transformation of MCDT

The following presents algorithms for transforming a MCDT into a family of discretized decision tables. The basic idea was introduced in [18]. For each decision value i, objects in the indexed block $B(i, i)$ are used to define discretization or grouping (if a criterion is ordinal) intervals for transforming values in the MCDT.

procedure Transform_MCDT
inputs: a MCDT with criteria $q_1, \ldots, q_m$ and decision criterion *d* with a finite number *K* of decision classes;
outputs: a family of *K* discretized decision tables;
begin
    Generate_IndexedBlockTable from the MCDT;
    for i = 1 to K **do**
        Generate Discretized_DT using blocks B(i, i);
end; //Transform_MCDT

procedure Generate_IndexedBlockTable
inputs: a MCDT with criteria $q_1, \ldots, q_m$ and decision criterion *d* with a finite number *K* of decision classes;
outputs: an indexed block table IBT for the MCDT;
begin
for k=1 to m **do**
    for i=1 to K **do**

```
          for j=i to K do
                Compute inconsistent intervals Iqk(i, j);
for i=1 to K do
      for j=i to K do
      begin
          I(i, j) := Iq1(i, j);
          for k=2 to m do
                I(i, j) := I(i, j) ∩ Iqk(i, j); //combine inconsistent intervals
      end;
for i=1 to K do
      for j=i to K do
          Generate block B(i, j) := [I(i, j)]i ∪ [I(i, j)]j;
end; //Generate_IndexedBlockTable

procedure Generate_Discretized_DT
inputs: a MCDT with criteria q1, ..., qm and a decision criterion d with a finite num-
          ber K of decision classes and
          an indexed block table IBT generated from the MCDT;
outputs: a discretized decision table;
begin
      for each criterion qj do
          begin
              L := {v | f(qj, x) = v, x in B(i, i) }; //unique criteria values for
                                                        // objects in B(i, i)
              Sort L in increasing order;
              Generate intervals Iqi = { [min, v1], (v1, v2], ..., (v|L|, max)] };
          end;
      for each object y in MCDT do
          begin
              Discretize criteria values of the object y by intervals in Iqi;
              if y is in B(i, i)
              then Label decision value of y with 1
              else Label decision value of y with 0;
          end;
end; // Generate_Discretized_DT
```

4 Generate Rules from Transformed Tables

To generate rules from the family of transformed tables, we will generate a decision table by taking the union of all the discretized tables. From rough set theory, it is clear that decision classes in the resulting table are definable, since they are generated from consistent indexed blocks. In our experiments, we have used the ROSE2 tool [14, 15] to find reducts, then rules were generated by using BLEM2 [1]. The algorithm is formulated in the following.

```
procedure Generate_Rules_From_MCDT
inputs: a MCDT with criteria q1, ..., qm and decision criterion d with a finite number
          K of decision classes;
```

```
outputs: a minimal set of rules with support, strength, certainty, and coverage factors;
begin
    Tansform_MCDT into a family of discretized decision tables;
    Union the tables into one table;
    Find one reduct using the ROSE2 tool;
    Generate rules from the reduct using the BLEM2;
end; // Generate_Rules_From_MCDT
```

5 Conclusions

In this paper we introduced a new methodology for learning rules from a multiple criteria decision table based on the concept of dominance-based rough sets. The dominance approximation space is represented by indexed blocks, which are used to transform a MCDT into a decision table that can be processed by existing learning tools. Our experiments showed that when the family of indexed blocks forms a partition on the universe of objects in a MCDT, only certain rules are generated. In general, some of the rules learned may be uncertain.

References

1. Chan, C.-C., Santhosh, S.: BLEM2: Learning Bayes' rules from examples using rough sets. In: Proc. NAFIPS 2003, 22nd Int. Conf. of the North American Fuzzy Information Proc-essing Society, Chicago, Illinois, July 24–26, 2003, pp. 187–190 (2003)
2. Greco, S., Matarazzo, B., Slowinski, R.: Rough approximation of a preference relation by dominance relations. European Journal of Operational Research 117(1), 63–83 (1999); ICS Research Report 16/96, Warsaw University of Technology, Warsaw (1996)
3. Greco, S., Matarazzo, B., Slowinski, R.: A new rough set approach to evaluation of bank-ruptcy risk. In: Zopounidis, C. (ed.) Operational Tools in the Management of Financial Risks, pp. 121–136. Kluwer Academic Publishers, Dordrecht (1998)
4. Greco, S., Matarazzo, B., Slowinski, R.: The use of rough sets and fuzzy sets in MCDM. In: Gal, T., Stewart, T., Hanne, T. (eds.) Advances in Multiple Criteria Decisions Making, p. 14. Kluwer Academic Publishers, Dordrecht (1999)
5. Greco, S., Matarazzo, B., Slowinski, R.: Rough sets theory for multicriteria decision analysis. European Journal of Operational Research 129(1), 1–47 (2001)
6. Greco, S., Matarazzo, B., Slowinski, R.: Rough sets methodology for sorting problems in presence of multiple attributes and criteria. European Journal of Operational Research 138(2), 247–259 (2002)
7. Greco, S., Matarazzo, B., Slowinski, R., Stefanowski, J.: An algorithm for induction of de-cision rules consistent with the dominance principle. In: Ziarko, W.P., Yao, Y. (eds.) RSCTC 2000. LNCS (LNAI), vol. 2005, pp. 304–313. Springer, Heidelberg (2001)
8. Greco, S., Matarazzo, B., Slowinski, R., Stefanowski, J., Zurawski, M.: Incremental versus non-incremental rule induction for multicriteria classification. In: Peters, J.F., Skowron, A., Dubois, D., Grzymała-Busse, J.W., Inuiguchi, M., Polkowski, L. (eds.) Transactions on Rough Sets II. LNCS, vol. 3135, pp. 33–53. Springer, Heidelberg (2004)

9. Pawlak, Z.: Rough sets: basic notion. International Journal of Computer and Information Science 11(15), 344–356 (1982)
10. Pawlak, Z.: Rough sets and decision tables. In: Skowron, A. (ed.) SCT 1984. LNCS, vol. 208, pp. 186–196. Springer, Heidelberg (1985)
11. Pawlak, Z., Grzymala-Busse, J., Slowinski, R., Ziarko, W.: Rough sets. Communication of ACM 38(11), 89–95 (1995)
12. Pawlak, Z.: Flow graphs and decision algorithms. In: Wang, G., Liu, Q., Yao, Y., Skowron, A. (eds.) RSFDGrC 2003. LNCS, vol. 2639, pp. 1–10. Springer, Heidelberg (2003)
13. Pawlak, Z.: Flow graphs and intelligent data analysis. Fundamenta Informaticae 64, 369–377 (2005)
14. Predki, B., Slowinski, R., Stefanowski, J., Susmaga, R., Wilk, S.: ROSE - Software Implementation of the Rough Set Theory. In: Polkowski, L., Skowron, A. (eds.) RSCTC 1998. LNCS, vol. 1424, pp. 605–608. Springer, Heidelberg (1998)
15. Predki, B., Wilk, S.: Rough Set Based Data Exploration Using ROSE System. In: Raś, Z.W., Skowron, A. (eds.) ISMIS 1999. LNCS (LNAI), vol. 1609, pp. 172–180. Springer, Heidelberg (1999)
16. Roy, B.: Methodologie Multicritere d'Aide a la Decision. Economica, Paris (1985)
17. Chan, C.-C., Tzeng, G.-H.: Dominance-based rough sets using indexed blocks as granules. In: Wang, G., Li, T., Grzymala-Busse, J.W., Miao, D., Skowron, A., Yao, Y. (eds.) RSKT 2008. LNCS (LNAI), vol. 5009, pp. 244–251. Springer, Heidelberg (2008)
18. Chan, C.-C.: Approximate dominance-based rough sets using equivalence granules. In: Zurada, J.M., Yen, G.G., Wang, J. (eds.) Computational Intelligence: Research Frontiers. LNCS, vol. 5050, pp. 2433–2438. Springer, Heidelberg (2008)
19. Chan, C.-C., Tzeng, G.-H.: Computing approximations of dominance-based rough sets by bit-vector encodings. In: Chan, C.-C., Grzymala-Busse, J.W., Ziarko, W.P. (eds.) RSCTC 2008. LNCS (LNAI), vol. 5306, pp. 131–141. Springer, Heidelberg (2008)

A Mobile Location Algorithm with Least Range and Clustering Techniques for NLoS Environments

Cha-Hwa Lin, Chien-Chih Wang, and Chih-Hung Tsai

Abstract. We propose an efficient location algorithm which can mitigate the influence of NLOS error. Based on the geometric relationship between known positions of the base stations, the theorem of "Fermat Point" is utilized to collect the candidate positions (CPs) of the mobile station. Then, a set of weighting parameters are computed using a density-based clustering method. Finally, the location of mobile station is estimated by solving the optimal solution of the weighted objective function. Different distributions of NLOS error models are used to evaluate the performance of this method. Simulation results show that the performance of the least range measure (LRM) algorithm is slightly better than density-based clustering algorithm (DCA), and superior to the range based linear lines of position algorithm (LLOP) and range scaling algorithm (RSA) on location accuracy under different NLOS environments. The simulation results also satisfy the location accuracy demand of Enhanced 911 (E-911).

1 Introduction

With the great growth on the demand of location based services (LBS), the location estimation of a mobile station (MS) has gained considerable attention of the researchers in recent years, especially for vital emergency, personal safety services, and commercial applications such as location sensitive billing, fleet management, and intelligent transportation systems (ITS).

The most popular technologies of wireless location [1, 2, 3, 9] include signal strength (SS), time of arrival (TOA), time difference of arrival (TDOA), and angle of arrival of the signal (AOA). In these approaches the MS whose position is being tracked interacts with several base stations (BSs). The method of location estimation solution will be introduced and mainly in view of location method based on TOA in this paper. However, the accuracy of mobile location schemes

Cha-Hwa Lin, Chien-Chih Wang, and Chih-Hung Tsai
National Sun Yat-sen University, Department of Computer Science and Engineering
Kaohsiung 80424, Taiwan
e-mail: chlin@cse.nsysu.edu.tw

B.-C. Chien and T.-P. Hong (Eds.): Opportunities and Challenges, SCI 214, pp. 247–252.
springerlink.com

depends on the propagation conditions of the wireless channels. If line of sight (LOS) exists between the MS and BSs, high location accuracy can be achieved. But, in most of the cases the MS is not in LOS with a BS, i.e., the signal faces obstacles as it propagates between the MS and a BS. Thus, the signal needs more time to reach the BS which will be directly added to the range measurement as an extra error which is called non-line-of-sight (NLOS) error [8]. This extra added error is usually large and will cause the MS location estimate to be far from the true location. Let the measured distance from MS to BS is denoted as l and the true distance is R, the NLOS condition can be expressed as $l = R +$ NLOS error.

Several approaches have been proposed to detect, remove, or account for the biases due to NLOS in order to obtain accurate location estimates. The typical TOA based geometrical techniques for NLOS mitigation considered in the literature are linear lines of position (LLOP) algorithm [4] and range scaling algorithm (RSA) [7]. LLOP algorithm tries to convert the nonlinear intersection problem to several linear ones. The geometrical interpretation is presented in which straight lines of position (LOP), rather than the circular LOP. It not only can be programmed and quickly executed but also can get a reliable result than Taylor Series method. This approach can mitigate the NLOS error as well as the measurement noise, but it needs at least four BSs to achieve a favorable result, and its performance highly depends on the relative locations of the MS and BSs. The RSA method involves the requirement of solving an optimization problem based on a nonlinear objective function. The inefficiency incurred by the algorithm may not be feasible to be applied in practical systems. RSA indeed improved the location accuracy for NLOS environments, but still can not satisfy the location accuracy demand of E-911 [5]. Recently, a new location algorithm named density-based clustering algorithm (DCA) [6] estimates MS location by solving the optimal solution of the objective function based on the high density cluster. The location accuracy of DCA has been greatly improved over previous algorithms (i.e., LLOP and RSA).

In this paper, we propose a novel algorithm named least range measure (LRM) algorithm that attempts to determine the optimal ranges given range measurements from MS to three BSs with NLOS errors. Based on the geometry of the range circles and the known positions of BSs, the theorem of "Fermat Point" is utilized to predict the candidate positions of the mobile station. The remainder of this paper is organized as follows. The proposed algorithm is detailed in Section 2. The simulation results are given in Section 3, followed by some concluding remarks in Section 4.

2 Least Range Measure Algorithm

In this section, a TOA-based technique is introduced that attempts to determine the optimal ranges given range measurements from the MS to three BSs with NLOS errors. Based on the geometrical relationships of the range circles and the known positions of three BSs, the theorem of "Fermat Point" is utilized to reconstruct a candidate position for the mobile station. The MS location is then estimated by solving the optimal solution of a weighted objective function.

The measured range measurement equation between the MS and the ith BS in a NLOS environment can be obtained by including the NLOS error, η_i, as

$$l_i = R_i + n_{R_i} + \eta_i, i = 1, 2, 3 \tag{2.1}$$

The measurement noise, n_{R_i}, is a zero-mean Gaussian random value with relatively small standard deviation and its effect is negligible if the measured ranges are averaged over a few seconds. It is also negligible compared to the magnitude of the NLOS error. Given the range measurement, l_i and neglecting the standard measurement error, it can be inferred that the range of possible values for η_i is between 0 and l_i.

Then true range measurement equation between the MS and the ith BS can be described as

$$R_i = \sqrt{(x - x_i)^2 + (y - y_i)^2}, i = 1, 2, 3 \tag{2.2}$$

where (x_i, y_i) is the ith BS location and (x, y) is the true location of MS. So the relation between the measured ranges l_i and the true ranges R_i can be written as

$$R_i = \delta_i l_i, i = 1, 2, 3 \tag{2.3}$$

Since NLOS error is a positive bias that causes the measured ranges to be greater than the true ranges. Hence, the value of δ_i is bounded by $0 < \delta_i \le 1$. If and only if there is no NLOS error, δ_i equals to (2.1). Squaring the range in (2.2) and substituting into (2.3) results in

$$(x - x_i)^2 + (y - y_i)^2 = \delta_i^2 l_i^2, i = 1, 2, 3 \tag{2.4}$$

If we make the reasonable assumption that, at any instant, not more than one BS is LOS, it is clear that each pair of overlapping range circles intersect at two distinct points, so the three circles form an area where the MS locates. (Fig. 1) shows the area as UVW, where BS_1, BS_2, BS_3 indicate the three BSs. Since MS must locate in the intersected area UVW, we take the objective function [7] to be the sum of the square of the distances from the MS location to point U, V and W,

$$f(x, y) = (x - U_x)^2 + (y - U_y)^2 + (x - V_x)^2 + (y - V_y)^2 + (x - W_x)^2 + (y - W_y)^2$$

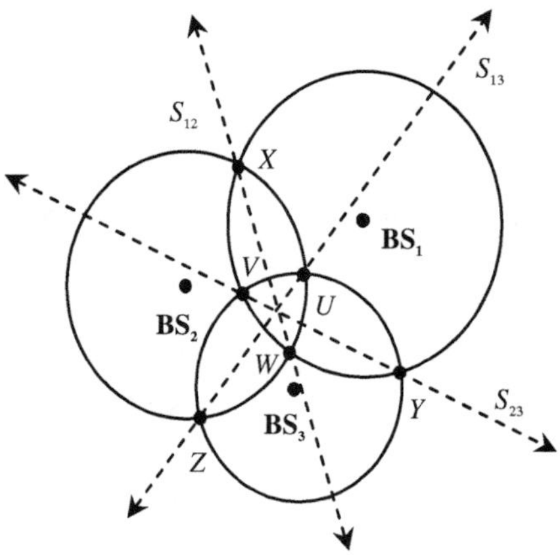

Fig. 1 Linear equations S_{12}, S_{13}, and S_{23} are the lines which pass through the intersection points of any two range circles

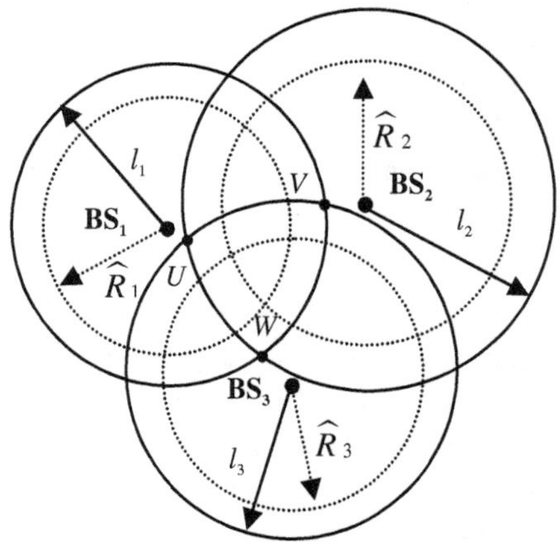

Fig. 2 The overlapping region of the three dotted-line circles formed by the optimal estimates $\widehat{R}_i$ is the smallest region with densest candidate positions by clustering

where the coordinates of U, V, and W are (U_x, U_y), (V_x, V_y), and (W_x, W_y), respectively. However, the extra propagation distance of the NLOS path directly corresponds to an overestimate of range between an MS and a BS. The phenomenon could be alleviated by adjusting the weights of the terms in the objective function. The objective function is modified as

$$f'(x,y) = \mu_1[(x-U_x)^2+(y-U_y)^2]+\mu_2[(x-V_x)^2+(y-V_y)^2]+\mu_3[(x-W_x)^2+(y-W_y)^2]$$

$$\mu_i = \widehat{R}_i \; / \; l_i$$

where μ_i = the weight of the ith term, $0 < \mu_i \leq 1$, $\widehat{R}_i$ = the optimal estimates for true ranges R_i, l_i = the measured ranges, and i = 1, 2, and 3. The weights μ_i can be selected to reflect the reliability of the measurements for each BS-MS pair. The optimal estimates $\widehat{R}_i$ for true ranges R_i are determined by the dotted-line circles (Fig. 2). Since the candidate positions (CPs) are all the possible locations for the mobile station, the location estimation problem can be formulated as a nonlinear optimization problem. The optimal candidate position which minimizes the weighted objective function is location estimation of the MS by the proposed LRM algorithm. Thus, the estimation for the optimal MS location will be

$$Optimal\ MS\ Location = \arg\min_{x,\,y} f'(x,y)\ .$$

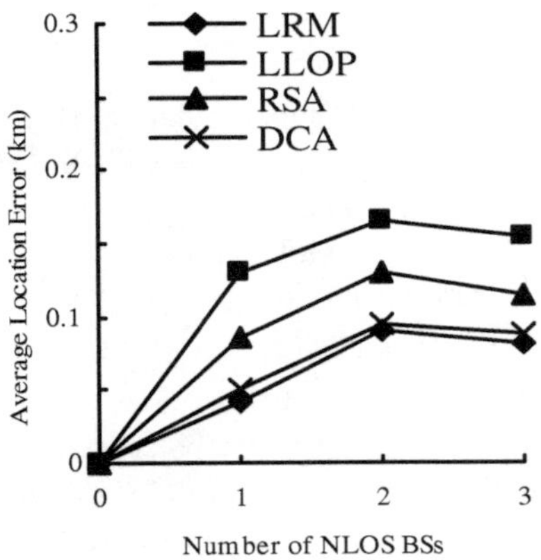

Fig. 3 The average location error for LRM, LLOP, RSA, and DCA algorithm versus the number of NLOS BSs with CDSM error model

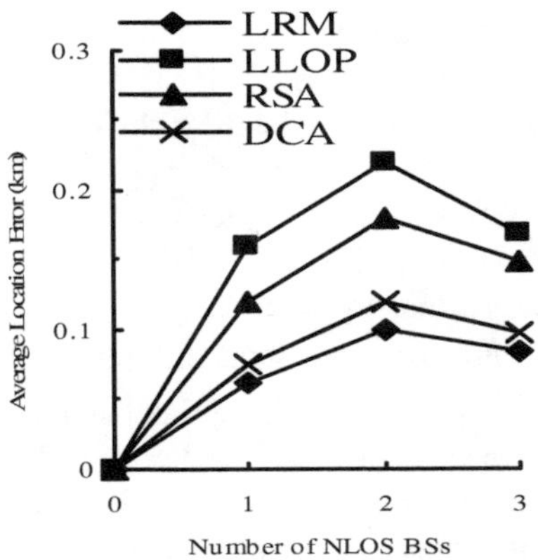

Fig. 4 The average location error for LRM, LLOP, RSA, and DCA algorithm versus the number of NLOS BSs with uniform error model

3 Simulation Results and Discussions

The performance of the proposed location algorithm was examined with regular hexagon cell layout as shown in with the MS position chosen randomly according to a uniformly distributed function within the area covered by the triangle formed by three BSs, BS_1, BS_2, and BS_3. The NLOS range errors were modeled as positive random variables having support over [0,0.4] km, generated according to different probability density functions, such as CDSM, normal distribution, and uniform distribution. All units are expressed in kilometer (km).

The coordinates of the three BSs are BS_1: (0.866, 1.5), BS_2: (0, 0), and BS_3: (1.732, 0), forming an equilateral triangle located at center of hexagonal cells with radius 1 km. The simulation results are averaged over 1000 test runs, each time using a different NLOS error and MS position. The performance of the LRM algorithm is compared with the LLOP [4] algorithm, RSA [7], and DCA [6]. The TOA measurements used in the LLOP, RSA, and DCA algorithms were the same as that used for LRM algorithm.

The experiments are implemented to investigate how the average location error is affected by the number of BSs that do not have a LOS path to the MS when LRM is employed. The serving BS is assumed to be LOS with the MS, except for the case when all BSs are NLOS. This performance also was compared with LLOP, RSA, and DCA for CDSM and uniform error model as shown in Fig. 3 and Fig. 4. As expected, it is observed that the average location error increases with number of NLOS BSs. However, the average location error decreases slightly when all BSs are NLOS as compared to the case when two BSs are NLOS. Simulation results show that the performance of the LRM is slightly better than DCA and superior to the range based LLOP and RSA regardless of the number of NLOS BSs.

4 Conclusions

In this paper, a TOA-based technique is introduced that attempts to determine the optimal MS location given range measurements from MS to three BSs with NLOS errors. Based on the geometry of the range circles and the known positions of BSs, the theorem of "Fermat Point" is utilized to collect the candidate positions of the mobile station. In order to omit the non-significant calculations, additional constraints are added on range scaling parameters α, β, and γ. A set of weighting parameters related to the optimal and measured range circles are computed using a density-based clustering method. The MS location is then estimated by solving the optimal solution of the weighted objective function. Simulation results show that the performance of the LRM is slightly better than DCA and superior to the range based LLOP and RSA in terms of location accuracy. The location error of LRM is close to 0.075 km for 67% of the time, and less than 0.14 km for 95% of the time. The results of the experiment satisfy the location accuracy demand of E-911. The proposed algorithm contributions are supported by simulations and analysis that demonstrate the performance improvement over previous algorithms.

References

1. Caffery, J., Stuber, G.: Overview of Radiolocation in CDMA Cellular Systems. IEEE Communications Magazine 36, 38–45 (1998)
2. Caffery, J., Stuber, G.: Subscriber Location in CDMA Cellular Networks. IEEE Transactions on Vehicular Technology 47(2), 406–416 (1998)
3. Caffery, J.: Wireless Location in CDMA Cellular Radio Systems. Kluwer, Massachusetts (1999)
4. Caffery, J.: A New Approach to the Geometry of TOA Location. In: Proceedings of the IEEE Vehicular Technology Conference, pp. 1943–1949 (2000)
5. FCC, Revision of the Commissions Rules to Insure Compatibility with Enhanced 911 Emergency Calling Systems. Technical Report, RM-8143. Washington, DC: U.S. Federal Communications Commission (1996)
6. Lin, C.-H., Cheng, J.-Y., Wu, C.-N.: Mobile Location Estimation by Density-Based Clustering for NLoS Environments. In: Proceedings of the 20th International Conference on Advanced Information Networking and Applications, vol. 1, pp. 295–300 (2006)
7. Venkatraman, S., Caffery, J., You, H.-R.: A Novel ToA Location Algorithm Using LoS Range Estimation for NLoS Environments. IEEE Transaction on Vehicular Technology 53, 1515–1524 (2004)
8. Wylie, M.P., Holtzman, J.: The Non-Line of Sight Problem in Mobile Location Estimation. In: Proceedings of the IEEE International Conference on Universal Personal Communications, vol. 2, pp. 827–831 (1996)
9. Zhao, Y.: Standardization of Mobile Phone Positioning for 3G Systems. IEEE Communications Magazine 40, 108–116 (2002)

Learning Interaction Structure Using a Hierarchy of Dynamical Systems

Yasser Mohammad and Toyoaki Nishida

Abstract. The IAM (Interaction Adaptation Manager) algorithm was recently proposed to learn the optimal parameters of a hierarchical dynamical system incrementally through interacting with other agents given that the structure of the system is known (the number of processes in each layer and their interconnections) and that the agent knows how to interact in all roles except the one it is learning (e.g. an agent learning to listen should know how to speak). This paper presents an algorithm for learning the structure of a hierarchical dynamical system representing the interaction protocol at various timescales and using multiple modalities relaxing these two constraint. The proposed system was tested in a simulation environment in which rich human-like agents are interacting and showed accurate recognition of the interaction structure using few training examples. The learned structure showed acceptable performance that allowed subsequent application of the adaptation algorithm to converge to a good solution using as few as 15 interactions.

1 Background

Researchers in nonverbal communication and social studies have discovered many forms of entrainment and nonverbal synchronization behaviors in human-human interactions including body alignment, prosody entrainment and various kinds of verbal entrainment [1]. The gestural dance theory proposed by [2] hypothesizes that nonverbal bodily communication shows synchrony effects between the interacting humans in time scales of 40 milliseconds or less. Although this hypothesis

Yasser Mohammad
Department of Intelligence Science and Technology,
Graduate School of Informatics, Kyoto University, Japan
e-mail: yasser@ii.ist.i.kyoto-u.ac.jp

Toyoaki Nishida
Department of Intelligence Science and Technology,
Graduate School of Informatics, Kyoto University, Japan
e-mail: nishida@i.kyoto-u.ac.jp

B.-C. Chien and T.-P. Hong (Eds.): Opportunities and Challenges, SCI 214, pp. 253–258.
springerlink.com

is not confirmed, fine scale synchrony in body movements during interactions [1] (specially during smooth turn taking) challenges the possibility of using disembodied techniques for interactive agents that utilize not only verbal but nonverbal communication channels.

Many researchers in HRI and ECA communities have tried to build interactive agents that can sustain levels of synchrony with humans comparable to the human-human case [3]. For example [4] studied the body alignment of the robot to a human instructor during an explanation of the route to a target and found significant difference between a fixed robot and a robot that correctly aligns its body to the partner. In this example, and other systems based on the situated modules architecture, the designer have to decide the kinds of behaviors needed and the detailed relation between them. This means that the interaction structure must be known beforehand. By the interaction structure we mean the kinds of synchronization needed, their time scales, and the structure of the processes needed to achieve them up to an unknown parameter vector.

Some researchers tried to alleviate this problem and enable the robot to learn the structure of the interaction in specific interaction situations (e.g. [5] [6]). In [7], the authors developed the L_iEICA architecture (Embodied Interactive Control Architecture) to handle the limitations of the currently available systems concerning management of nonverbal interaction protocols in human-like interactions. The system showed fast learning where agents could construct a valid protocol representation utilizing as low as nine interactions. The algorithm used in this system was called IAM (Interaction Adaptation Manager).

One limitation of this system was that the agent learning a specific role in an interaction (e.g. listening to explanations) needs to know how to behave in all the other roles in that interaction (e.g. speaking). Another limitation was that the system can only learn the protocol given that it is fully specified up to an unknown parameter vector (e.g. the number of control layers, the number of processes in every layer, as well as the internal structure of every process must be known).

In this paper we try to alleviate these two restrictions of the IAM by enabling the agent to learn the structure of the interaction protocol by *watching* other agents interacting. The only information need by the algorithm proposed in this paper is a set of basic interactive acts (BIAs) that represent the motor plans meaningful to the current interaction.

2 The Architecture

The architecture used in this paper is a layered control architecture called L_iEICA [7]. This architecture is designed to control the nonverbal communication channels of the robot (e.g. gaze control, proximities, body alignment, etc) to provide human-like nonverbal behavior that is considered to be important in many HRI applications (e.g. companion robots, rehabilitation robots, etc).

Within each layer a set of processes provide the competencies needed to synchronize the behavior of the agent with the behavior of its partner(s) at a specific level of abstraction. The synchronization protocol implemented within each layer is called the *within-layer protocol*. Each layer controls the layer under it using a *higher-layer protocol*. This paper focuses on how the agent can learn the structure of this architecture up to an unknown parameter vector for each process which can then be learned by the algorithm described in [7].

Fig. 1 shows a simplified version of the architecture. Every process in the system is implemented as a dynamical system. The first processes that respond to the behavior of the partner are the Perspective Taking Processes (PTPs) responsible of estimating the responses of the sensors of the partner. The actual behavior of the robot is generated by a set of Forward Basic Interactive Acts (FBIAs) that represent the basic motor plans that the robot/agent can use to behave in a specific role in the interaction. Every FBIA has an activation level (corresponding to its *intentionality* in L_iEICA vocabulary [8]) that determines its influence on the final behavior of the robot. For every FBIA there is a Reverse Basic Interactive Act (RBIA) that can use the outputs of the PTPs corresponding to one partner to estimate the activation level of this FBIA in the partner (remember that the architecture assumes that the behavior of the partner can be modeled by the competencies of the agent/robot itself). The FBIAs and RBIAs constitute the first layer of control in the robot (Interaction Control Layer 0). Higher layers of control are constituted of Forward Interaction Control Processes (FPICs) and Reverse Interaction Control Processes (RPICs) that control the activation levels of the processes in the preceding layer.

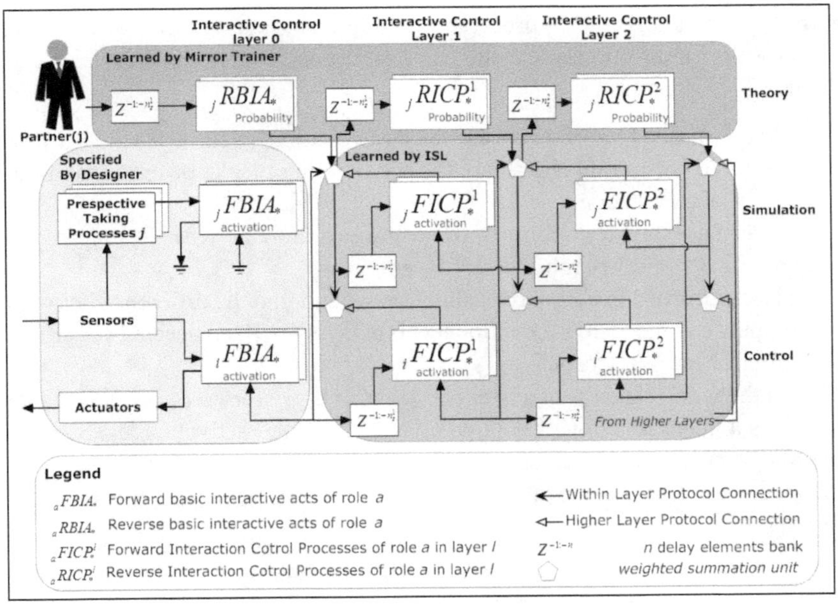

Fig. 1 The relation between the inputs and outputs of processes in different control layers during online interaction. The agent is assumed to have role i (e.g. listener) while its partner is having role j (e.g. speaking)

3 The Interaction Structure Learner

The Interaction Structure Learner is invoked by a set of training examples and a set of verification examples. Every example (c) is a record of the sensory information (^{c}s) and body motion information ($^{c}_{i}m$) of every agent i during a specific interaction c of the type to be learned.

The robot designer has to provide the PTPs, the FBIAs and the RBIAs related to the interaction at hand and the goal of the interaction structure learner is to learn the rest of the control architecture of the robot. FBIAs can be learned using constrained motif discovery as described in [9], while RBIAs can be learned using the mirror trainer [7].

The outputs of the interaction structure learner are: the number of layers needed to represent the interaction (l_{max}), the number of ICPs for every role in each layer ($_{r}n_{l}$), and parameter initialization of every FICP ($_{r}FICP^{l}_{j}$) in every layer l for every role in the interaction r and the corresponding RICPs ($_{r}RICP^{l}_{j}$) [learned by the Mirror Trainer presented in [7]]. These initial values of the parameters can then be adapted while the agent is interacting in various roles using IAM as described in [7].

The ISL algorithm involves the following steps:

1. Project sensor readings to the perspective of every partner in the interaction using the PTPs.
2. Convert the outputs of the PTPs of all interactions into activation levels of the FBIAs using the RBIAs.
3. Learn the interaction control processes starting from layer l incrementally one layer at a time until one layer contains at most one process for every agent as follows:

    a. Apply constrained motif discovery for the activation levels of layer $l-1$ processes to get $_{i}n^{l}$ motifs characterizing the building blocks of the behavior at this level of abstraction.
    b. For layer l use the occurrences of these motifs to train $_{i}n^{l}$ RBFNNs representing the FICPs of every role i in this layer.
    c. Apply this learned FICPs to the validation set and find the difference between its outputs and the action activation levels at layer $l-1$ and accept every FICP if this error level is acceptable.
    d. Learn the within-layer protocol connections of every forward process in layer $l-1$ as a linear function of the one time step delayed activation levels of the partner representing processes in layer $l-1$.
    e. Invoke the mirror trainer to learn the RICPs corresponded to the accepted FICPs

4 Evaluation

As a proof of concept for the proposed algorithm, a simulation study was conducted to measure the capacity of the agent to learn how to control the gaze direction during listening and speaking by *watching* interactions between other agents who already have learned these roles based on human-human interaction data (fully designed agents) [7]. A simulation study rather than a real world human-agent interaction was selected because it allows us to quantitatively measure the accuracy of the system in learning a *known* predefined structure. Ten different fully designed agents were implemented that differ in the details of how they conduct instruction and how they respond to it while one agent used their interactions to learn the structure of the instructing-listening interaction. Details of the internal design of each fully designed agent are omitted due to lack of space. For more details refer to [7].The FBIAs were implemented as augmented state machines. A zero mean Gaussian random process error signal was added to the final join angles used to simulate the pose of the agent. 100 interactions between fully designed agents were collected and divided into 75 training examples and 25 verification examples. Each interaction lasted between 10 and 15 minutes. Fifty different training/verification sets were selected using a random permutation of 20, 40, 60, 80 and 100% of the data. To measure effect of environmental noise on the algorithm, random uniform noise signals of range $\pm\delta$ (where δ=0,2,4,6,8cm) were applied to the sensory information presented to the system. This leads to a total of 250 training sessions.

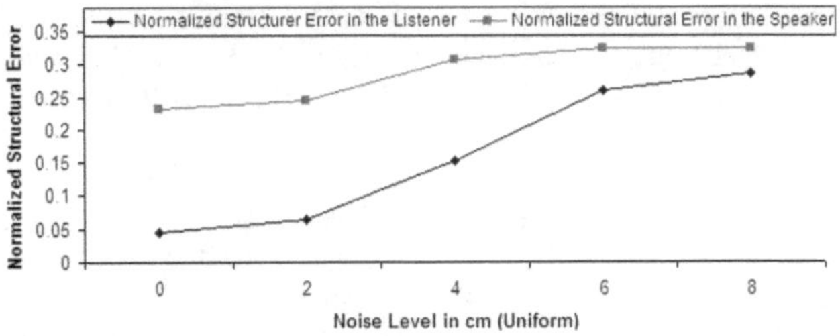

Fig. 2 Structural Error as a function of the noise level

For the case with no noise, the system was able to learn a consistent structure in all cases using a training set of 30 interactions or more(i.e. 2 layers for the listener with 5 processes in layer one and one process in layer 2 and one layer for the instructor with five processes in it). The fact that the system did not discover any higher level protocol for the instructor reflects the asymmetry of the interaction as the behavior of the listener is completely determined by the instructor's behavior while the behavior of the instructor depends also on the needs of the explanation script. In the case of 15 training samples, the system learned the correct structure for the listening process in all cases but discovered only four of the five interactions in layer 1 for the instructor in two cases.

In general noise levels less than 4cm did not affect the structure learned by the algorithm except for the smallest training set (15 interactions).

5 Conclusion

In this paper, a novel algorithm for learning a hierarchical architecture of dynamical systems representing the structure of complex human-like nonverbal interactions. The system was tested using a simulated environment in which humanoid agents trained using actual human data are interacting and showed accurate recognition of the interaction structure in low noise levels with a small number of training examples (15 scenarios).

References

1. Argyle, M.: Bodily Communication. Routledge, London (2001)
2. Condon, W.S., Ogston, W.D.: Sound Film Analysis of Normal and Pathological Behavior Patterns. Journal of Nervous and Mental Disease 143, 338–347 (1966)
3. Ono, T., Kanda, T., Imai, M., Ishiguro, H.: Embodied communications between humans and robots emerging from entrained gestures. In: IEEE International Symposium on Computational Intelligence in Robotics and Automation, vol. 2(2), pp. 558–563 (2003)
4. Kanda, T., Kamasima, M., Imai, M., Ono, T., Sakamoto, D., Ishiguro, H., Anzai, Y.: A humanoid robot that pretends to listen to route guidance from a human. Autonomous Robots 22(1), 87–100 (2007)
5. Ogata, T., Sugano, S., Tani, J.: Open-end human robot interaction from the dynamical systems perspective: mutual adaptation and incremental learning. In: Orchard, B., Yang, C., Ali, M. (eds.) IEA/AIE 2004. LNCS, vol. 3029, pp. 435–444. Springer, Heidelberg (2004), `http://dx.doi.org/10.1007/b97304`
6. Imai, M., Kawasima, H., Honda, Y.: Generating Behavioral Protocol for Human-Robot Physical-Contact Interaction. In: IEEE ICRA 2005, pp. 229–234 (2005)
7. Mohammad, Y., Nishida, T.: Towards Combining Autonomy and Interactivity for Social Robots. In: AI and Society: Special Issue about SID 2007 (in press) (2008)
8. Mohammad, Y., Nishida, T.: A Cross-Platform Robotic Architecture for Autonomous Interactive Robots. In: Nguyen, N.T., Borzemski, L., Grzech, A., Ali, M. (eds.) IEA/AIE 2008. LNCS, vol. 5027, pp. 108–118. Springer, Heidelberg (2008)
9. Mohammad, Y., Nishida, T.: Constrained Morif Discovery. In: Third International Workshop on Data-Mining and Statistical Science, pp. 16–19 (2008)

Fuzzy Assessment for Survey Defuzzified by Signed Distance Method

Lily Lin and Huey-Ming Lee

Abstract. In this study, we propose a model to do fuzzy aggregative assessment defuzzified by signed distance method. The proposed fuzzy assessment method on sampling survey analysis is easily to assess the sampling survey and make the aggregative evaluation.

1 Introduction

Traditionally, we compute statistics with sample data by questionnaires according to the thinking of binary logic. But, this kind of result may lead to an unreasonable bias since the human thinking is full with fuzzy and uncertain. Fuzzy sets theory was introduced by Zadeh [4] to deal with problem in which vagueness is present, linguistic value can be used for approximate reasoning within the framework of fuzzy set theory [5] to effectively handle the ambiguity involved in the data evaluation and the vague property of linguistic expression, and normal triangular fuzzy numbers are used to characterize the fuzzy values of quantitative data and linguistic terms used in approximate reasoning.

There usually exists two different methods, both multiple-item and single-item choices, while using linguistic variables as rating item. We use mark or unmark to determine the choice for each item, i.e., the marked item is represented by 1, while the other unmark item is represented by 0. Generally speaking, the linguistic variable possesses the vague nature. Therefore, Lin and Lee [1, 2] applied a value m which belongs to the interval of [0, 1] to represent the reliability or membership grade in the fuzzy sense of marking item.

In this study, we propose a model to do aggregative assessment via questionnaires and fuzzy composition rule of inference based on signed distance method.

Lily Lin
Department of International Business, China University of Technology
56, Sec. 3, Hsing-Lung Road, Taipei 116, Taiwan
e-mail: lily@cute.edu.tw

Huey-Ming Lee
Department of Information Management, Chinese Culture University, Taiwan
55, Hwa-Kung Road, Yang-Ming-San, Taipei(11114), Taiwan
e-mail: hmlee@faculty.pccu.edu.tw

B.-C. Chien and T.-P. Hong (Eds.): Opportunities and Challenges, SCI 214, pp. 259–264.
springerlink.com

The proposed fuzzy assessment method on sampling survey analysis is easily to assess the sampling survey and do the aggregative evaluation.

2 Preliminaries

For the proposed algorithm, all pertinent definitions of fuzzy sets are given below [3, 5, 6].

Let $\tilde{A} = (p, q, r)$ be a triangular fuzzy number, then the α level set is

$$A(\alpha) = \{x | \mu_{\tilde{A}}(x) \geq \alpha\} = [A_L(\alpha), A_R(\alpha)], 0 \leq \alpha \leq 1 \quad (1)$$

where

$$A_L(\alpha) = a + (b-a)\alpha, A_R(\alpha) = c - (c-b)\alpha \quad (2)$$

Definition 1. The signed distance [3]: We define $d_0(a,0) = a,\ for\ a, 0 \in R$

Definition 2. ([3]) Let $\tilde{D}$ be a fuzzy set, we define the signed distance of $\tilde{D}$ measured from $\tilde{0}$ as

$$d(\tilde{D},\tilde{0}) = \frac{1}{2}\int_0^1 [D_L(\alpha) + D_R(\alpha)]d\alpha \quad (3)$$

Let $\tilde{A} = (p, q, r)$, $p<q<r$ be a triangular fuzzy number, then, from Definition 2, we have

$$d(\tilde{A},\tilde{0}) = \frac{1}{4}(p + 2q + r) \quad (4)$$

3 Fuzzy Aggregative Assessment Based by Signed Distance Method

In most cases, questionnaire of sampling survey exists many topics and questions, let's say, main items and sub-items. For instance, one specific questionnaire regarding satisfactory level may include main survey items such as satisfactory level for product, service and price etc., also sub-items may exist under each main item. We can define them as follows:

Main items: $B_1,\ B_2, ...,\ B_r$

with weight: $b_1, b_2, ..., b_r$, respectively

subject to: $0 \leq b_j \leq 1$, $j = 1, 2, \ldots, r$ and $\sum_{j=1}^{r} b_j = 1$

Sub-items: $B_{j1},\ B_{j2}, ...,\ B_{jm_j}$ under main items B_j, $j = 1, 2, \ldots, r$;

with weight: $b_{j1}, b_{j2}, ..., b_{jm_j}$, respectively

subject to:

$$0 \le b_{jt} \le 1,\ j = 1, 2, \ldots, r; t = 1, 2, \ldots, m_j \ \text{ and } \sum_{t=1}^{m_j} b_{jt} = 1$$

Owing to the questionary of the sample survey is in fuzzy sense, we can not evaluate the aggregative assessment by the traditional statistics. Let L_q (for q=1, 2,..., *m*) be the *m* different linguistic variables as criteria of questionnaire, expressed in fuzzy language such as very low, low, medium, high, very high, etc. We consider that the fuzzy linguistics $L_1, L_2, \ldots, L_m$ with the corresponding series of fuzzy numbers $\tilde{L}_1, \tilde{L}_2, \ldots, \tilde{L}_m$, where

$$\tilde{L}_q = ((q-1)(\frac{100}{n+1}), q(\frac{100}{n+1}), (q+1)(\frac{100}{n+1})) \tag{5}$$

for $q=1, 2, \ldots, m$. Suppose the evaluators assess the aggregative grades for some one main item B_j with sub-items $B_{j1}, B_{j2}, \ldots, B_{jm_j}$ of the company. We propose the contents of the *j*th (for *j*=1, 2, …, *r*) main item assessment form as shown in Table 1.

From Table 1, we let

$$n_{jq} = \sum_{t=1}^{m} n_{jqt},\ j = 1, 2, \ldots, r;\ q = 1, 2, \ldots, m_j \tag{6}$$

We let $(n_{jqt} / n_{jq})\tilde{L}_q$ be the weighted triangular fuzzy number of $\tilde{L}_q$ respective to B_{jq_j}. Let $L=\{L_1, L_2, \ldots, L_m\}$ be the set of the *m* fuzzy linguistics. We can form a fuzzy relation on B_jxL with triangular fuzzy number elements as follows:

$$\tilde{R}_j = \begin{bmatrix} (n_{j11}/n_{j1})\tilde{L}_1 & (n_{j12}/n_{j1})\tilde{L}_2 & \ldots(n_{j1m}/n_{j1})\tilde{L}_m \\ (n_{j21}/n_{j2})\tilde{L}_1 & (n_{j22}/n_{j2})\tilde{L}_2 & \ldots(n_{j2m}/n_{j2})\tilde{L}_m \\ \cdot & \cdot & \cdot \\ \cdot & \cdot & \cdot \\ (n_{jm_j1}/n_{jm_j})\tilde{L}_1 & (n_{jm_j2}/n_{jm_j})\tilde{L}_2 & \ldots(n_{jm_jm}/n_{jm_j})\tilde{L}_m \end{bmatrix}$$

Then, the first stage aggregative assessment for the attribute $B=\{B_{j1}, B_{j2}, \ldots, B_{jmj)}\}$ is as follows:

$$(\tilde{a}_{j1}, \tilde{a}_{j2}, \ldots, \tilde{a}_{jm}) = (b_{j1}, b_{j2}, \ldots, b_{jm_j}) \bullet \tilde{R}_j \tag{7}$$

where

$$\tilde{a}_{jq} = (b_{j1} \cdot \frac{n_{j1q}}{n_{j1}})\tilde{L}_q \oplus (b_{j2} \cdot \frac{n_{j2q}}{n_{j2}})\tilde{L}_q \oplus \ldots \oplus (b_{jm_j} \cdot \frac{m_{jm_jq}}{n_{jm_j}})\tilde{L}_q$$

Let B={B_1, B_2, …, B_r}, then the second stage aggregative composition inference is as follows:

$$(\tilde{g}_1,\tilde{g}_2,...\tilde{g}_m)=(b_1,b_2,...,b_r)\circ\begin{bmatrix}\tilde{a}_{11}\ \tilde{a}_{12}\ ...\ \tilde{a}_{1m}\\ \tilde{a}_{21}\ \tilde{a}_{22}\ ...\ \tilde{a}_{2m}\\ .\quad .\quad .\\ \tilde{a}_{r1}\ \tilde{a}_{r2}\ ...\ \tilde{a}_{rm}\end{bmatrix}$$

Then, we have the following proposition:

Proposition 1

(1^0) For each $j\in\{1, 2,..., r\}$, and for the fuzzy linguistic language $\tilde{L}_q$ (q=1, 2, …, m), the aggregative assessment of the main item B_j is

$$d(\tilde{a}_{jq},\tilde{0})=\frac{1}{4}\sum_{k=1}^{m_j}b_{jk}\frac{n_{jkq}}{n_{jk}}(t_{q-1}+2t_q+t_{q+1}) \tag{8}$$

(2^0) For each $j\in\{1, 2,..., m\}$, the integrated assessment of the main item B_j is

$$P_2^{(j)}=\sum_{q=1}^{m}d(\tilde{a}_{jq},\tilde{0})$$

$$=\frac{1}{4}\sum_{k=1}^{m_j}b_{jk}\sum_{q=1}^{m}\frac{n_{jkq}}{n_{jk}}(t_{q-1}+2t_q+t_{q+1}) \tag{9}$$

(3^0) For the fuzzy linguistic language $\tilde{L}_q$ (q=1, 2, …, m), the Second stage aggregative assessment is

$$d(\tilde{g}_q,\tilde{0})=\frac{1}{4}\sum_{j=1}^{r}b_j\sum_{k=1}^{m_j}b_{jk}\frac{n_{jkq}}{n_{jk}}(t_{q-1}+2t_q+t_{q+1}) \tag{10}$$

(4^0) The integrated of the assessment is

$$P_2=\sum_{q=1}^{m}d(\tilde{g}_q,\tilde{0})$$

$$=\frac{1}{4}\sum_{j=1}^{r}b_j\sum_{k=1}^{m_j}b_{jk}\sum_{q=1}^{m}\frac{n_{jkq}}{n_{jk}}(t_{q-1}+2t_q+t_{q+1}) \tag{11}$$

Table 1 Contents of the proposed assessment form

Main-Item	Item - weight	Sub-item	Linguistic variables	Numbers of answer	Triangular fuzzy number of the linguistic language
B_j	a_j	B_{j1}	L_1	n_{j11}	$\tilde{L}_1 = (0, t_1, t_2)$
			L_2	n_{j12}	$\tilde{L}_2 = (t_1, t_2, t_3)$
		$\vdots$	$\vdots$	$\vdots$	$\vdots$
			L_m	n_{j1m}	$\tilde{L}_m = (t_{m-1}, t_m, 100)$
		B_{j2}	L_1	n_{j21}	$\tilde{L}_1 = (0, t_1, t_2)$
			L_2	n_{j22}	$\tilde{L}_2 = (t_1, t_2, t_3)$
			$\vdots$	$\vdots$	$\vdots$
			L_m	n_{j2m}	$\tilde{L}_m = (t_{m-1}, t_m, 100)$
		$\vdots$	$\vdots$	$\vdots$	$\vdots$
		B_{jm_j}	L_1	n_{jm_j1}	$\tilde{L}_1 = (0, t_1, t_2)$
			L_2	n_{jm_j2}	$\tilde{L}_2 = (t_1, t_2, t_3)$
			$\vdots$	$\vdots$	$\vdots$
			L_m	n_{jm_jm}	$\tilde{L}_m = (t_{m-1}, t_m, 100)$

4 Conclusion

In this study, we propose a model to do assessment analysis for sampling survey with the linear order character of fuzzy linguistics by signed distance method. Since the proposed model described in this study is to measure the group evaluation, the final value is more objective and unbiased than just one evaluator's assessment. Moreover, if there is only one evaluator existing, the proposed model is also appropriate to assess.

Acknowledgments. The authors gratefully acknowledge the helpful comments and suggestions of the reviewers, which have improved the presentation.

References

1. Lin, L., Lee, H.-M.: A New Assessment Model for Global Facility Site Selection. Int. J. Innov. Comp. Inf. Control 4, 1141–1150 (2008)
2. Lin, L., Lee, H.-M.: Fuzzy Assessment Method on Sampling Survey Analysis. Expert Syst. Appl. 36, 5955–5961 (2009)

3. Yao, J.-S., Wu, K.: Ranking fuzzy numbers based on decomposition principle and signed distance. Fuzzy Sets Syst. 116, 275–288 (2000)
4. Zadeh, L.A.: Fuzzy Sets. Information and Control 8, 338–353 (1965)
5. Zadeh, L.A.: The Concept of a Linguistic Variable and its Application to Approximate Reasoning. Inf. Sci. 8, 199–249 (I), 301–357 (II), 9, 43–58 (III) (1975)
6. Zimmermann, H.-J.: Fuzzy Set Theory and Its Applications. Kluwer Academic Publishers, Dordrecht (1991)

Multi-Mobile Agent Based Virtual Alliance Formation (MAVAF) Platform

Chang Yu Chiang and Chuan Jun Su

Abstract. With the advance of the Internet, today's interconnected environment creates a world of global competition featured time pressures, complexity, and rapid changes and also makes a new relationship the Virtual Alliance possible. In this research, we propose an information infrastructure MAVAF for effective global sourcing to conjoin various organizations and to strengthen business competence. The MAVAF comprises mobile agent enabled platforms and partner selection strategy patterns. With the proposed infrastructure, the emerging concept of Virtual Alliance can possibly be effectively realized and achieved.

1 Introduction

With the advance of the Internet, today's interconnected environment creates a world of global competition with time pressure, complexity, and rapid changes and leads the products to have special characteristics different from past, such as global marketing, short time-to-market, short life cycle, and high customization. With these conditions, each individual entity can respond to the oncoming features of this global competitive environment with global collaboration with other partners via the Internet.

With the tendency towards global collaboration, we intend to address the issue of a virtual alliance (VA)/virtual team formation based on multi- and mobile agent technology, named MAVAF (Multi- Mobile Agent based Virtual Alliance Formation). MAVAF enables the ability to search potential partners around the vast Internet automatically that would provide the great help for the VA initiators to organize their virtual alliance efficiently and rapidly.

1.1 Agent Technology

An agent is a software entity that continuously performs tasks given by a user within a particular restricted environment autonomously. The autonomy characteristic of a software agent distinguishes it from general software programs [3].

Chang Yu Chiang and Chuan Jun Su
Department of Industrial Engineering and Management, Yuan Ze University, Taiwan
135 Far East Road, Chungli, Taoyuan, Taiwan
e-mail: s978905@mail.yzu.edu.tw, iecjsu@saturn.yzu.edu.tw

B.-C. Chien and T.-P. Hong (Eds.): Opportunities and Challenges, SCI 214, pp. 265–270.
springerlink.com

A mobile agent is a particular class of agent with the ability during execution to migrate from one host to another where it can resume its execution. It has been suggested that mobile agent technology, amongst other things, can help to reduce network traffic and to overcome network latencies [1].

Java Agent DEvelopment Framework (JADE) [4] is a framework that facilitates the development of agent applications in compliance with the FIPA (The Foundation of Intelligent Physical Agents) [2], an IEEE standards, specifications for interoperable intelligent multi-agent systems. In JADE, a runtime environment is called a container as it can contain several agents. A set of active containers is called a platform. In a platform, there is a special main container that must be always active, and all other normal containers that must register with the main container as they start [5].

1.2 Virtual Alliance

The "*virtual alliance*" in this research is a flexible, far reaching, information intensive relationship and belongs to the Value Alliance classification of the virtual organization. The main characteristics of a virtual alliance are that it is a temporary network of alliances with a limited lifetime where the partners are distributed geographically and collaborate electronically; it is goal-oriented and commitment-based; and it is supported by communication and information flow and the partners share their skills, costs and profits [6].

A VA goes through four distinct phases during its life cycle, identification, formation, operation, and termination [7]. In this paper, we focus on the formation phase of the VA's life cycle and propose an information system framework to support VA formation that initiators can efficiently and rapidly find the suitable partners and organize their VA team. Few researches focus on this issue before and even no ready applications can support now. The research that proposed an agent-based model to support the virtual alliance formation is the more complete research in recent years. A high level multi-agent based framework model was proposed [6], which can support the VA initiator finding partners and organizing VA team conveniently and automatically.

In our opinion, with the advantages of the mobile agent, we propose a multi- and mobile agent enabled information infrastructure, MAVAF, to support VA formation and implement MAVAF by using JADE agent platform for more approximating to the actual application layer but not a high level model.

2 System Architecture

The MAVAF is composed of five types of architectural components as depicted in Fig. 1: Knowledge-based data Server, Provider Agent (PA), User Agent (UA) Coordinator Agent (CA), and Registrant Agent (RA).

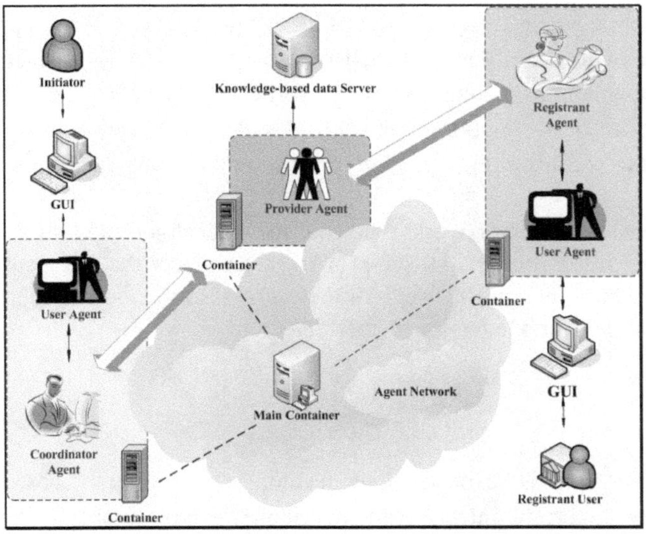

Fig. 1 MAVAF architecture

The PA is a stationary agent and operates at a higher level of trust and mediates access resources from mobile agents to the knowledge-based data server. In MAVAF, the PA plays an important role to dynamically interface with the knowledge-based data server. Through PA, mobile agent can obtain the resources in the knowledge-based data server. The PA would take the place of the web sites to bridge the users to the databases.

The UA is a type of stationary agent that serves as a bridge to interface with a host's computer and applications. It acts as a mediator for users and applications and invokes application services when receiving users' requests. The UA collects a request from GUI, delegates the tasks to the mobile agents, CA or RA, to invoke internal services and presents the final results from mobile agents to the user.

The CA is a mobile agent which acts as a commissioner to explore the possible partners in each data container. The CA is invoked when an initiator submits a request to form a virtual alliance via the UA. The CA migrates/travels to every data containers, communicates with the PA for searching the suitable potential partners in accordance with the user requests, and returns the searched results to the user.

Besides, the CA also has an analysis engine that is responsible for evaluating the explored potential individuals' capability through a capability evaluation function. It is designed to help the initiator conveniently and rapidly choose the ideal partners. In this research, we use a sample example to carry out the capability evaluation: We define some capability indicators and design evaluation patterns and give scores for manufacturer, seller, and distributor.

The RA is also a mobile agent, which is responsible for the access control, update, and management of user profiles on behalf of a user. The RA is invoked when a user intends to maintain his/her personal profile via the UA. The RA carrying the user maintained information migrates to the data container, communicates with the PA for reconfiguring the user profile, and returns the tasked results to the user.

The Knowledge-based data Server works with the Provider Agent and stores the user's information, called profile. A user profile encapsulates the interests, skills, experiences, capacities, etc. of the user. All the users' information and their capabilities are stored in the Knowledge-based data Server.

3 Usage Scenarios

The objective of this research is to design and develop a mobile agent based information platform to support the intensive and distributed nature of virtual alliance formation that allows the initiators to select prospects using a PDA, laptop, or desktop computer. Under this circumstance, two types of users are distinguished in the proposed system: Initiators and Registrants who are interested in participating challenge projects over the Internet.

3.1 Scenario 1: Initiators' Perspective – Exploring Team Members

Mr. John Smith has come up with a brilliant product idea, which can potentially be a hot item in a niche market. Consequently, he is inspired to get the product to the market as early as possible. Through his PC or PDA, he specifies the capability requirements for the team members using UA's GUI and makes a request of exploring potential candidates (see Step1 of Fig. 2). Upon receiving the request, the UA launches and delegates the job to a mobile agent, the CA via a local container (see Step2 of Fig. 2). Mr. Smith may then switch off his device and proceed with his daily work. In the mean time, the CA will surf in the logical mobile agent networks and work with available PA, which are static in data containers to identify potentially qualified team members. When a data container with potential people is found, the CA will send a request notification to the PA for retrieving the person's profile data (see Step3 of Fig. 2). If the PA approved the request notification, the PA will retrieve the data using SQL request and subsequently deliver to the CA (see Step4, 5, 6 of Fig. 2). The CA will continue pursuing the delegated tasks until it does all nodes of the network. Once the CA completes the trip, it displays the data acquired on the application GUI when Mr. Smith is back on line with his device (see Step7, 8 of Fig. 2).

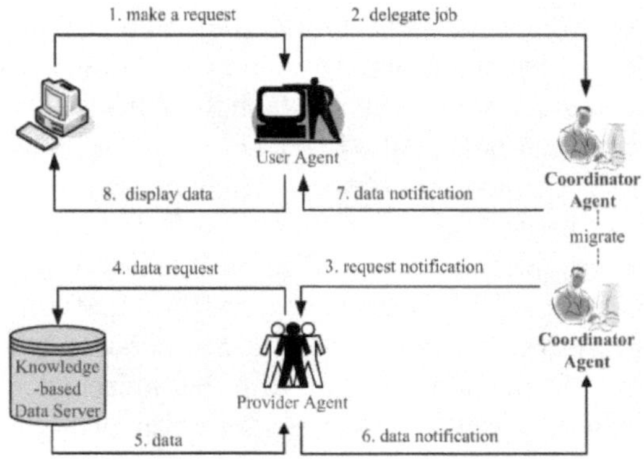

Fig. 2 Implementation scenario from alliance initiator's perspective

3.2 Scenario 2: Registrants' Perspective

Lisa is an experienced production engineer. She is interested in participating innovative projects and working with people from all over the world. When she learnt about the MAVAF platform from the Internet, Lisa downloads a UA and a RA from the MAVAF homepage and registers at MAVAF for creating account identifications (ID). With her PC or PDA, she creates her profile and sends it to the MAVAF by using the RA (see Step1, 2 of Fig. 3). The profile may be updated at any time she feels necessary with the UA and RA by the same process (see Step3, 4, 5, 6, 7, 8 of Fig. 3). As a registered participant, Lisa becomes a potential candidate to join a team for an exciting project. A confirmation and negotiation session; i.e., video conferencing for discussing the potential collaboration will then be initiated by MAVAF between the initiator and Lisa.

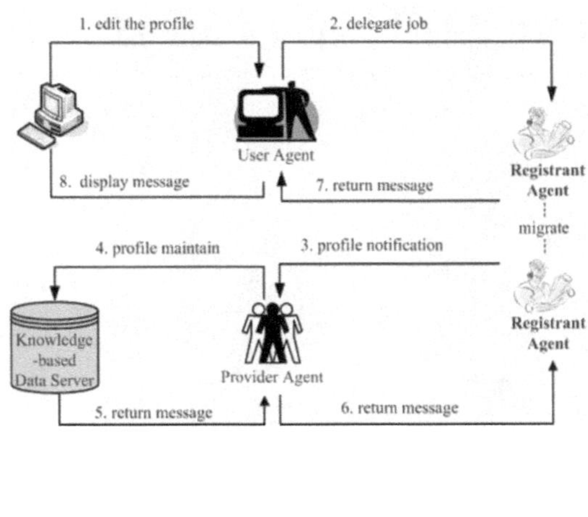

Fig. 3 Implementation scenario from the registrants' perspective

4 Conclusion

In this paper, we have presented a mobile multi- agent enabled information infrastructure MAVAF based on the JADE platform, which encapsulates potentials to reduce the VA formation cycle time for the VA initiators. Multi-agent systems have a great potential to encourage the process efficiency of the VA formation. Despite the appealing nature of MAVAF, the works has undergone for improving the system are: 1) Security improvement: the information transmitted between agents can be classified and sensitive. The JADE equips with a security plug-in, which enables a set of security features and provides the base technology for such agent-based application as MAVAF. However, more robust and comprehensive security mechanisms need to be incorporated in MAVAF. 2) The MAVAF was built on top of an open and scalable platform JADE, which makes the system extensible. Coordinator Agent incorporating with partner selection and the capability evaluation function can be independently developed and dynamically plugged in to the system. As an essential extension of this work, robust partner selection and evaluation strategies are explored to enhance the effectiveness of decision potential partner searching and integrated into the MAVAF. The overall system performance can subsequently be strengthened.

References

1. Danny B, L., Oshima, M.: Programming and Deploying Java Mobile Agents with Aglets. Addison-Wesley, Reading (1998)
2. FIPA, `http://www.fipa.org`
3. Hewitt, C.E.: Viewing Control Structures as Patterns of Passing Messages. Journal of Artificial Intelligence 8(6), 323–364 (1997)
4. JADE, `http://jade.tilab.com`
5. JADE Programmer's Guide, `http://jade.tilab.com/doc/programmersguide.pdf`
6. Petersen, S., Rao, J., Matskin, M.: Virtual Enterprise Formation supported by Agents and Web Services, Agent and Web Service Technologies in Virtual Enterprises. Idea Group Publishing (2007)
7. Strader, T.J., Lin, F.R., Shaw, M.J.: Information infrastructure for electronic virtual organization management. International Journal of Decision Support Systems 23, 75–94 (1998)

On Supporting Cross-Platform Statistical Data Analysis Using JADE

Chien-Ho Wu, Yuehjen E. Shao, Jeng-Fu Liu, and Tsair-Yuan Chang

Abstract. Data collected by information systems for decision support in a modern business is very often distributed over different databases on different computing platforms. This fact impedes the practice of cross-functional data analysis. To address this problem, use of data warehouses of various forms has been suggested. However, this approach may not be cost-effective and is time-consuming. Nevertheless, it also requires a lot of efforts on maintaining the data warehouses. In this research we are evaluating the feasibility of applying agent technology to collect and integrate, on a regular and ad hoc basis, data distributed over various computing platforms to facilitate statistical data analysis. The prototype designed for the evaluation is implemented in the JADE development environment.

1 Introduction

In the trend of globalization, businesses are inevitably forced to face severe foreign and domestic competitions. In order to survive the never-ending harsh challenges emerging from all over the world, businesses have to maintain a high degree of operational efficiency and achieve excellent quality of decision effectiveness. And all these require the support of information systems to present accurate and valid information to the managers on a regular and ad hoc basis.

In a modern business data collected for decision support by information systems are very often stored on networked databases over different computing platforms. Accordingly, use of data warehouses has been suggested to facilitate cross-functional data analysis. This approach to support cross-functional data analysis is feasible. However, it requires a fortune of hardware and software investment, and may be time-consuming. Furthermore, maintenance of the data warehouses is also a pain in the neck.

Chien-Ho Wu, Yuehjen E. Shao, and Jeng-Fu Liu
Graduate Institute of Applied Statistics, Fu-Jen Catholic University, Taiwan
e-mail: {stat2016, stat1003, stat1011}@mail.fju.edu.tw

Tsair-Yuan Chang
Department of Information Management, Ming-Chuan University, Taiwan
e-mail: tychang@mail.mcu.edu.tw

B.-C. Chien and T.-P. Hong (Eds.): Opportunities and Challenges, SCI 214, pp. 271–277.
springerlink.com

In this research, we are trying to facilitate the task of cross-functional data analysis by using agent technology. By using the agent technology, we have a better chance of collecting just-enough data on either regular or ad hoc basis, thus adding more flexibility to the practice of data analysis. To evaluate the feasibility of the approach, we are implementing a prototype with the JADE development packages.

2 JADE-The Development Environment

In general, an agent is an active assistant that acts or is capable of acting or is empowered to act for another person or thing [6]. An agent is autonomous, reactive, sociable, capable of learning and mobile [8, 12, 14]. A multi-agent system is a system in which agents are organized communicating with each other and share resources to complete computation tasks [4, 15]. There are many tools available to develop multi-agent systems [11]. JADE (Java agent development framework) distributed by Telecom Italia is one of those tools that can provide a computation environment for developing multi-agent systems. It is completely implemented in Java language and is FIPA-compliant with excellent interoperability [1, 5].

As pointed out by Bellifemine [1, 2, 3] a JADE platform consists of agent containers that can be distributed over the network. Agent containers provide the JADE run-time and all the services needed for hosting and executing agents. The main container is the bootstrap point of a JADE platform and all other containers must join a main container by registering with it. Since JADE is de facto platform-independent and can bring about a high degree of interoperability between agents, the development framework provided by JADE is considered very suitable for implementing applications that require distributing computation tasks over the network.

3 The System Architecture for Distributed Statistical Data Analysis

3.1 The Flow of Statistical Data Analysis

The main purpose for a business to conduct statistical data analysis is to generate information for decision purposes such as maintaining strategic competitiveness and continuous survival. The quality of statistical data analysis is of no doubt a major factor for an effective decision.

Since business processes of an organization are very often computerized, data required for statistical data analysis are usually distributed over databases on different computing platforms. To conduct statistical data analysis in such a scenario, a statistical data analyst generally involves the following flow of activities:

1. Clarify, with the support from the management concerned, what sort of information is required of a business decision.
2. Identify the data and the analysis method to produce the required information.
3. Collect, cleanse, validate, and integrate data from various sources.
4. Analyse the cleansed data using the chosen methods.
5. Generate analysis reports

In practice, step 3 of the above procedure is very often taking an awful lot of time and efforts, and can have a direct impact on the quality of a statistical data analysis.

How IT can support the execution of business processes is always extracting attentions from modern organizations. The current focus of this research is to explore the feasibility of addressing the problem of collecting and integrating data from various sources using technology that can support the idea of distributed data processing. The success of this approach may reduce the heavy reliance on building data warehouses.

3.2 The System Architecture

We believe that agent technology is a very good candidate to address the difficulties in collecting and integrating data in a computing environment where data required for statistical analysis are distributed over networked databases. Our belief is based on the fact that agents can be deployed as needed dynamically and can be empowered with intelligence. Furthermore, systems developed using agent-technology can be conveniently scaled by adding agents of various functions to or removing agents from the system, despite the pitfalls that may come with multi-agent systems [13].

The proposed approach to conducting statistical data analysis is supported by a multi-agent system implemented with JADE. In general, there are two types of nodes, i.e. main node and slave node, in the system architecture. The main node is the control node for statistical data analysis. The slave node is the remote site that hosts the decision-related data. The conceptual structure of the main node and slave node are shown in Figure 1 and Figure 2 respectively in the next page.

The information catalog in the main node is designed to assist the analyst in locating decision-related data for analysis. Besides the information catalog, there are 4 types of agents:

1. Workflow control agent: this agent is the core of our proposed approach. It is the agent that provides the data analyst with a GUI that can dynamically deploy data acquisition agents and activate data analysis agents.
2. Data acquisition agent: agents of this type are activated by the workflow control agent and may in turn deploy one or more data collection agents to remote sites to collect data. The data acquisition agent coordinates with slave nodes to make sure that the data required for analysis will be successfully retrieved from remote sites.

3. Data analysis agent: these are the agents in charge of various types of statistical analysis and produce reports for decision support. The laborious task of data analysis is done, to a large degree, at the main node.
4. Status monitoring agent: This is the agent at remote site that monitors the status of a deployed data collection agent. This agent coordinates with main node to ensure the success completion of the data collection agent.

Despite that the proposed agent-based approach is very promising, we need to face the fact that cross-functional data access has to tackle the challenge of disputes over data ownership and security.

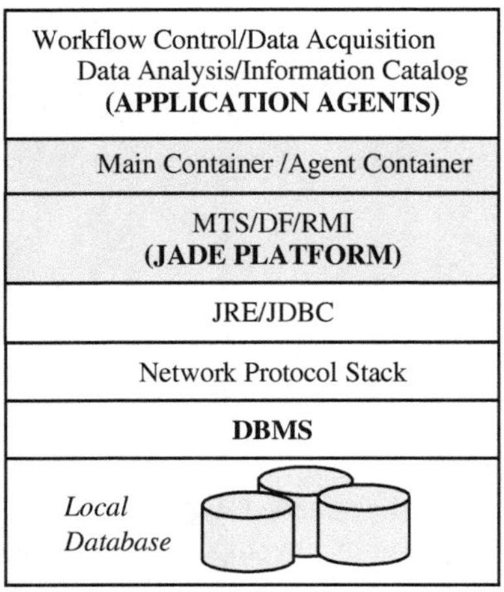

Fig. 1 The composition of the main node

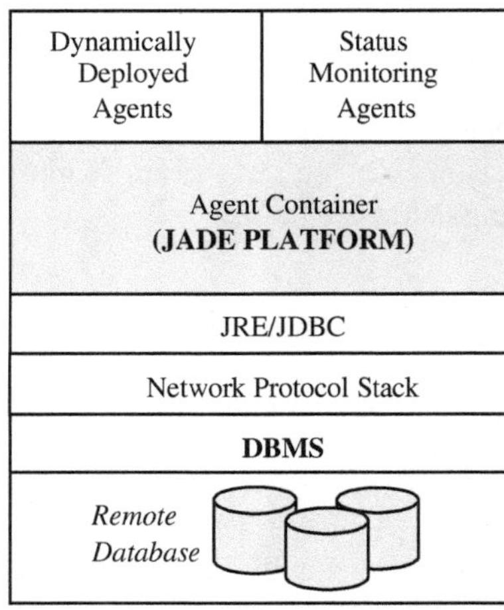

Fig. 2 The composition of the slave node

3.3 The Workflow Control of Distributed Data Analysis with Multi-agent System Support

With the support of the multi-agent system, the workflow of distributed statistical data analysis, prescribed by the behaviour of workflow control agent, will go as follows:

1. Locate and tailor decision-related data with the support of the information catalog and the decision maker at stake.
2. Activate data acquisition agent and deploy data collection agents to remote sites hosting decision-related data. Data acquisition agents will coordinate with status monitoring agents at remote sites to ensure that data will be successfully retrieved and integrated.
3. Activate data analysis agents for statistical data analysis and report generation.

This flow of activity control, though not completely automated, still can facilitate the work of data analysis.

4 The Prototype

During this stage of our research, we have implemented a simplified prototype to early test our idea of supporting distributed statistical data analysis using agent technology. The attention is on implementing data collection and data analysis agents for now. The basic idea is that, through the GUI for agents, data analysts can locate and retrieve the raw data elements for analysis. Then the analysts can activate analysis agents to analyse the retrieved data. The analysis agent implemented at this stage can replace missing responses with values calculated by using either Approximate Bayesian Bootstrap (ABB) or Ratio Imputation [7, 9, 10]. These two imputation methods are briefly described as follows:

I. Approximate Bayesian Bootstrap (ABB)

Suppose that a target variable Y has a total of n observations of which m values are missing and $r=(n-m)$ values are observed. The ABB method works as follows:

1. Draw randomly r values, denoted $y_1, y_2, \ldots, y_r$, with replacement from the r observed values to create Y^*_{obs}.
2. Draw m values from Y^*_{obs} as imputed values for the m missing values in the target variable.

In general, ABB imputation method works well for within-class imputations if the missing mechanism depends only on the variables used to construct the imputation classes.

II. Ratio Imputation

If a variable X is known to be closely relate to the target variable Y. The ratio imputation method uses the following formula to generate the numeric imputed value y^*_{hi} for the i_{th} missing response in the h_{th} imputation class.

$$y^*_{hi} = \frac{\overline{y}_{rh}}{\overline{x}_{rh}} x_{hi}$$

where $\overline{y}_{rh}$ and $\overline{x}_{rh}$ are the *mean values* of the responses for the h_{th} imputation class. The ratio imputation method can provide very good imputations if the target variable Y mainly depends on the highly correlated auxiliary variable X.

After dealing with the missing values the analysis agent can generate reports based on the results from descriptive analysis. The use case diagram for the prototype is shown as Figure 3 in the following. Figure 4 is the GUI for the data collection agent and Figure 5 is the GUI for data analysis agent. Figure 6 presents one of the results from descriptive analysis.

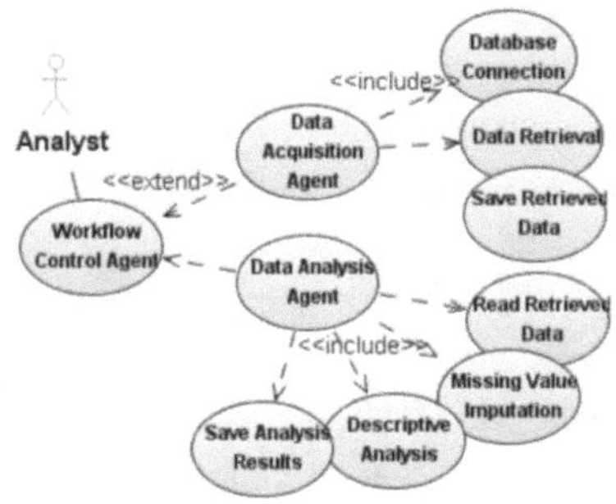

Fig. 3 The use case diagram for the prototype

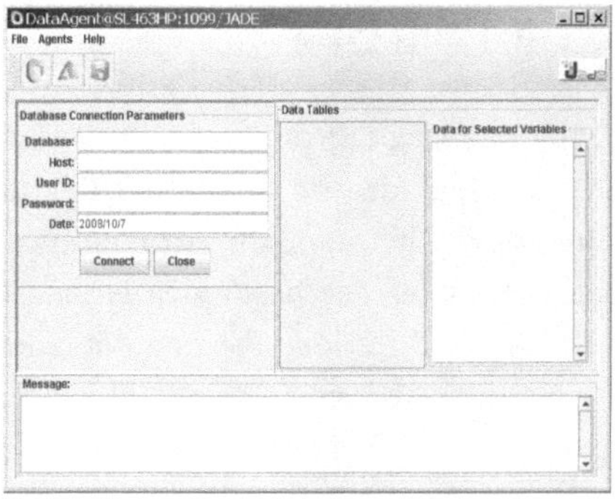

Fig. 4 The GUI for data collection agent

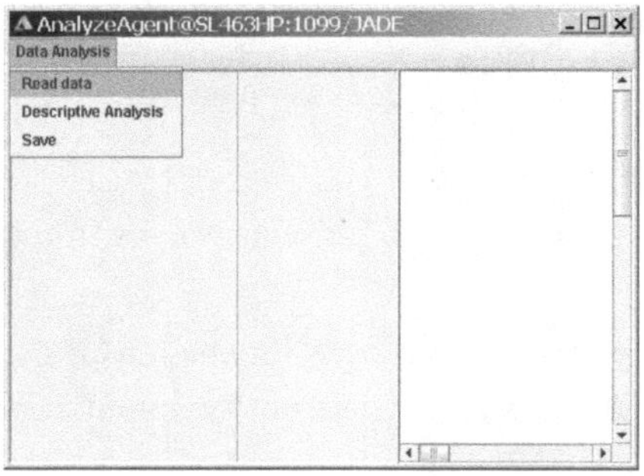

Fig. 5 The GUI for data analysis agent

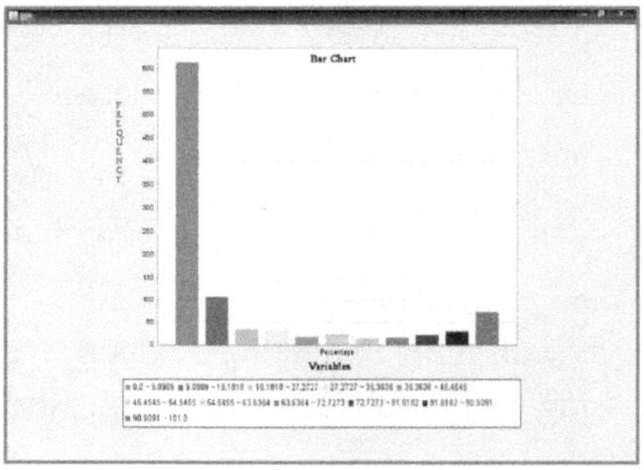

Fig. 6 The exemplar analysis result

5 Final Remarks

In modern organizations, data required for statistical data analysis are very often distributed over networked databases. Although data warehouses are suggested by many to facilitate statistical data analysis, the reality is that building of data warehouses may not be cost-effective and requires a lot of efforts on maintaining the data warehouses.

In this application-oriented research, we have proposed an agent-based approach to support the idea of distributed statistical data analysis. In our approach, agents are dynamically deployed to remote sites to collect data on the fly, and the collected data can then be analysed by data analysis agents. This approach can be more economic and more flexible in terms of collecting data on the fly and providing specialized statistical analysis functions. The early implementation of the prototype using JADE has shown the feasibility of the approach.

In the future, there is a need to more fully design and implement the proposed approach. For example we need to lay out the detailed specification of each type of agent, including coordination and integration schemes, and test out the limitations on the dynamic deployment of data collection agents. We also need to work out a better way of building the information catalog. Furthermore, we may extend the functions of data analysis agents to make our approach more practical and appealing.

References

1. Bellifemine, F., Caire, G., Greenwood, D.: Developing Multi Agent Systems with JADE. John Wiley & Sons, Chichester (2007)
2. Bellifemine, F., Caire, G., Trucco, T., Rimassa, G.: Jade Administrator's Guide. CSELT S.p.A, Italy (2007)
3. Bellifemine, F., Caire, G., Trucco, T., Rimassa, G.: Jade Programmer's Guide. CSELT S.p.A, Italy (2007)
4. Ferber, J.: Multi-Agent Systems: An Introduction to distributed Artificial intelligence. Addison-Wesley, New York (1999)
5. FIPA: FIPA Agent Management Specification, Geneva, Switzerland (2004), `http://www.fipa.org/specs/fipa00023/SC00023K.pdf`
6. Guralink, D.B.: Webster's New World Dictionary of the American Language. Grand Central Publishing, New York (1987)
7. Little, R.J.A., Rubin, D.B.: Statistical Analysis with Missing Data. John Wiley & Sons, New York (1987)
8. Maes, P.: Agents that Reduce Work and Information Overload. Communications of the ACM 71(7), 31–40 (1994)
9. Rubin, D.B.: Multiple Imputation for Non-response in Surveys. John Wiley & Sons, New York (1987)
10. Schafer, J.L.: Analysis of Incomplete Multivariate Data. Chapman and Hall, London (1997)
11. Shen, W., Norrie, D.H., Barthes, J.-P.: Multi-Agent Systems for Concurrent Intelligent Design and Manufacturing. Taylor & Francis, London (2001)
12. Wooldridge, M.J., Jennings, N.R.: Intelligent agents Theory and practice. The Knowledge Engineering Review 10(2), 115–152 (1995)
13. Wooldridge, M., Jennings, N.R.: Pitfalls of Agent-Oriented Development. In: 2nd Int. Conf. on Autonomous Agents (Agents 1998), Minneapolis, USA (1998)
14. Wooldridge, M., Sycara, K., Jennings, N.: A Roadmap of Agent Research and Development. In: Autonomous Agents and Multi-agent Systems, pp. 275–306. Kluwer Academic Publishers, Boston (1998)
15. Zambonelli, F., Jennings, N., Omicini, A., Wooldridge, M.: Agent-Oriented Software Engineering for Internet Applications. In: Omicini, A., Zambonelli, F., Klusch, M., Tolksdorf, R. (eds.) Coordination of Internet Agents-Models, Technologies, and Applications, pp. 326–346. Springer, Berlin (2001)

Building Agents That Learn by Observing Other Agents Performing a Task: A Sequential Pattern Mining Approach

Philippe Fournier-Viger, Roger Nkambou, Engelbert Mephu Nguifo, and Usef Faghihi

Abstract. In this paper, we propose to build agents that learn by observing other agents performing a task by extracting frequent temporal patterns from their behavior. We propose a learning mechanism consisting of three phases: (1) recording other agents' behavior, (2) mining temporal patterns from this data and (3) utilizing the resulting knowledge. We illustrate this approach with a tutoring system for training learners to robotized arm manipulation where we have integrated a tutoring agent that observes humans performing a task to learn it. The agent then exploits this knowledge to provide assistance to learners.

1 Introduction

For a virtual agent, observing the behavior of other agents constitutes an excellent opportunity for learning. But implementing such a learning mechanism in a virtual agent is a tough challenge. Researchers have made various proposals for integrating learning-by-observation in agents for example by providing mechanism for learning production rules or decision trees (see for example [11] for a review). However, many of them rely on strong assumptions. For example, [11] proposes a framework for learning production rules from recorded human behavior. In this approach, a human has to teach an agent by performing a task. But this approach is tightly linked to a very specific conception of intelligence, as humans performing a demonstration are required to specify their actions as complex operators organized in a hierarchy and having goal conditions, and they have to explicitly state their goals during the demonstrations. Contrarily to this view, we here address the problem of learning-by-observation by considering that an agent

---

Philippe Fournier-Viger, Roger Nkambou, and Usef Faghihi
Department of Computer Science, Université du Québec à Montréal, Montréal, Canada

Engelbert Mephu Nguifo
LIMOS-CNRS and Department of Mathematics and Computer Science,
Université Blaise Pascal, Clermont-Ferrand, France

B.-C. Chien and T.-P. Hong (Eds.): Opportunities and Challenges, SCI 214, pp. 279–284.
springerlink.com

cannot make any assumption on the nature of the decision-making processes of other agents. Thus we consider that an agent can only perceive actions of other agent, without any additional information.

Because the amount of data to be recorded and analyzed for learning practical knowledge can be huge, we suggest using data mining algorithms to analyze the behavior of other agents and extract useful knowledge. Our hypothesis is that learning mechanisms based on the discovery of temporal patterns in other agents' behavior would allow implementing effective procedural learning in virtual agents. This paper is organized as follows. First, it describes an algorithm for mining temporal patterns and discusses related works in agent learning. Then, the paper describes how this algorithm is integrated in a tutoring agent deployed in a real application. Finally, the paper presents an experiment and conclusions.

2 Mining Temporal Patterns from Sequences of Events

According to [1], there are four kinds of patterns that can be mined from time series-data. These are trends, similar sequences, sequential patterns and periodical patterns. In this work we chose to mine sequential pattern [2], as we are interested in finding relationships between occurrences of sequential events. To mine sequential patterns, several efficient algorithms have been proposed (e.g. [2]). But to our knowledge, only a few works have been published on using sequential pattern mining for agent learning. For example, [4] proposed to implement sequential pattern mining in a robot playing soccer. In this case, sequential patterns are then used to derive prediction rules about what actions or situations might occur if some preconditions are satisfied. This is different from the form of learning that we consider in this paper, which is learning-by-observation. For this work, we chose a sequential pattern mining algorithm that we have developed [3] as it provides several more features than classical sequential pattern algorithms, such as accepting symbols with numeric values, eliminating redundancy and handling time constraints and contextual information.

The algorithm takes as input a database D of sequences of events. An event $X=(i_1, i_2, \ldots i_n)$ contains a set of items $i_1, i_2, \ldots i_n$, that are considered simultaneous, and where each item can be annotated with an integer value. Formally, a sequence is denoted $s = <(t_1,X_1), (t_2,X_2),\ldots, (t_n,X_n)>$, where each event X_k is associated to a timestamp t_k indicating the time of the event. For example, the sequence S1 of figure 1 (left) contains two events. It indicates that item *a* appeared with a value of 2 at time 0 and was followed by items *b* and *c* with a value of 0 and 4 respectively at time 1. An events sequence $s_a = <(ta_1,A_1), (ta_2,A_2),\ldots, (ta_n,A_n)>$ is said to be contained in another events sequence $s_b = <(tb_1,B_1), (tb_2,B_2),\ldots, (tb_n,B_m)>$, if there exists integers $1 \leq k1 < k2 < \ldots < kn \leq m$ such that $A_1 \subseteq B_{k1}$, $A_2 \subseteq B_{k2}, \ldots, A_n \subseteq B_{kn}$, and that $tb_{kj} - tb_{k1}$ is equal to $ta_j - ta_1$ for each $j \in\{1\ldots m\}$. The relative support of a sequence s_a in a database D is defined as the percentage of sequences $s \subseteq D$ that contains s_a, and is denoted by $supD(s_a)$. The problem of mining frequent sequences is to find all the sequences s_a such that $supD(s_a) \geq minsup$ for a sequence database D, given a support threshold *minsup*, and optional time constraints. The optional time constraints are the minimum and maximum time intervals required between the head and tail of a sequence and the minimum and maximum time intervals required between two adjacent events of a sequence.

As an example, figure 1 illustrates a database of 6 sequences (left) and the corresponding patterns found for a *minsup* of 33% (right). Consider pattern M5. This pattern appears in sequence S4 and S5, respectively. It has thus a support of 33% (2 out of 6 sequences). Now consider patterns M1 and M2. Because the item *a* appears in sequence S1, S2, S3 and S4 with values 2, 2, 5 and 6 respectively the algorithm separated these values in two groups to create patterns M1 and M2 instead of creating a single pattern with a support of 66 %. For each of these groups, the median (2 and 5) was kept as an indication of the values grouped. This clustering of similar values only occurs when the support is higher or equal to 2 * *minsup* (see [3]).

ID	Sequences
S1	<(0,a{2}), (1,bc{4})>
S2	<(0,a{2}), (1,c{5})>
S3	<(0,a{5}), (1,c{6})>
S4	< (0,f), (1, g),(2,a{6}e)>
S5	<(0, f b{3}), (1,h),(2,ef) >
S6	<(0,b{2}), (1,d)>

->

ID	Mined Sequences	Supp.
M1	<(0,a{2})>	33 %
M2	<(0,a{5})>	33 %
M3	<(0,a{2}), (1, c{5})>	33 %
M4	<(0,c{5})>	50 %
M5	<(0,f), (2, e)>	33 %
M6	...	...

Fig. 1 A database of 6 Sequences (left) and mined sequences (right)

3 A Tutoring Agent That Learn by Observing Humans' Behavior

We now describe how we integrated our algorithm in an agent so that it can learn by observing other agents. We present this agent in the context of RomanTutor [5] (fig. 2), a virtual learning environment for learning how to operate the Canadarm2 robotic arm on the international space station.

The main learning activity in RomanTutor is to move the arm from one configuration to another. This is a complex task, as the arm has 7 joints and the user must chose at any time the 3 best cameras for viewing the environment from around 12 cameras on the space station, and adjust their parameters. We have integrated a tutoring agent in RomanTutor to provide assistance to learners during this task. However, there are a very large number of possibilities for moving the arm from one position to another, and because one must also consider the safety of the maneuvers, it is very difficult to define a task model for generating the moves that a human would execute [6]. For this reason, instead of providing domain knowledge to the agent, we have implemented a learning mechanism which allows the agent to learn by observing the behavior of other agents performing the task, which in this case are humans. The agent then uses this knowledge to provide assistance to learners. The three next subsections describe the three operation phases of the learning mechanism as they are implemented in the virtual agent.

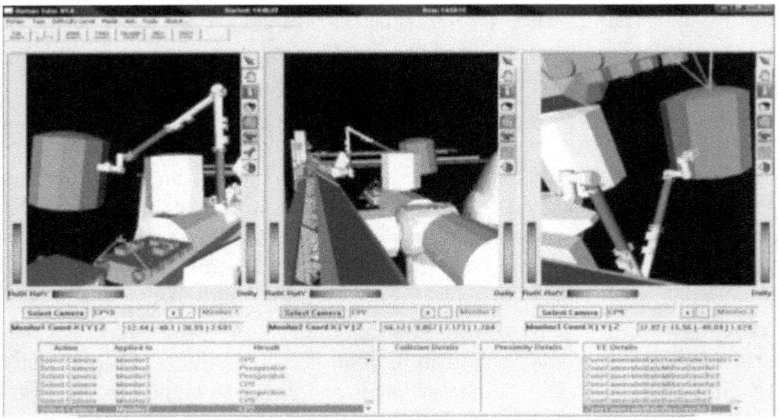

Fig. 2 The RomanTutor user interface

In the *observing phase*, the virtual agent observes and records the behavior of users that attempt an arm manipulation exercise (moving the arm from an initial configuration to a goal configuration). For each attempt, a sequence of events is created in a database. In this context, an event is a set of actions (items) that are considered unordered temporally. We defined 112 primitive actions that can be recorded in RomanTutor, which are (1) selecting a camera, (2) performing an increase or decrease of the pan/tilt/zoom of a camera and (3) applying a rotation value to an arm joint.

But in a tutoring system context, it would be also useful to annotate sequences with contextual information such as success information and the expertise level of a user, to then mine patterns containing this information. Our solution to this issue is to take advantage of an extra feature of our algorithm (based on [7]), which is to add dimensional information to sequences. A database having a set of dimensions D={D1, D2,… Dn} is called an MD-Database. Each sequence of a MD-Database (an MD-Sequence) possesses a symbolic value for each dimension. This set of value is called an MD-Pattern and is denoted {d1, d2… dn}. In the context of our virtual agent, we have defined two dimension "success" and "expertise level", which are added manually to each sequence recorded. The left part of figure 3 shows an example of an MD-Database having these two dimensions. As an example, the MD-Sequence B1 has the MD-Pattern {"true", "novice"} for the dimensions "success" and "expertise level". The symbol "*", which means any values, can also be used in an MD-Pattern. This symbol subsumes all other dimension values. An MD-Pattern $Px=\{dx_1, dx_2\ldots dx_n\}$ is said to be contained in another MD-Pattern $Py=\{dy_1, dy_2\ldots dy_m\}$ if $dx_1 \subseteq dy_1$, $dx_2 \subseteq dy_2$, …, $dx_n \subseteq dy_n$. The problem of mining frequent sequences with dimensional information is to find all MD-Sequence appearing in an MD-Database with a support higher or equal to *minsup*. As an example, right part of figure 3 shows some patterns that can be extracted from the MD-Database of figure 3, with a *minsup* of 2 sequences.

In the *learning phase,* the virtual agent applies the algorithm to extract frequent sequences, which build its domain knowledge. For mining patterns, we setup the algorithm to mine only sequence of size 2 or greater, as sequence shorter would not be useful in a tutoring context. Furthermore, we chose to mine sequences with a maximum time interval between two adjacent events of 2. The benefits of accepting a gap of 2 is that it eliminates some "noisy" (non-frequent) learners' actions, but at the same time it does not allow a larger gap size that could make the patterns less useful for tracking a learner's actions.

ID	Dimensions	Sequences
B1	true, novice	<(0,a),(1,bc)>
B2	true, expert	<(0,d) >
B3	false, novice	<(0,a),(1,bc)>
B4	false, interm.	<(0,a),(1,c), (2,d)>
B5	true, novice	<(0,d), (1,c)>
B6	true, expert	<(0,c), (1,d)

->

Dimensions	Sequences
*, novice,	<(0,a)>
*, *	<(0,a)>
*, novice	<(0,a), (1,b)>
true, *	<(0,d)>
true, novice	<(0,c)>
true, expert	<(0,d)>

Fig. 3 An example of sequential pattern mining with contextual information

In the third phase, the *application phase,* the virtual agent provides assistance to the learner by using the knowledge learned in the *learning phase*. To recognize a learner's plan, the virtual agent proceeds as follow. The first actions of the learner are compared with the first action of each frequent pattern. If the actions do not match for a pattern, the system discards the pattern. Each time the learner makes an action, the system repeats the same process. It compares the actions done so far by the learner with the remaining patterns. If at any given moment a user action does not match with any patterns, the algorithm ignores the last user action or the current action to match for each pattern. This makes the plan recognizing process more flexible and has shown to improve its effectiveness. One utility of the plan recognizing process for actions/problem states is to assess the expertise level of the learner (novice, intermediate or expert) by looking at the patterns applied.

The plan recognizing algorithm also allows the agent to guide the learner. It allows determining possible actions from the current situation. This functionality is triggered when the student selects "What should I do next?" in the interface menu. The algorithm returns the set of possible actions with the associated patterns. The tutoring service then selects the action among this set that is associated with the pattern that has the highest support and that is the most appropriate for the estimated expertise level of the learner. In the cases where no actions can be identified, the virtual agent uses a path planner [5] to generate an approximate solution.

We conducted a preliminary experiment in RomanTutor with two exercises to qualitatively evaluate the virtual agent's capability to provide assistance. We asked 12 users to record plans for these exercises. The average length of plans was around 20 actions. From this data, the virtual agent extracted sequential patterns with the algorithm. In a subsequent work session, we asked the users to evaluate the tutoring services provided by the virtual agent. On the whole, users agreed that the assistance provided was helpful. We also observed that the virtual agent often correctly inferred the estimated expertise level of learners.

4 Conclusion

We presented the idea of building agents that learn from the behavior of other agents by recording behaviors and finding temporal patterns from this data. This is different from other approaches for learning by observation as it doesn't make any assumption on the internal processes of other agents. To demonstrate our approach, we presented an agent that record the behavior of humans performing a task and then use a sequential pattern mining algorithm to extract patterns. The agent then uses this knowledge base to provide assistance to humans. First experiments have demonstrated that the agent can effectively learn complex procedural tasks and that it can use this knowledge to provide useful tutoring services. In future work, we will measure empirically how the virtual agent influences the learning of students. Because the input format for the learning algorithm is simple, the same approach could be applied for learning other tasks and in agents that are not tutoring agents. In future work, we will compare the learning mechanisms that we proposed with other agent learning mechanisms and investigate the possibility of mining other types of temporal patterns such as trends from agents' behavior.

Acknowledgments. The authors thank the Canadian Space Agency, the FQRNT and the NSERC for their logistic and financial support. The authors also thank members of the GDAC/PLANIART teams who participated in the development of RomanTutor.

References

1. Han, J., Kamber, M.: Data mining: concepts and techniques. Morgan Kaufmann Publ., San Franc. (2000)
2. Agrawal, R., Srikant, R.: Mining Sequential Patterns. In: Proc. ICDE, pp. 3–14 (1995)
3. Fournier-Viger, P., Nkambou, R., Mephu Nguifo, E.: A Knowledge Discovery Framework for Learning Task Models from User Interactions in Intelligent Tutoring Systems. In: Proc. MICAI 2008, pp. 765–778 (2008)
4. Lattner, A.D., Miene, A., Visser, U., Herzog, O.: Sequential Pattern Mining for Situation and Behavior Prediction in Simulated Robotic Soccer. In: Bredenfeld, A., Jacoff, A., Noda, I., Takahashi, Y. (eds.) RoboCup 2005. LNCS, vol. 4020, pp. 118–129. Springer, Heidelberg (2006)
5. Kabanza, F., Nkambou, R., Belghith, K.: Path-planning for Autonomous Training on Robot Manipulators in Space. In: Proc. IJCAI 2005 (2005)
6. Fournier-Viger, P., Nkambou, R., Mayers, A.: Evaluating Spatial Representations and Skills in a Simulator-Based Tutoring System. IEEE Trans. Learning Techn. 1(1), 63–74 (2008)
7. Pinto, H., et al.: Multi-Dimensional Sequential Pattern Mining. In: Proc. CIKM 2001, pp. 81–88 (2001)
8. van Lent, M., Laird, J.E.: Learning procedural knowledge through observation. In: Proc. K-CAP 2001, pp. 179–186 (2001)

A Multi-agent System Architecture for Personal Support during Demanding Tasks

Tibor Bosse, Rob Duell, Mark Hoogendoorn, Michel Klein,
Rianne van Lambalgen, Andy van der Mee, Rogier Oorburg,
Alexei Sharpanskykh, Jan Treur, and Michael de Vos

Abstract. Task performance of humans that act under demanding circumstances may vary over time, depending on the characteristics of human, task and environment. To increase the effectiveness and efficiency of task performance, personalised assistance may be provided, in the form of automated personal assistant agents that constantly monitor the task execution and well-being of the human, and intervene when a problem is detected. This paper proposes a generic design for a multi-agent system architecture including such personal assistant agents, which can be deployed in a variety of domains.

1 Introduction

Systems supporting humans during execution of demanding tasks often need to fulfil two important requirements: 1) they need to be personalised to a specific human and his or her cognitive and task performance characteristics, and 2) they need to incorporate dynamical models for the analysis of the functioning of the human in a given task. For example, human task performance can degrade over time due to available resources being exceeded [6], which may lead to a reduction in attention and situation awareness [2, 3, 7]. By dynamical models involving the internal (exhaustion level, work pressure level) states of the human this can be predicted. As another example, it can be analysed whether the human remains healthy during the processes of task execution. Intelligent personal assistants proposed to support humans during the execution of tasks (see e.g. [4, 5]) depend on models that represent the state of the human and his or her tasks at particular

Tibor Bosse, Mark Hoogendoorn, Michel Klein, Rianne van Lambalgen
Alexei Sharpanskykh, and Jan Treur
Vrije Universiteit Amsterdam, Department of Artificial Intelligence
De Boelelaan 1081, 1081 HV Amsterdam, The Netherlands
e-mail: {tbosse, mhoogen, mcaklein, rm.van.lambalgen, sharp, treur}@few.vu.nl

Rob Duell, Andy van der Mee, Rogier Oorburg, and Michael de Vos
Force Vision Lab, Barbara Strozzilaan 362a, 1083 HN Amsterdam, The Netherlands
e-mail: {rob, andy, rogier, michael}@forcevisionlab.nl

B.-C. Chien and T.-P. Hong (Eds.): Opportunities and Challenges, SCI 214, pp. 285–290.
springerlink.com

time points, for example, a model addresses the cognitive load of the human (see e.g. [8]). To provide more widely applicable assistance it is important to have a larger set of models from which an appropriate instance can be chosen depending on circumstances. This paper presents a generic design for a multi-agent system architecture including personal assistant agents. The personal assistant includes generic constructs that allow for self-configuration by loading domain-specific models and thus altering its own functionality. These domain-specific models also address the dynamics of states over time. A personal assistant agent can use these models to monitor and analyse the current state of the human and select the best intervention method (if needed) in the specific domain and task.

This paper is organised as follows. The multi-agent system architecture is described in Section 2. Model maintenance and state maintenance agents are discussed in Section 3. Further, the functions of the self-maintaining personal assistant agent are considered in Section 4. Finally, Section 5 concludes the paper.

2 The Agent-Based System Architecture

The developed conceptual component-based architecture comprises a number of essential components:

- *Process*: ensures request/provision of data from/to different components.
- *Reflection*: exercises control/monitoring over the functioning of the whole system. In particular, by performing meta-reasoning using human, task, system and environmental characteristics this component activates analysis methods.
- *Library of specifications* contains specifications of analysis methods, workflow, cognitive and dialogue models.
- *Storage of execution information* is used for storage and retrieval of information about the human, the world, the execution of workflows, dialogues and systems.

Given the essential components that have been identified above, the system has been modelled by a multi-agent system architecture consisting of the following types of agents:

- *Self-maintaining personal assistant agent* (SMPA) that supports a human during the execution of a task;
- *Model maintenance agent* (MMA) that contains a library of models used for the configuration of SMPA's.
- *State maintenance agent* (SMA) that maintains characteristics, states and histories of other agents, of the world and of the executions of tasks.
- *Mental operations agent* (MOA) that represents the mental part of the human.
- *Task execution support agent* (TESA) used by the human as an (active) tool during the execution of a task.

These agents are depicted graphically in Figure 1. The agents are represented as squares with small boxes on both sides (input and output). Another component of the multi-agent system architecture is the physical world that comprises all material (or physical) objects including the body of the human. See the complete

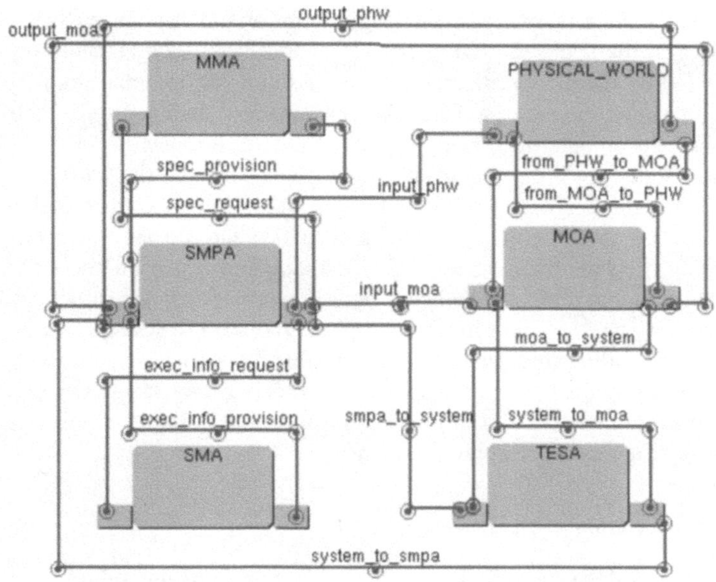

Fig. 1 Overview of the multi-agent system architecture

specification in [9] for more details. The personal assistants have two modes of functioning: (1) *the self-maintenance mode*, in which they are able to reason about the model specifications required to perform their tasks and to achieve their goals, to request these models and to load them (altering their functionality); (2) *the monitoring and guidance mode*, in which they perform monitoring and guidance of the human to whom they are related.

In the self-maintenance mode communication takes place between personal assistant agents and model maintenance agents. In the monitoring and guidance mode personal assistant agents communicate with state maintenance agents, the mental operations agent, and task execution support agents. Furthermore, they interact with the physical world by performing observations (e.g., of the human's state). The mental part of a human represented by a mental operations agent is connected to the human's physical body, which can act in the physical world.

3 Maintenance Agents

Two types of maintenance agents are included in the multi-agent system architecture: model maintenance agents (MMA) and state maintenance agents (SMA).

The model maintenance agent contains a library of models that can be used by self-maintaining personal assistant agents to perform their tasks. Models of four types are maintained in the library: monitoring and guidance task models, cognitive models, workflow models, and dialogue models. Models are provided by the model maintenance agent to self-maintaining personal assistant agents upon request. To facilitate the model acquisition process, each maintained model is annotated by particular parameters. The ontology used for the annotation is assumed to be known to the agent-requester. In the general case, such an ontology may be also provided by the model maintenance agent to an self-maintaining personal assistant agent upon request. In Table 1 some of the parameters and their possible values used to annotate the different types of models are listed. The models maintained in model maintenance agents may be specified using different knowledge representation languages. However, it is important to ensure that a model provided to a self-maintaining personal assistant agent can also be interpreted by this agent. The state maintenance agent maintains information about the characteristics, states and histories of the agent types *mental operations agent* and *task execution support agent*, of the physical world, of the workflows and of dialogues related to them. Information about states and histories (i.e., sequences of states) is stored in a time-indexed format using the predicate at(prop, time), where a state property is specified by the first argument and the time point at which this property holds is specified by the second argument.

Table 1 Cognitive and Workflow Model Parameters

Parameter	Some possible values	Parameter	Some possible values
Cognitive models		**Workflow models**	
Name	string value	Name	string value
Cognitive processes	reasoning, consciousness, perception	Task type	string value
States	stress, motivation level, fatigue level	Task executor capabilities	excellent analytic skills, quick typing, domain-related knowledge
Agent type	robot, human, animal	Task executor traits	openness, extraversion, neuroticism
Related physical parts	frontal lobe, parietal lobe, temporal lobe	Minimum/maximum duration	integer value / integer value
Characteristics	qualitative/quantitative; stochastic; statistical	Consumable resources	building materials

Such information is gathered and provided to the state maintenance agent by self-maintaining personal assistant agents, which may also use this information in their analysis. Information for which a self-maintaining personal assistant agent has no immediate need after being stored in the state maintenance agent can be removed from the assistant agent's memory. When stored information is required by a self-maintaining personal assistant agent, it can be requested from the state maintenance agent. An information request includes the identification of the element (i.e., mental operations agent, task execution support agent, physical world, a workflow, a dialogue), the aspect (i.e., characteristic, state, history) and the time interval for which information should be provided.

4 Self-maintaining Personal Assistant Agent

For each human that needs to be supported during the task execution, a self-maintaining personal assistant agent is created. Initially, the personal assistant agent contains generic components only. The configuration of the personal assistant agent is performed based on an organisational role that needs to be supported by the agent, on the characteristics of a human who is assigned to this role, and on the goals defined for the personal assistant agent.

The human is assigned a role of being responsible for a package of tasks, which is provided to the personal assistant agent. For the whole task package, as well as for each task separately a set of goals and norms related to the execution of the task(s) may be defined. To determine the characteristics of the human responsible for the execution of these tasks, the personal assistant agent sends a request to the state maintenance agent. If the human is known to the state maintenance agent, his/her known professional, cognitive, psychological and physical characteristics are provided to the personal assistant agent. Otherwise, the state maintenance agent returns to the personal assistant agent the default profile (i.e., a standard set of characteristics).

For the personal assistant agent a set of prioritised general goals is defined, which it strives to achieve. Some of these goals are related to the quality of the task execution, others concern the human's well-being. Based on information about the human and the assigned tasks, some of these goals may be refined and instantiated into more specific, operational goals. Next, the personal assistant agent will configure itself by identifying the suitable monitoring and guidance task model(s) that need(s) to be requested from the model maintenance agent. As soon as it possesses these models, it can perform monitoring and guidance of the human, according to four sub-processes: *monitoring* (i.e., determining which observation foci are needed), *analysis* (detecting potential problems and their causes), *plan determination* (determining plans to remedy these problems), and *plan execution preparation* (refining these plans by relating them to specific actions to execute).

More details about the functioning of the self-maintaining personal assistant agent, as well as a prototype implementation (and an example simulation trace) can be found in [1].

5 Conclusion

In this paper, a multi-agent system architecture for personal support during task execution has been proposed. This architecture includes self-maintaining personal assistant agents with a generic design. Such agents possess self-configuration abilities, which enable them to dynamically load domain-specific models, thereby specialising these agents for the execution of particular tasks in particular domains. Using these models and information about the assigned goals and tasks, the personal assistant agent performs monitoring and analysis of the behaviour of the supported human in his/her environment. In case a known problem is detected, the agent tries to identify and execute an appropriate intervention action. The fact that the architecture is generic differentiates the approach from other personal assistants such as presented in [4, 5]. The proposed self-maintaining personal assistant agent has an advantage of being relatively lightweight, as it only maintains and processes those models that are actually needed for the performance of the tasks. It can therefore run upon for instance a PDA or cell phone. To provide the required functionality for personal assistant agents, the proposed architecture includes model maintenance and state maintenance agents.

References

1. Bosse, T., Duell, R., Hoogendoorn, M., Klein, M.C.A., Lambalgen, R., van Mee, A., van der Oorburg, R., Sharpanskykh, A., Treur, J., de Vos, M.: A Generic Personal Assistant Agent for Support in Demanding Tasks. In: Proc. of the Fourth International Conference on Augmented Cognition and 13th Int. Conference on Human-Computer Interaction, HCI 2009. LNCS. Springer, Heidelberg (to appear, 2009)
2. Endsley, M.R.: The role of situation awareness in naturalistic decision making. In: Zsambok, C., Klein, G. (eds.) Naturalistic decision making, pp. 269–284. Erlbaum, Mahwah (1997)
3. Endsley, M.R.: Theoretical underpinnings of situation awareness. In: Endsley, M.R., Garland, D.J. (eds.) Situation awareness analysis and measurement, pp. 1–21. Erlbaum, Mahwah (2000)
4. Modi, P.J., Veloso, M., Smith, S.F., Oh, J.: CMRadar: A Personal Assistant Agent for Calendar Management. In: Bresciani, P., Giorgini, P., Henderson-Sellers, B., Low, G., Winikoff, M. (eds.) AOIS 2004. LNCS, vol. 3508, pp. 169–181. Springer, Heidelberg (2005)
5. Myers, K., Berry, P., Blythe, J., Conley, K., Gervasio, M., McGuinness, D.L., Morley, D., Pfeffer, A., Pollack, M., Tambe, M.: An Intelligent Personal Assistant for Task and Time Management. AI Magazine Summer 2007, 47–61 (2007)
6. Posner, M.I., Boies, S.J.: Components of attention. Psychological Bulletin 78, 391–408 (1971)
7. Wickens, C.D.: Situation awareness and workload in aviation. Current Directions in Psych. Science 11, 128–133 (2002)
8. Wilson, G.F., Russell, C.A.: Performance enhancement in an uninhabited air vehicle task using psychophysiologically determined adaptive aiding. Human Factors 49(6), 1005–1018 (2007)
9. `http://www.cs.vu.nl/~wai/PersonalAssistant/`

The Selective Traveling Salesman Problem with Regular Working Time Windows

Hu Qin, Andrew Lim, and Dongsheng Xu*

Abstract. Most literature on variations of traveling salesman problem (TSP) assumes that the worker is continuously available. However, this assumption may not be valid in some real applications due to the human's natural needs. In this work, we study a selective TSP with regular working time windows in which a set of jobs at dispersed locations requires a specific length of time for completion from the worker. In the whole working time horizon, the worker can only travel or process jobs during the working time windows and the rest of time is used for taking breaks. The objective is to maximize the sum of the job processing time. By conducting extensive experiments, we show that our devised tabu search is a good approach to solve this problem.

1 Introduction

The selective traveling salesman problem with regular working time windows (STSP-RWTW) is defined on a graph $G = (V, E)$ where $V = \{0, 1, \ldots, n, n+1\}$ is the vertex set and $E = \{(i, j) : i, j \in V, i \neq j\}$ is the arc set. Vertex 0 and $n+1$ represent the exit from a depot and the entrance to the depot, respectively. Let s_i be the processing time required by the job situated at vertex $i \in V$ (with $s_0 = 0$ and

Hu Qin and Andrew Lim
Department of Management Sciences, City University of Hong Kong,
Tat Chee Ave, Kowloon Tong, Hong Kong
e-mail: hunqin3@student.cityu.edu.hk, lim.andrew@cityu.edu.hk

Dongsheng Xu
School of Business, Sun Yat-Sen University, Guangzhou, P.R. China
e-mail: xudsh@mail.sysu.edu.cn

* Corresponding author.

B.-C. Chien and T.-P. Hong (Eds.): Opportunities and Challenges, SCI 214, pp. 291–296.
springerlink.com

$s_{n+1} = 0$) and let distance $d_{i,j}$ be associated with arc $(i, j) \in E$. One worker travels vertices and processes jobs within some consecutive days, but he can only travel or process jobs during the regular working time (e.g., from $10:00am$ to $8:00am$). After one regular working time period, the worker must take a break and because only can the vertices supply accommodation, he can not stop traveling at any point on the arc connecting two vertices. Moreover, if a job can not be completed before the break time point, it must be restarted rather than continued. The objective is to determine a directed tour for the worker, visiting each vertex at most once and maximizing the total job processing time.

The classical TSP and its variations have been widely studied by researchers over the last 50 years or so and the book edited by G.Gutin and A.P.Punnen [1] collects many of the known results on modeling and solving them. To our best knowledge, the STSP-RWTW is a new TSP variation on which little has been published to date. If we view breaks as one type of jobs and have determined their locations, the problem can be transformed to a special TSPTW [2]. If we can keep the worker traveling or processing jobs continuously without any break, the problem can be reduced to the selective TSP [3, 4], the orienteering problem [5] or traveling salesman problem with profits [6]. If we let d_{ij} be the setup time required for processing job j immediately after job i, the STSP-RWTW becomes a problem similar to the single machine scheduling problem with periodic maintenance and sequence-dependent setup time. Naderi *et al.* [7] recently published a paper considering the job shop scheduling with periodic preventive maintenance and sequence-dependent setup time, whereas they do not explicitly describe issues like whether the setup process can be interrupted by maintenance break, or whether the setup process can be conducted immediately before the maintenance break.

2 A Mixed Integer Programming Formulation

H and B represent the span of regular working time and breaking time, respectively. We assume the service time of each job and travel time between any two vertices are not larger than H. To formulate the problem, we transform the network graph $G = (V,E)$ to $G' = (V',E')$ by splitting each vertex $i \in V/\{0,n+1\}$ into two vertices, i_+ and i_-, where i_+ denotes the point of reaching vertex i and i_- represents the point of finishing the job i. Vertex 0 and $n+1$ still denote the starting and ending points. For arcs $(i, j), (0, j), (i,n+1) \in E$, we add corresponding arcs $(i_-, j_+), (0, j_+), (i_-,n+1)$ with the same length to the graph G'. Dummy arc (i_+,i_-) mimicking the duration of processing job i is also added into G' and the length and time needed associated with arc (i_+,i_-) are zero and s_i. As a consequent, the worker can choose to take a break at any vertex in $V' = V'_+ \cup V'_- \cup \{0,n+1\}$, where $V'_+ = \{1_+,\dots,n_+\}$ and $V'_- = \{1_-,\dots,n_-\}$. The notations used in our model are defined as follows:

Parameters

t_{i_+,i_-} — $t_{i_+,i_-} = s_i$, service time for job i, i.e., the time associated with arc (i_+, i_-);

t_{i_-,j_+} — $t_{i_-,j_+} = d_{i_-,j_+}$, travel time from job i to j, i.e., the time associated arc (i_-, j_+);

(a_l, b_l) — $(a_1,b_1),(a_2,b_2),\ldots,(a_\tau,b_\tau)$ are τ regular working time windows, where $a_1 = 0$, $b_1 = H$, $a_{l+1} = a_l + B + H$ and $b_{l+1} = b_l + B + H$;

Variables

$x_{i,j}$ — $x_{i,j} = 1$ if and only if the worker goes to vertex j immediately after finishing vertex i, where $(i, j) \in E'$; 0 otherwise;

$v_{i,l}$ — $v_{i,l} = 1$ if i is visited in day l, where $i \in V'$; 0 otherwise;

$u_{i,l}$ — $u_{i,l} = 1$ if the worker takes a break at vertex i in day l, i.e., vertex i is the last one reached in day l, where $i \in V'$; 0 otherwise;

p_i — the time point at which the worker reaches vertex i, where $i \in V'$.

With above notations, we have the following mixed integer programming (MIP) model for our STSP-RWTW problem:

$$\textbf{MIP} \quad \max \sum_{i\in V/\{0,n+1\}} s_i \sum_{l=1}^{\tau} v_{i_+,l} \tag{1}$$

$$\text{s.t.} \quad \sum_{(i_-,j_+)\in E'} x_{i_-,j_+} \le x_{j_+,j_-},\ for\ j \in V/\{0,n+1\} \tag{2}$$

$$\sum_{(0,j_+)\in E'} x_{0,j_+} = \sum_{(i_-,n+1)\in E'} x_{i_-,n+1} = 1 \tag{3}$$

$$\sum_{(j_-,i_+)\in E'} x_{j_-,i_+} = \sum_{(i_-,j_+)\in E'} x_{i_-,j_+} = \sum_{l=1}^{\tau} v_{i_+,l} = \sum_{l=1}^{\tau} v_{i_-,l} \le 1, \quad i \in V/\{0,n+1\} \tag{4}$$

$$\sum_{i\in V'} u_{i,l} \le 1,\ for\ 1 \le l \le \tau \tag{5}$$

$$\sum_{i\in V'} v_{i,l} \le M \sum_{i\in V'} u_{i,l},\ for\ 1 \le l \le \tau \tag{6}$$

$$u_{i,l} \le v_{i,l},\ for\ i \in V',\ 1 \le l \le \tau \tag{7}$$

$$p_i + t_{i,j} - M(1 - x_{i,j}) \le p_j,\ for\ (i,j) \in E' \tag{8}$$

$$a_{l+1}u_{i,l} + t_{i,j} - M(1 - x_{i,j}) \le p_j,\ for\ (i,j) \in E' \tag{9}$$

$$a_l - (1 - v_{i,l})M \le p_i \le b_l + (1 - v_{i,l})M,\ for\ i \in V',\ 1 \le l \le \tau \tag{10}$$

$$x_{i,j}, u_{i,t}, v_{i,t} \in \{0,1\},\ for\ i,j \in V',\ 1 \le l \le \tau$$

$$p_i \ge 0,\ for\ i \in V'$$

where M is a big number in this *MIP* model. The objective function (1) maximizes the total processing time of jobs visited since $\sum_{l=1}^{\tau} v_{i_+,l} = 1$ guarantees the job i must be processed. Constraints (2) ensure that vertex i_+, i_- must be visited successively. The tour must start and end at vertex 0 and $n+1$ respectively, so constraints (3)

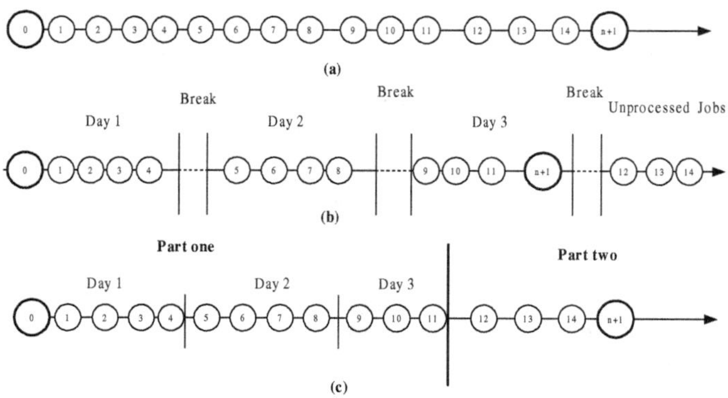

Fig. 1 Example of a feasible solution with 14 jobs and 3 working days

applies. Constraints (4) ensure that each vertex can be visited at most once during τ days and if job i is processed, vertex i_+ has at most one incoming arc and i_- also has at most one outgoing arc. Constraints (5)-(7) jointly represent the logic relationship between $u_{i,l}$ and $v_{i,l}$. Sufficient time is guaranteed for the worker traveling from vertex i to vertex j by constraints (8). If i is the last vertex in day l, the earliest departure time to next vertex is the beginning of next day, which is specified by constraints (9). Constraints (10) ensure that the time point reaching vertex i must lies in the working hours. Obviously, since the STSP-RWTW is a variation of TSP problem, it is also $\mathcal{NP}$ hard.

3 Solution Procedure–Tabu Search

Apply nearest neighbor heuristic (NNH) to construct a job chain, which is shown in Figure 1 (a). Then, use multiple working time windows to segment this chain into several pieces and insert vertex $n+1$ to the last day at appropriate place. The resulting tour shown in Figure 1 (b) can be regarded as the initial feasible solution in which the jobs processed during working days are called on-tour vertices and others are indicated as off-tour vertices.

From any feasible solution, we can construct one job chain like the one in Figure 1 (c). Given the the job chain derived from current feasible solution, the moves we used for our problem are SWAP and ADD.

ADD Adding job i processed in day τ' to one of the other working days or adding unprocessed job i to one of the working days. Remove job i, where the two vertices adjacent to i on the chain are connected together, and then add it to one working day. Suppose we try to add job i in day τ_1 and i_1, i_2 represent the first and last job processed in τ_1. If job i is inserted between vertex i_1 and i_2, we minimize the length of the chain segment between i_1 and i_2 by performing some TSP algorithms, i.e., 2-Opt or 3-Opt algorithm. After this, we may get a new part one and may need to reconstruct part two. Then, a new job chain may be brought into existence and by segmenting this job chain using working time windows and inserting the vertex

$n+1$ to the last day, we can easily obtain another feasible solution. If we add job i to day τ_1 as the first job, we insert i between i_1 and i_1's preceding vertex and then minimize the length of the chain segment between i and i_2 such that we may get a new job chain from which a new feasible solution can be generated by using the same approach in previous case. Similarly, we can add job i to day τ_1 as the last job.

SWAP Swapping two on-tour jobs or two jobs from different chain parts. After swapping, we can get a new job chain with which we can obtain a feasible solution by conducting the same operations used in ADD move.

During the tabu search process, it is always to execute the move resulting in the largest objective value, even if that move makes the current solution worse. The tabu list consists of vertices links, which are deleted in previous moves. Aspiration criterion is associated with the objective value found. A tabu move can be declared acceptable only it yields a solution better than the previously best-known solution. Once the tabu is overridden by aspiration criterion, we empty the tabu list to avoid the deadlock from occurring in next iteration.

4 Numerical Experiments and Conclusion

The testing instances we generated were ones with the distance being the ordinary Euclidean distance, whose $d_{i,j}$ is the distance between two of n random points in a planer rectangle $[0,100] \times [0,100]$. We had five types of instance size, (n, τ), which were (6, 1), (10, 1), (15, 2), (40, 5), (80, 10), each of which has 5 generated instances. The length of working time window is 300 and the processing time of each job is a uniformly distributed random number on $[10,20]$. For our tabu search, we specified: tabu tenure equaled $n^2/10$ and the maximum number of moves with no improved solution found was $3n$. We implemented all the algorithms with Java and ran all experiments on a notebook with 1.83GHz Intel Core 2 Duo and 1.00 GB RAM. We applied the branch-and-cut search schemes, which was provided by CPLEX 9.0, on our *MIP* model. Table 4 presents the results of all 25 instances ran by *MIP*, nearest neighbor heuristic (NNH) and tabu search (TS). Computation times reported here are in CPU seconds on this computer.

In table 4, the results associated with *MIP* indicate that the performance of CPLEX is such bad that it even can not handle small-size instances with only more than 10 jobs. On the contrary, our tabu search produced good-quality solutions within reasonable time. For the instances with 6 jobs, the *MIP* model and tabu search both performed well, solving the instances optimally in little time. As the problem size increased to 10 jobs, the *MIP* model still gained optimal solution, but the time spent increased explosively, whereas our tabu search found all the optimal solutions almost in no time. For the instances with more than 10 jobs, although CPLEX did not reach feasible solutions within 5 hours, our tabu search still found better solutions compared with the solutions achieved by the simple nearest neighbor heuristics. In summary, tabu search has good performance among all the various test instances in practice.

Table 1 Experimental results

		MIP		NNH		TS		$\frac{\text{(TS-MIP)}}{\text{TS}}$	$\frac{\text{(TS-NNH)}}{\text{TS}}$
		Obj	Time(sec)	Obj	Time(sec)	Obj	Time(sec)		
6 jobs 1 day	1	107.46	2.74	59.51	0.00	107.46	0.08	0.0%	44.7%
	2	70.34	4.19	57.67	0.00	70.34	0.05	0.0%	18.0%
	3	94.37	2.75	94.37	0.00	94.37	0.05	0.0%	0.0%
	4	71.58	3.23	48.26	0.00	71.58	0.05	0.0%	32.6%
	5	110.89	4.14	85.77	0.00	106.77	0.08	0.0%	19.7%
10 jobs 1 day	1	110.89	10542.48	104.67	0.00	110.89	0.13	0.0%	5.6%
	2	92.23	8936.02	70.49	0.00	92.23	0.08	0.0%	23.6%
	3	108.50	16246.88	44.95	0.00	108.50	0.13	0.0%	58.6%
	4	120.98	7933.69	98.24	0.00	120.98	0.11	0.0%	18.8%
	5	120.83	14602.44	107.16	0.00	120.83	0.14	0.0%	11.3%
15 jobs 2 days	1	N/A	18000.00	243.10	0.00	320.35	0.58	N/A	24.1%
	2	N/A	18000.00	302.85	0.00	328.19	0.56	N/A	7.7%
	3	N/A	18000.00	249.79	0.00	276.37	0.50	N/A	9.6%
	4	N/A	18000.00	288.70	0.00	323.11	0.59	N/A	10.7%
	5	N/A	18000.00	244.42	0.00	331.73	0.42	N/A	26.3%
40 jobs 5 days	1	N/A	18000.00	1032.35	0.00	1066.69	19.47	N/A	3.2%
	2	N/A	18000.00	873.98	0.00	1011.24	24.00	N/A	13.6%
	3	N/A	18000.00	951.27	0.00	1059.06	28.56	N/A	10.2%
	4	N/A	18000.00	916.48	0.00	1034.65	22.08	N/A	11.4%
	5	N/A	18000.00	895.60	0.00	948.62	40.53	N/A	5.6%
80 jobs 10 days	1	N/A	18000.00	2093.34	0.02	2182.39	462.91	N/A	4.1%
	2	N/A	18000.00	2070.52	0.00	2145.77	436.41	N/A	3.5%
	3	N/A	18000.00	2077.97	0.00	2214.99	496.97	N/A	6.2%
	4	N/A	18000.00	2034.56	0.00	2158.19	391.11	N/A	5.7%
	5	N/A	18000.00	2041.43	0.02	2152.62	532.70	N/A	5.2%

In this paper, we formulated and studied a selective traveling salesman problem with regular working time windows, which is a new variation of TSP. We developed a tabu search based solution procedure and implemented tabu search, nearest neighbor heuristics and CPLEX based branch-and-cut search. The extensive experiments showed that our tabu search behaved the best.

References

1. Gutin, G., Punnen, A.P.: The Traveling Salesman Problem and Its Variations. Kluwer Academic Publishers, Dordrecht (2002)
2. Dumas, Y., Desrosiers, J., Gelinas, E., Solomon, M.M.: An optimal algorithm for the traveling salesman problem with time windows. Operations Research 43(2), 367–371 (1995)
3. Gendreau, M., Laporte, G., Semet, F.: A branch-and-cut algorithm for the undirected selective traveling salesman problem. Networks 32(4), 263–273 (1998)
4. Laporte, G., Martello, S.: The selective travelling salesman problem. Discrete Applied Mathematics 26(2-3), 193–207 (1990)
5. Fischetti, M., González, J.J.S., Toth, P.: Solving the orienteering problem through branchand-cut. INFORMS Journal on Computing 10(2), 133–148 (1998)
6. Feillet, D., Dejax, P., Gendreau, M.: Traveling salesman problems with profits. Transportation Science 39(2), 188–205 (2005)
7. Naderi, B., Zandieh, M., Fatemi Ghomi, S.M.T.: Scheduling sequence-dependent setup time job shops with preventive maintenance. The International Journal of Advanced Manufacturing Technology (accepted) (2008)

Scheduling for Dedicated Machine Constraint Using the Shortest Paths in Graphs*

Huy Nguyen Anh Pham, Arthur Shr, Peter P. Chen, and Alan Liu

Abstract. The dedicated machine constraint (DMC) in semiconductor manufacturing is one of the new challenges of scheduling problems. The constraint set by the process engineer is due to the natural bias between the photolithography machines. Previous studies either did not take this constraint into account or the proposed heuristic approach might not result in an efficient cost. This paper proposes a new framework based on the graph theory to deal with scheduling for the DMC. Finding the shortest paths in the graph will minimize the overall production cost in an efficient time. Experiments were provided to validate the work.

1 Introduction

The production process in the fabrication consists of four basic parts: wafer fabrication, wafer probe, assembly, and final test [3]. Wafer fabrication is one of the most complex parts, consisting of several hundred steps in a certain order. In this part, photolithography (photo) steps transfer the patterns of each layer of the IC into wafers, and then these wafers will be processed in the other stations bearing on the patterns. The most challenging task of photo steps is the alignment of all the photo layers. Therefore, the yield of IC products relies heavily on photo machines.

Natural bias between photo machines can result in the precision of alignment for patterns between photo layers. To prevent the impact caused by natural bias

Huy Nguyen Anh Pham, Arthur Shr, and Peter P. Chen
Department of Computer Science, 298 Coates Hall, Louisiana State University, Baton Rouge, LA 70803, U.S.A
e-mail: {hpham15,mshr1,pchen}@lsu.edu

Alan Liu
Department of Electrical Engineering, 168 University Rd, National Chung Cheng University, Chia-Yi 621, Taiwan, R.O.C.
e-mail: aliu@ee.ccu.edu.tw

* This research is partially supported by the U.S. National Science Foundation grant No. IIS-0326387, U.S. AFOSR grant: FA9550-05-1-0454, and National Science Council of Taiwan grants No. 97-2221-E-194-048-MY3. This research is also partially supported by the research grant from Louisiana Board of Regents and from LSU on the "Center for Secure Cyberspace" project.

B.-C. Chien and T.-P. Hong (Eds.): Opportunities and Challenges, SCI 214, pp. 297–302.
springerlink.com

between photo machines, process engineer introduces the dedicated machine constraint (DMC) to semiconductor manufacturing systems. The DMC is defined as follows: if wafers are scheduled to one of the photo machines at the first photo step, they must be assigned to the same machine at the rest of the photo steps. After the DMC is brought in, unexpected abnormal events or breakdown of a photo machine could cause a pile-up of many wafers waiting for it and the load among photo machines might become unbalanced. This will significantly increase cycle time. This issue leads the main contributor to the complexity and uncertainty of semiconductor manufacturing.

A study proposed a load balance allocation function between photo machines and applied the function to a dynamic programming method to schedule photo machines [4]. Vargas-Villamil [6] proposed a three-layer hierarchical approach for inventory control and production optimization of the semiconductor manufacturing system. It reduced the effort and frequency of the control decisions of photo machines. A mixed-integer programming model for optimally scheduling machines was proposed by Kimmes [2]. Akçali et al [1] admitted that a dedicated machine assignment assured better alignment for separate layers of wafers and proposed a machine dedication policy. Studying solutions for scheduling the DMC were done by applying a heuristic Load Balancing scheduling method [5]. These studies neither consider the DMC nor optimize the total production cost.

In reality, wafers of a load unbalancing factory would be switched from the highly congested machines to the idle machines. However, this may take much time and relies on experienced engineers to manually handle alignment problems with a different situation off-line. In this paper, we present a graph framework for the scheduling with DMC. The framework will first form a graph based on the current scenario represented by a task matrix called a Resource Schedule and Execution Matrix (RSEM). It will minimize the total production cost by finding the shortest paths in the graph.

2 A Graph Framework

A semiconductor factory needs to produce N wafers W_1, W_2,…, W_N. The problem would consider a production process where each wafer W_i consists of S steps. The S steps consist of photo steps W_{i,j_1}, W_{i,j_2},…, W_{i,j_v} for $1 \le j_1, j_2, \ldots, j_v \le S$, used for aligning layers by M photo machines. Other steps are called non-photo steps. The DMC for photo steps defines that if photo step W_{i,j_l} is running on machine k, then the photo steps $W_{i,j_{l+1}}$,…,W_{i,j_v} must run on the same machine. This scenario as shown in Fig. 1 can be represented by a RSEM defined in [5].

At the observation time T, assume that machine k, used for photo step $W_{i,j}$, is seriously broken down in $BD_{j,k}$ units, for k = 1, 2,..., M. Breakdown times can cause either serious problems inside the machine or maintenance of the machine. We also assume that at time T, the breakdown time for further steps is available. Therefore, the goal for the scheduling problem is set to find a minimization of the production cost for N wafers in a semiconductor manufacturing system with the DMC.

$$\text{RSEM } A:\quad \begin{array}{c} \\ \text{Wafer} \end{array}\begin{array}{c|cccc|} & \multicolumn{4}{c}{\text{Step}} \\ W_1 & & 1 & 1 & \\ W_2 & 2 & 2 & & \\ W_3 & 1 & & 1 & 1 \\ W_4 & 3 & & & 3 \end{array}$$

$$\text{Matrix } BD:\quad \text{Machine}\ \begin{array}{c} \\ M_1 \\ M_2 \\ M_3 \end{array}\begin{array}{c} \text{Step} \\ \begin{bmatrix} 0 & 10{,}000 & 100 & 700 \\ 0 & 0 & 0 & 0 \\ 0 & 0 & 700 & 0 \end{bmatrix} \end{array}$$

$$\text{RSEM } B:\quad \begin{array}{c} \\ \text{Wafer} \end{array}\begin{array}{c|cccc|} & \multicolumn{4}{c}{\text{Step}} \\ W_1 & & 2 & 2 & \\ W_2 & 2 & 2 & & \\ W_3 & 1 & & 1 & 1 \\ W_4 & 3 & & & 3 \end{array}$$

Fig. 1 RSEM *B* is derived from RSEM *A* by using the graph framework. RSEM *A* shows that four wafers W_1 to W_4 pass the four steps to complete the production process. Non-photo steps are represented by empty items, while each photo step is depicted by a photo machine name. The first wafer has two photo steps in the 2nd and 3rd steps using machine 1 (M_1). The second wafer has two photo steps in the 1st and 2nd steps. These photo steps use M_2. The same definition exists in the other wafers. Matrix *BD* shows that if the third step is the photo step and one of the wafers uses M_1, then that wafer hits a breakdown in 100 units. If M_3 is used, then the wafer hits a breakdown in 700 units. A similar manner exists for M_2

The framework forms a graph from RSEM *A* and breakdown matrix *BD* with the three following assumptions. (1) *The production time for each photo step is the same*. This assumption is due to the fact that the production time for each wafer is a long process time (one to two months depending on the technology types). Thus, the difference of the production times at photo steps can be ignored. The production time for each photo step is then added with a setup time when assigning the photo step to a new machine and a queue time for when the wafer is waiting in the queue of the new machine. (2) *The queue time can be ignored.* Once $W_{i,j}$ is assigned to a new machine, then it is always set as the highest priority in the queue of that machine. (3) *The production procedure between wafers is independent.* The third assumption resulted in the second assumption. The basic idea to keep the DMC is that for all the photo steps, except for the first photo step, the system would be penalized with a very high cost for the overhead time when they are switching to a new machine. Under these three assumptions, the graph framework as shown in Fig. 2 is presented as follows:

Input: RSEM *A* of size $N \times S$ and matrix *BD* of size $S \times M$.

Step 1: Construct graph *G*= <*V*, *H*> based on RSEM *A* and matrix *BD*.
Step 2: Find the shortest paths in *G* from the start nodes to the end node.
Step 3: Create a new RSEM *B* by using the shortest paths in *G*.

Output: RSEM *B* with the optimal cost.

Fig. 2 In Step 1, the graph *G* consists of *N* sub-graphs: $G_1, G_2, \ldots, G_N$. Since weights for the edges in *G* are positive, we can find the shortest paths in *G* from start nodes "W_i" to end node "*E*" in Step 2 by applying Dijkstra's algorithm [7]. A new RSEM *B* is created in Step 3 by using the name of nodes along the shortest paths. The shortest path for each start node is maintaining the DMC for each wafer

In Step 1, each G_i represented for W_i has maximum $S\times M+1$ nodes as follows:

- A start node named by "W_i" for each wafer W_i.
- Intermediate nodes. If step $W_{i,j}$ is a photo step (i.e., $A(i, j) \neq 0$), then there are intermediate nodes named by "$W_{i,j,k}$" for each combination of wafer W_i running at step $W_{i,j}$ and using machine k.
- An end node named by "E" for G.

Each G_i has maximum $(S\text{-}1)\times M^2+2\times M$ edges with the edges and their weights:

- Edges from a start node "W_i" to node "$W_{i,j,k}$" if j is the first column satisfying $A(i, j)\neq 0$ in RSEM A. The weight for these edges is equal to α units (where α, called the production time, is a symbolic parameter and its value is a positive constant), if $W_{i,j}$ is using machine k (i.e., $A(i, j)=k$). Otherwise, the weight is equal to $\alpha+\beta$ (where β, called the setup time, is a symbolic parameter and its value is a positive and very big constant). Setting a very big value for β shows that the approach will penalize a high cost if $W_{i,j}$ is assigned to a new machine. The value for α should be much less than β.
- Edges between intermediate nodes "$W_{i,j,k}$" and "$W_{i,j',k'}$" if j is less than j' and j' is the next column after the column j satisfying $A(i, j')\neq 0$ in RSEM A. The weight for the edges from "$W_{i,j,k}$" to "$W_{i,j',k'}$" is equal to $\alpha+BD(k, j)$ if k is equal to k'. Otherwise, the weight is equal to $\alpha+\beta+BD(k, j)$.
- Edges from intermediate node "$W_{i,j,k}$" to "E" if j is the last column satisfying $A(i, j)\neq 0$ in RSEM A. The weight for these edges equals $\alpha+BD(k, j)$.

If the path of a wafer chooses other nodes having a different k with the previous nodes, then the path is penalized by the big value β. Hence, the path will not be the shortest path. The complexity of Dijkstra's algorithm for finding the shortest paths in G is $O(S^2\times M^2\times N^2)$.

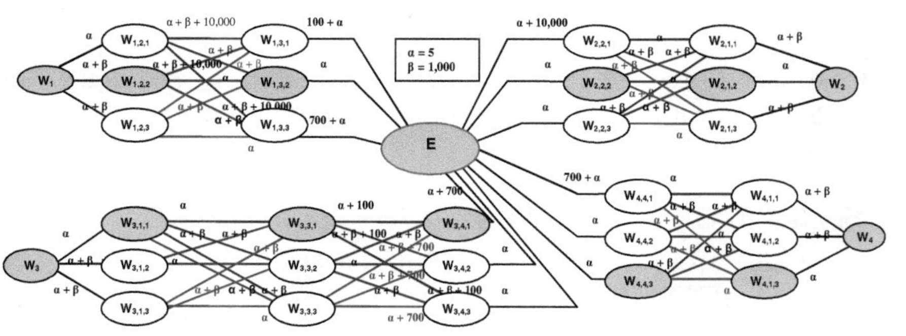

Fig. 3 The shortest paths depicted by the dark nodes on the graph G from the start nodes named by W_1, W_2, W_3, and W_4 are found by using Dijkstra's algorithm. The shortest path from the start nodes W_1 with the cost equal to 1,015 units consists of nodes W_1, $W_{1,2,2}$, $W_{1,3,2}$, and E. This shortest path shows that wafer W_1 is assigned to machine 2, while wafers W_2, W_3, and W_4 are to use the old machines. The new RSEM B is shown as in Fig. 1

3 Computational Studies

We use an example to demonstrate our graph approach as shown in Fig. 3. The Graph is using RSEM *A* and matrix *BD* depicted in Fig. 1 with 4 sub-graphs. The values for α and β are assumed equal to 5 and 1,000 units, respectively. The total production cost for RSEM *B* is 1,865 units, while the total production cost for RSEM *A* is 10,965 units. The graph framework saved about 83.0% of the production cost. The following provides more experiments.

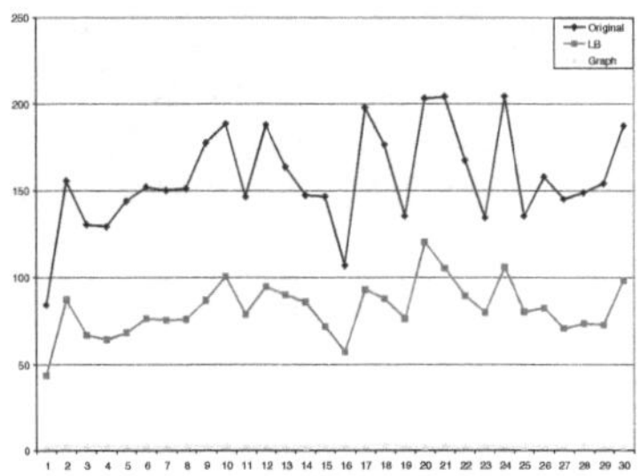

(a) 100 wafers, 10 steps, and 3 machines.

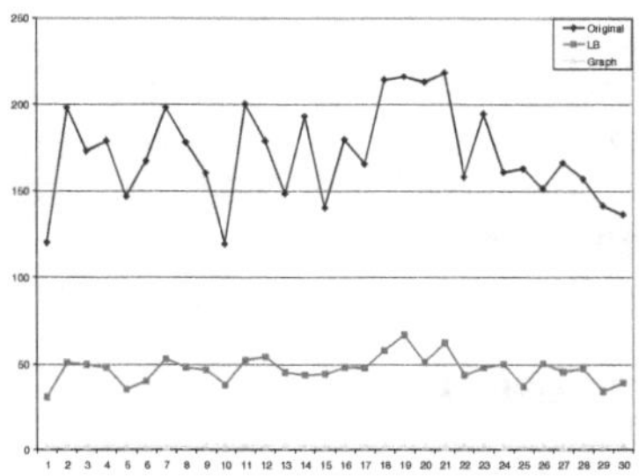

(b) 100 wafers, 10 steps, and 6 machines

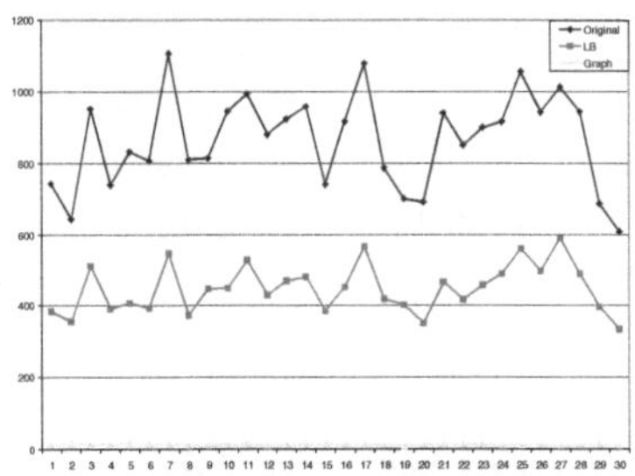

(c) 500 wafers, 10 steps, and 3 machines

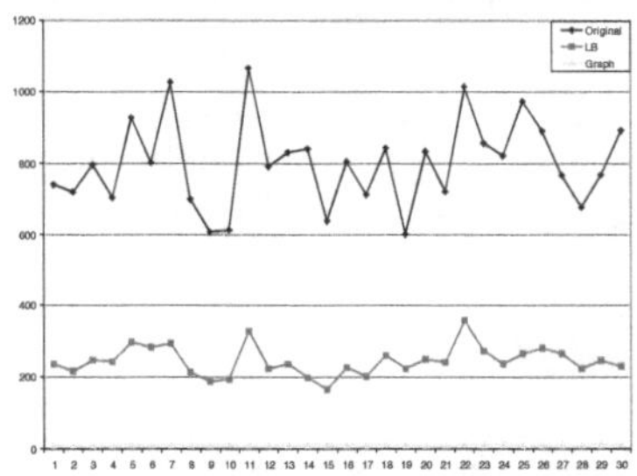

(d) 500 wafers, 10 steps, and 6 machines

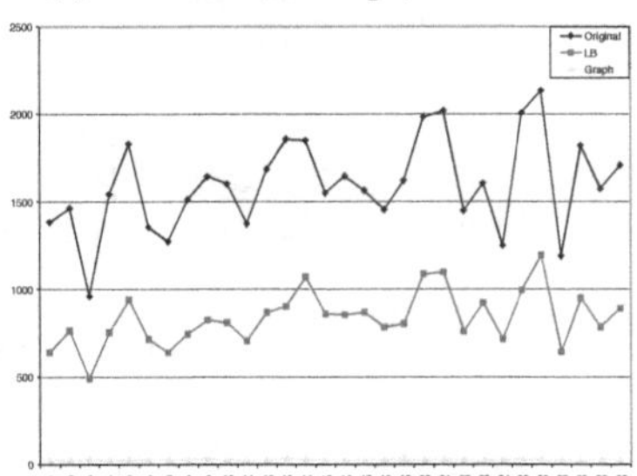

(e) 1000 wafers, 10 steps, and 3 machines

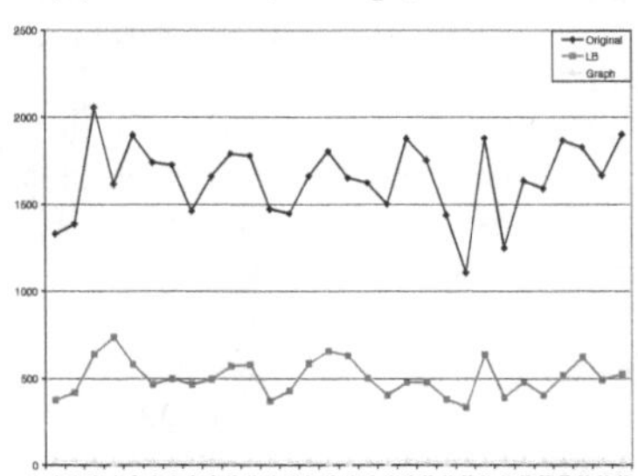

(f) 1000 wafers, 10 steps, and 6 machines

Fig. 4 These six test beds are depicted in (a)-(f), respectively. In each figure, the X-axis represents 30 examples, while the left Y-axis shows the total production costs obtained from the original, the LB, and the graph approaches divided by 100,000 units. (a)-(f) shows that the curves for the graph approach are always below the others. The average production costs obtained from the graph framework are better than the original and LB approaches by about 98.6% and 59.15%, respectively

Different numbers of wafers and machines were used to evaluate the performance for the graph framework. The RSEMs were created by respectively using 100, 500, and 1,000 wafers with 10 steps for 3 and 6 machines. This selection was due to the fact that in reality a typical factory with 8-inch wafers produced could handle these scenarios. The breakdown matrices were randomly assigned by values using the Normal distribution in the range [0, 10,000] units. There were six test beds with 100, 500, and 1,000 wafers and ran 30 examples for each test bed. The values for α and β were the same as the first example.

The experimental methodology consists of two phases. The first phase is to run the test beds with the different number of wafers and machines on both the graph framework and the LB approach [6]. The LB approach attempts to schedule wafers with the DMC according to the balance factors among machines. In the second phase, the total production costs obtained from the original (i.e., scheduling wafers as the input RSEM), the graph, and the LB approaches would then be compared together.

4 Conclusion

We have presented the graph framework for the scheduling issue of the DMC and have compared its results with the LB approach. The experiments conclude that formulating the DMC by using a graph form can provide an efficient approach to minimize the production cost with an appropriate complexity. The assumptions for the graph framework were only from empirical observation and some of the assumptions, such as the queue time, should also be considered in future work.

References

1. Akçali, E., Nemoto, K., Uzsoy, R.: Cycle-time improvements for photolithography process in semiconductor manufacturing. IEEE T Semi Manuf. 14, 48–56 (2001)
2. Kimms, A.: Multi-level, Single-machine lot sizing and Scheduling. Euro. J. OR 89, 86–99 (1996)
3. Li, H., Ramírez-Hernández, J.A., Fernandez, E., et al.: A framework for standard modular simulation: application to semiconductor wafer fabrication. NISTIR 7236 (2005)
4. Miwa, T., Nishihara, N., Yamamoto, K.: Automated Stepper Load Balance Allocation System. IEEE T Semi. Manuf. 18, 510–516 (2005)
5. Shr, A.M.D., Liu, A., Chen, P.P.: A Heuristic Load Balancing Scheduling Approach to the Dedicated Machine Constraint. Int'l. J. AI Tools 17, 339–353 (2008)
6. Vargas-Villamil, F.D., Rivera, D.E., Kempf, K.G.: A Hierarchical Approach to Production Control of Reentrant Semiconductor Manufacturing Lines. IEEE T Cntl. Sys. Tech. 11, 578–587 (2003)
7. Dijkstra, E.W.: A note on two problems in connexion with graphs. Num. Math. 1, 269–271 (1959)

Comparing Path Length by Boundary Following Fast Matching Method and Bug Algorithms for Path Planning

Chia Hsun Chiang, Jing-Sin Liu, and Yo-Shing Chou

Abstract. Local or sensor-based path planning in an unknown environment is a challenging problem for mobile robots. In completely known grid environments, the shortest Euclidean-length path connecting a given start-goal point pair can be generated using the Fast Marching Method (FMM). In contrast, in unknown environments, path planning must be reactive (i.e. based on recent sensory information) and *boundary following (*BF) is one type of reactive path planning technique. In [1], a hybrid method called Boundary Following FMM (BFFMM) was proposed that could extend the applicability of FMM to work well in an unknown environment using only local sensory perceptions. Bug algorithms are one family of navigation algorithms that also utilize BF technique in planning paths in unknown environments. This paper compares the path length of the paths generated by BFFMM with those by Bug1 and Bug2, the earliest versions of the Bug family algorithms to quantitatively reveal the effects of environment uncertainties on length of paths generated by navigation methods using boundary following. By comparison, BFFMM is a modest mobile robot path planner that bridges the gap between unknown and known terrain.

1 Introduction

Depending on whether the prior knowledge of the whole environment is available, the problems of path planning may be divided into two classes: local or sensor-based and global. For global path planning, or similar path planning with complete information, it is assumed that the mobile robot knows a priori the environment,

Chia Hsun Chiang
Institute of Information Science, Academia Sinica
e-mail: fetishist@gmail.com

Jing-Sin Liu
Institute of Information Science, Academia Sinica
e-mail: liu@iis.sinica.edu.tw

Yo-Shing Chou
Institute of Information Science, Academia Sinica
e-mail: shing@iis.sinica.edu.tw

B.-C. Chien and T.-P. Hong (Eds.): Opportunities and Challenges, SCI 214, pp. 303–309.
springerlink.com

and the main concern of path planning is efficiently generating a collision-free and/or optimal path. Many different methods of navigating mobile robots achieving varying degrees of success in a variety of conditions/criteria of motion and environments have been developed [9], [6]. In contrast, for local path planning, or similar path planning with uncertainty, the mobile robot has no a priori information about the environment in which it is moving around—an environment likely populated by obstacles whose positions, shapes, and quantities are unknown. The mobile robot locally updates its knowledge of environment during runtime by exploring using local information about proximate surroundings provided by its on-board sensors, allowing for planning a local path. The greatest concern when navigating an unknown path is generating a usable path to the destination.

For global path planning, the fast marching method (FMM) proposed by Sethian [7] has received more attention and has been applied to mobile robot path planning in a grid environment. Through wave propagation to solve the Eikonal equation, FMM can generate paths that closely resemble the geodesic via use of wave propagation to solve the Eikonal equation. Although FMM is a powerful tool for solving the problem of geodesic, shortest path planning, FMM is still limited to completely known terrain. In [1] we extended the applicability of FMM assuming the capability of boundary following (BF) by mobile robots and perfect sensing ability by combining it with the BF and developed a hybrid approach called Boundary Following FMM (BFFMM), which could be applied well to an unknown terrain. Bug Algorithms, are famous local path planning algorithms in an unknown planar environment utilizing boundary following and were invented by Lumelsky [3] and later a variety of Bug-like algorithms were proposed [3]-[5], [8]. A report was released recently that compared the performance of various Bug algorithms [4]. However, comparisons with other BF based path planning algorithms have not been made, and their optimality aspects have not been addressed. In this paper, we will compare the path length of paths generated by the BFFMM algorithm and the Bug algorithms, specifically the Bug1 and Bug2.

The paper is organized as follows. Section 2 introduces the technique used for experimental comparisons of the algorithms. Section 3 presents the simulation results in different environments and also discusses the difference between the different algorithms. The final sections include the conclusion and a possible future direction of research.

2 Local Navigation Methods Based on Boundary Following

In this paper, we make the following assumptions about the robot to which the simulations apply: The robot moves in an unknown 2D polygonal environment that is arbitrarily populated with a set of static polygonal obstacles. The robot is modeled as a point by enlarging the obstacles to account for the robot size. Moreover, the robot is equipped with a sensory system capable of exploring the environment by measuring the distance and direction to an obstacle within the visibility range of its sensors.

2.1 Boundary Following FMM (BFFMM) [1]

The BFFMM may be divided into an exploring stage and a path planning stage. In the exploring stage, the aim of the robot is to build its own partial map of the entire terrain via exploration strategy. For approach, the robot is equipped with a compass to detect the direction orienting the robot to its goal point. In unknown terrain, only the visible parts of obstacles are represented by the sensory readings. A local obstacle avoidance system is used to avoid collision when obstacles are encountered. The robot records the coordinate of the hit point on which the robot bumps the obstacle when the distance between the robot and the obstacle is below a certain threshold value, which is necessary to be taken into account according to the sensor range. The robot gathers map information about its vicinity and records these data on a partial map. After acquiring enough data, the robot generates a partial map of specific regions of the terrain connecting the start point and goal point as a global resemblance of the entire terrain. Since our objective is for the robot to navigate itself to a goal point, this partial map is enough for geodesic planning by the fast marching method (FMM).

2.2 Bug Algorithms

Bug algorithms operate under some assumptions. First, the localization ability of the robot is perfect. Second, the sensors of the robot are ideal. Different Bug algorithms may require different types of sensors or different environment settings, but all Bug algorithms share the same basic property in that they can be divided into two parts: a goal pursuing mode and a boundary following mode. In goal pursuing mode, the robot in general moves directly toward to the goal until the robot hits an obstacle. When it encounters an obstacle, the robot switches from goal pursuing mode to boundary following mode and does not switch back to goal pursuing mode until some switching condition is satisfied. If the robot does not have a partial map generated that connects the start and destination points, the algorithm will recognize this situation, terminate its running process and report that the destination point is unreachable.

Fig. 1 Three single-obstacle test maps of size 200x200 for experiment. The blue circle is the start; the red circle is the goal. (a) the cavity map. (b) the Sankar map. (c) the zigzag map

3 Experiments and Results

The three environments described by the bitmap file shown in Fig. 1 were used for comparative study of path planning performance of Bug1, Bug2, and BFFMM. All the simulations are run with MATLAB on a PC with Intel Core 2 Duo 1.8 GHz microprocessor.

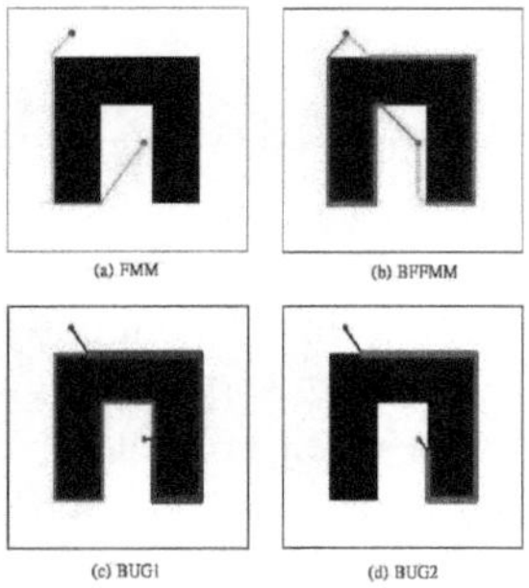

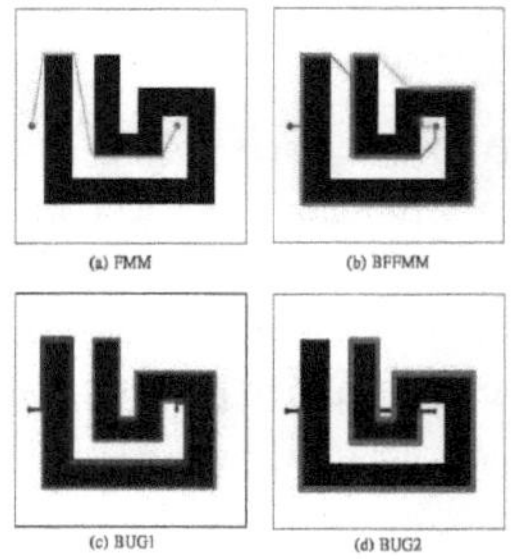

Fig. 2 Simulations on the Cavity map: path generated by (a) FMM (b) BFFMM (c) BUG1 (d) BUG2

Fig. 3 Simulations on the Sankar map: path generated by (a) FMM (b) BFFMM (c) BUG1 (d) BUG2

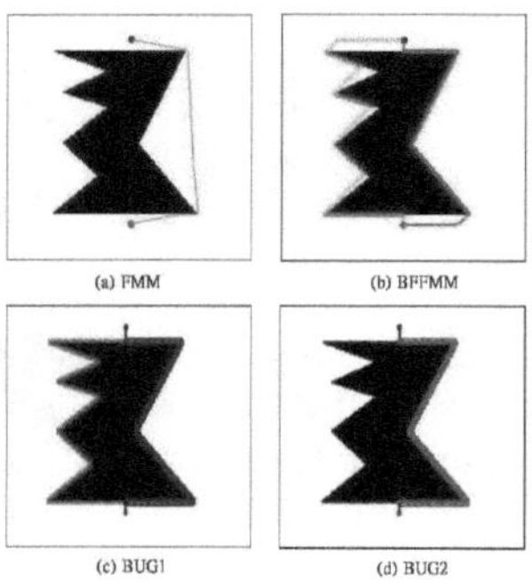

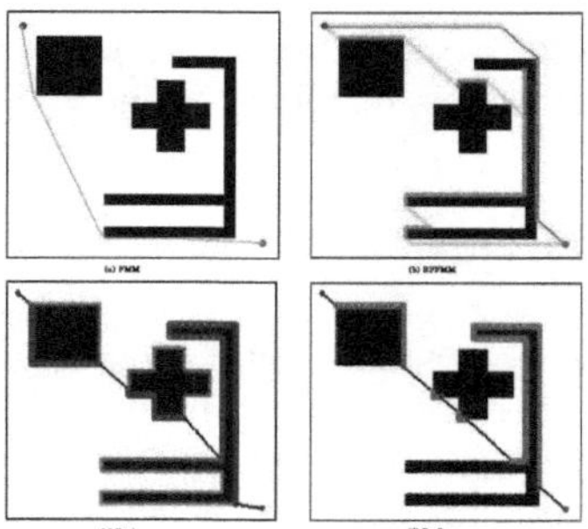

Fig. 4 Simulations on the Zigzag map: path generated by (a) FMM (b) BFFMM (c) BUG1 (d) BUG2

Fig. 5 Simulations on the multiple obstacles map: path generated by (a) FMM (b) BFFMM (c) BUG1 (d) BUG2

Table 1 The length of alternative paths generated by each algorithm in three single-obstacle test maps of size 200x200. The two numbers in the row of BFFMM represent the path length in the two stages of BFFMM respectively. The first number is the path length in exploring phase; the second one is the path length in path-planning phase

Algorithm	Path length on Cavity map	Path length on Sankar map	Path length on Zigzag map
FMM	245.62	261.46	227.74
BFFMM	635.14 / 311.11	858.48 / 279.36	633.07 / 257.19
BUG1	990	1438	903.6
BUG2	337	784	273.64

We conduct simulations with the assumption that the world is perfect, noise-free, and static. The results of the simulations are shown in Fig. 2-4 and summarized in Table 1.

As a reference for evaluation of path length, the path generated by FMM is the geodesic connecting the start and the destination points on the obstructed map, or the shortest path between the two points. We express the path length as percentage above optimal (PAO), which is defined as

$$\text{PAO}=\left(\frac{\text{generated path length}}{\text{path length generated by FMM}}-1\right)\cdot 100$$

to facilitate the comparison between different methods. The comparison is given in Table 2. From the results, the path length generated by Bug1 is the longest because the robot has to contour to the whole boundary curve of the encountered obstacle to determine the switching point. Bug2 generally produces shorter paths than Bug1 because Bug2 stops following the boundary of the obstacle before encircling it. Although the path generated by Bug 2 is shorter than Bug1, Bug2 does not know which side or which route would be better. BFFMM builds the partial map by contouring the boundary curves of obstacles, so the path length in exploring mode would approach the limit of Bug1.The total path length of BFFMM is close to that of Bug1, but the path length of the path planning phase is even shorter than that of Bug2. Besides, the robot could plan a shortest path for a new start-goal pair on the built partial map without resorting to the exploring phase again. Without following the boundary of the obstacles, the path length of FMM is not directly or explicitly influenced by the perimeters of the obstacles. Therefore, this comparison also quantitatively reveals the impact of environment uncertainties on length of paths generated by local navigation algorithms relying on contouring the boundary curve of the obstacle.

3.1 Effect of Multi-obstacles

To inspect the effect of multi-obstacles on the algorithms, we perform another simulation shown in Fig. 5 and Table 3. Through observation, we found that as the number of obstacles increased, the difference between the total path lengths generated by BFFMM and Bug1 clearly increased. This phenomenon resulted from the robot operating using Bug1 where it would encircle all the enroute obstacles encountered on the way towards the goal before the switching condition was satisfied. In contrast, when the robot is navigated by BFFMM, it could switch from boundary following mode earlier than when it operated using Bug1 and spent less effort on exploration. Overall, the BFFMM strikes a good balance between less exploration load and shorter path length even though the complexity of the environment increased.

Table 2 PAO of path generated by each algorithm

Algorithm	Cavity map	Sankar map	Zigzag map
BFFMM	159 / 27	228 / 7	1178 / 13
BUG1	303	450	297
BUG2	37	200	20

Table 3 PAO of path generated by each algorithm in multiple obstacles environment

Algorithm	Path length	Relative path length
FMM	306.33	1.00
BFFMM	947.10 / 333.64	3.09 / 1.09
BUG1	1843.53	6.02
BUG2	611.42	2.00

4 Conclusions

Comparisons of path lengths generated by the three local path planning algorithms utilizing boundary following BF techniques are presented. Although paths generated by these kinds of algorithms are far from optimal in path length, they are successful in navigating a mobile robot safely and reliably to a goal along the boundary curve of an obstacle even in situations where the sensor(s) is faulty. From simulations in test environments, we have observed that Bug1 algorithm produces the longest path length, while the BFFMM plans paths that are shorter than both the Bug1 and Bug2. Taking the path length of the exploring phase into account, the BFFMM generates paths with total length approaching those of Bug1 in single obstacle maps and may be even shorter in multiple obstacle maps. Furthermore, new paths could be re-planned by BFFMM easily for a variety of start and destination point pairs within already built partial maps. If the passage ways are known, we can save a lot of computational effort in pursuing destination points with changeable coordinates.

Acknowledgments. The research was supported by National Science Council under contract NSC 96-2221-E-001-018-MY2.

References

1. Chiang, C.H., Liu, J.S.: Boundary Following in Unknown Polygonal Environment Based on Fast Marching Method. In: IEEE International Conference on Advanced Robotics and its Social Impacts (2008)
2. Chiang, C.H., Chiang, P.J., Fei, J.C.C., Liu, J.S.: A Comparative Study of Implementing Fast Marching Method and A* Search for Mobile Robot Path Planning in Grid Environment: Effect of Map Resolution. In: IEEE Workshop on Advanced Robotics and its Social Impacts, pp. 7–12 (2007)
3. Lumelsky, V.J., Stepanov, A.A.: Dynamic Path Planning for a Mobile Automation with Limited Information on the Environment. IEEE Transactions on Automatic Control AC-31(11), 1058–1063 (1986)
4. Ng, J., Braunl, T.: Performance Comparison of Bug Navigation Algorithm. Journal of Intelligent and Robotic Systems 50, 73–84 (2007)
5. Langer, R.A., Coelho, L.S., Oliveira, G.H.C.: K-Bug, A new bug approach of mobile robot path planning. In: 16th IEEE International Conference on Control Application, Part of IEEE Multi-conference on Systems and Control, Singapore, pp. 403–408 (2007)

6. Rao, N.S.V., Shi, S.K.W., Ivengar, S.S.: Robot Navigation in Unknown Terrains: Introductory Survey of Non-Heuristic Algorithms. Technical Report, Oak Ridge National Laboratory (1999)
7. Sethian, J.A.: Level Set Methods. In: Evolving interfaces in geometry, fluid mechanics, computer vision, and materials science. Cambridge University Press, Cambridge (1999)
8. Kamon, I., Rivlin, E., Rimon, E.: A New Range-Sensor Based Globally Convergent Navigation Algorithm for Mobile Robots. In: Proceeding for the 1996 IEEE International Conference on Robotics and Automation, Minneapolis, Minnesota (1996)
9. LaValle, S.M.: Planning Algorithms. Cambridge University Press, Cambridge (2006)

A Capacitated Vehicle Routing Problem with Toll-by-Weight Rule

Chenghao Shen, Hu Qin*, and Andrew Lim

Abstract. Most literature on min-cost network flow problems such as shortest path problem (SPP), traveling salesman problem (TSP), vehicle routing problem (VRP), assumes that the rate of any arc is constant. However, this assumption may not be true for some real applications occurring in China due to the toll-by-weight rule. Different from most toll rules, toll-by-weight allows the charging rates on some arcs to vary with vehicle's total weight. Obviously, the min-cost path between two nodes may vary as the actual load of the vehicle changes when toll-by-weight rule is considered. In this work, we study a new variation of capacitated vehicle routing problem (CVRP) which considers the toll-by-weight rule, and the objective is to minimize the total transportation cost involved. To solve this problem, we implemented a simulated annealing (SA) algorithm and the results of extensive experiments showed the effectiveness of this algorithm.

1 Introduction

Most min-cost network flow problems in literature such as shortest path problem (SPP), traveling salesman problem (TSP), vehicle routing problem (VRP) assume that the rates associated with the network arcs are constant. However, this assumption may not always be true. For example, as of Jun 2007, about 20 provinces in China have put toll-by-weight rule into practice, which allows the charging rate to

Chenghao Shen and Andrew Lim
School of Computer Science and Engineering,
South China University of Technology, Guangzhou, P.R. China
e-mail: chenghao.shen@gmail.com

Hu Qin and Andrew Lim
Department of Management Sciences, City University of Hong Kong,
Tat Chee Ave, Kowloon Tong, Hong Kong
e-mail: huqin3@student.cityu.edu.hk, lim.andrew@cityu.edu.hk

* Corresponding author.

B.-C. Chien and T.-P. Hong (Eds.): Opportunities and Challenges, SCI 214, pp. 311–316.
springerlink.com

vary with vehicles' total weights. This example demonstrates that investigating network problems with non-constant arc rate has great practical value in China. In this paper, we study a capacitated vehicle routing problem (CVRP) with toll-by-weight rule (CVRPW), which can be described as follows: (1) The objective is to minimize the total transportation cost involved, which comprise two parts, vehicle cost and toll-by-weight. (2) There is only one depot and each vehicle starts from the depot, serves a set of customers and returns to the depot. The total demand of customers served by one vehicle must not exceed its capacity. (3) Both the capacity of the vehicle and the demands of all customers are measured by weight. (4) All vehicles have the same capacity and the number of vehicles is unlimited. (5) After serving one customer, the vehicle unloads customer's goods and the corresponding weight is deducted from its total weight. (6) All customers have to be served exactly once.

The charging rule is represented by expression (1). $C_v(d) = \alpha * d$ is the vehicle cost including expenditure of oil, driver and so on, where d denotes the length of the travel distance and α denotes the vehicle cost per kilometer. $T(w,d)$ is the cost resulted from toll-by-weight charged based on the vehicle's total weight w and the travel distance d. $T(w,d)$ may be different in different highways. To explain this point, we take two highways, Ronghuxi highway and Xiangjin highway in Hubei province as examples. Expression (2) is the toll-by-weight rule for Ronghuxi highway and expression (3) is for Xiangjin highway, which are both illustrated in Figure 1(a) by curves.

$$C(w,d) = C_v(d) + T(w,d) \tag{1}$$

$$T(w,d) = \begin{cases} 0.75d & \text{if } w < 5 \\ 0.15wd & \text{if } 5 \le w < 10 \\ (-\frac{1}{800}w^2 + \frac{7}{40}w - \frac{1}{8})d & \text{if } 10 \le w < 40 \\ (0.075w + 1.875)d & \text{if } w > 40 \end{cases} \tag{2}$$

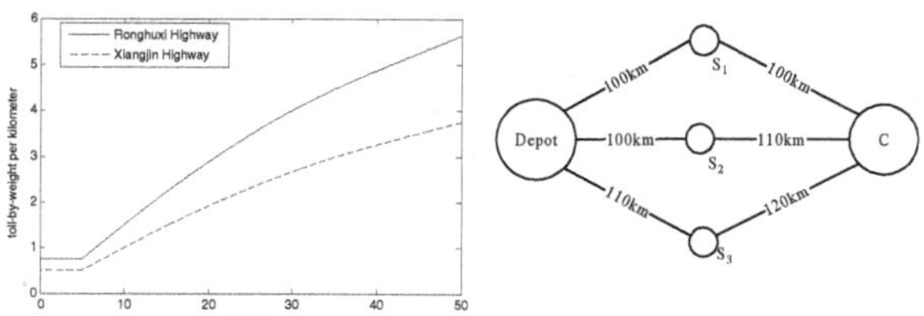

(a) Toll-by-weight per kilometer

(b) Three paths between Depot and Customer C

Highway	Toll-by-weight	Cost (a=5)		
		5t	20t	50t
Depot - S_1	Expression 3	575	788	1063
Depot - S_2	Expression 3	575	788	1063
Depot - S_3	Expression 4	605	761	963
S_1 - C	Expression 3	575	788	1063
S_2 - C	Expression 4	605	761	963
S_3 - C	Expression 4	660	830	1050

(c) The cost for the highway in different weight

Path	Cost(a=5)		
	5t	20t	50t
P_1: Depot -> S_1 -> C	**1150**	1576	2126
P_2: Depot -> S_2 -> C	1180	**1549**	2026
P_3: Depot -> S_3 -> C	1265	1591	**2013**

- When the demand is 5t, p1 is the best path
- When the demand is 20t, p2 is the best path
- When the demand is 50t, p3 is the best path

(d) The cost for the three paths in different weight

Fig. 1 Toll-by-Weight per kilometer

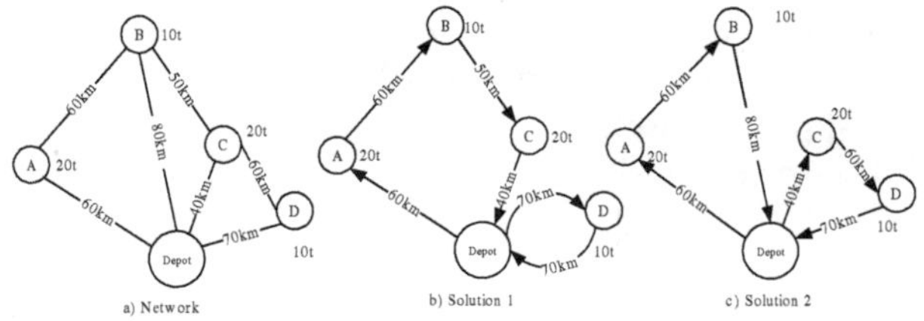

Solution	Distance	Cost
1	350	2659
2	370	2543

Note:
The vehicle cost is charged 5 per kilometer
The toll-by-weight is charged as expression (2)

Fig. 2 A simple highway network

$$T(w,d) = \begin{cases} 0.5d & \text{if } w < 5 \\ 0.1wd & \text{if } 5 \le w < 10 \\ (-\frac{1}{1200}w^2 + \frac{7}{60}w - \frac{1}{12})d & \text{if } 10 \le w < 40 \\ (0.05w + 1.25)d & \text{if } w > 40 \end{cases} \quad (3)$$

Let us consider a simple case shown in Figure 1(b) in which there is only one customer C and three nodes S_1 S_2 and S_3, which represent the crosspoints of different highways. We can easily find that between *Depot* and C, there are three paths p_1, p_2 and p_3 available. In the classical CVRP problem, the shortest path from *Depot* and customer C is p_1. However, when the toll-by-weight rule is considered, the best path from Depot to C varies according to the actual weight of the vehicle. Shown as Figure 1 (b) and (c), given the vehicle cost is 5 per kilometer, we can find when the customer demands change, the associated min-costs route also changes. Figure 2 shows an example with two vehicles and four customers. If we adopt constant rate, solution 1 is the better one and whereas if toll-by-weight rule is adopted, solution 2 becomes better in terms of the overall transportation cost.

Since the classical VRP is NP-hard, the CVRPW is obviously NP-hard. If only one vehicle and one customer are considered, our problem becomes a shortest path problem with the toll-by-weight rule, which has already been studied in [1]. If we fix the rates for all network arcs, the problem actually is a standard CVRP which has be extensively studied for many decades. The book edited by P. Toth and D. Vigo [2] collects many of the known methods and results regarding CVRP.

2 Calculating Cost between Two Nodes

In the algorithm we will introduced in next section, calculating the min-cost path of two nodes as long as the weight of the vehicle is changed is unrealistic since it may consume considerable computational time. To deal with this, we propose an approximate way described as follows: discretize the weight range at a given precision and calculate the min-cost path for each node pair and each value of the weight. For example, you can discretize the weight range [0, 3] into a discrete number set $A = \{0,1,2,3\}$. Then, calculate the min cost-path for all values in A and store these paths into one table. When you need the min-cost path for weight 1.2, which lies in [1, 2] and is close to 1, you can just use the min-cost path for weight 1. If the weight is 1.8, you can use the min-cost path for weight 2. If higher precision is required, more computational time will be spent as well.

```
Data: C is the list of customers to be inserted
Result: Solution S
create an empty solution S;
while C is not empty do
    randomly choose a customer c, and delete it from C ;
    minCost ← ∞, bestR ← null;
    foreach Route r in the current solution S do
        keep the order of the customers in r, try to insert c into r at a place with the lowest insertion cost e;
        if e < minCost then
            minCost ← e, bestR ← r
        end
    end
    create a new route r' to server customer c with cost e'.;
    if e' < minCost then
        add Route r' to current solution S;
    else insert c into bestR
end
```

Algorithm 1. The randomized insertion algorithm

```
Data: S the initial solution
Data: T0, Te the initial and termination temperature
Data: c the cooling ratio 0 < c < 1
Data: K the max successful operations at given temperature
Data: N the max attempts at given temperature
Result: best the best solution found
t ← T0; best ← S;
while t > Te do
    i ← 0, n ← 0;
    while i < K and n < N do
        randomly choose routes r1 and r2 from S;
        if r1 = r2 then apply the operation within route;
        else apply the operation between routes;
        if S is changed then
            cc ← the change of the total cost on S ;
            if cc < 0 then
                accept the change to S; try to update best, if S is better ;
            else
                μ ← cc/f(S) ;
                accept the changes with a probability e^(−μ/t), otherwise undo changes;
            end
        end
        if S is changed then i ← i + 1;
        n ← n + 1 ;
    end
    t ← t · c ;
end
```

Algorithm 2. The Simulated Annealing algorithm

3 Solution Procedures–Simulated Annealing Algorithm

To solve this problem, we have developed a tabu search (TS) algorithm and a simulated annealing (SA) algorithm. But from the results of the experiments we found that the SA dominated over TS in terms of solution quality and computation speed. Due to the space limitations, we only expose a part of the results. We use a randomized insertion (RI) algorithm to construct initial solutions. This insertion algorithm tries to insert customers one by one into the best position of the current solution until all the customers are served. Details of the insertion algorithm are exposed in Algorithm 1.

In our simulated annealing algorithm shown in Algorithm 2, we use some neighborhood operators called generalized n-op operators proposed by [3] for VRPTW, which are:(1) Move a sub-route from one route to a different random location on the same route; (2) Move a sub-route from one route to a random location on one of other routes; (3) Swap two disjoint sub-routes on the same route; (4) Swap a sub-route from one route and another sub-route from another route; (5) Reverse a sub-route in one route. The length of the sub-routes is randomly selected from the range $[0, M]$, where M is the maximal length of the sub-routes. According to some pilot experiments we observed that $M = 3$ or 4 could provide good results.

4 Experiments and Conclusion

As we did not find benchmark problems in the literature associated with our problem, we only tested the performance of our algorithm based on our self-generated instances. For all instances, data that they require are: charging rules, highway length and customer demand. We obtained Chinese highway network structure and associated distance data from [4] and then we randomly classified all highways into three types: close type highway, open type highway and free highway. According to the requirements in an official guiding document [5] issued by Chinese Communication Ministry, the toll-by-weight charging rules in three types of highways are specified as follows:

Open Type $T_o(w,d)$***:*** If $w \leq 20$, the vehicle is charged by β per kilometer per ton; if $20 < w \leq 40$, the former 20 ton is charged as $T_o(20,d)$, and the left weight $(w-20)$ is charged by δ per kilometer per ton, where δ is a rate that linearly decreases on weight from β to 0.8β ; if $w > 40$, the former 40 ton is charged as $T_o(40,d)$, and the left weight $(w-40)$ is charged by 0.8β per kilometer per ton.
Close Type $T_c(w,d)$***:*** If $w \leq 20$, the vehicle is charged by β per kilometer per ton; if $20 < w \leq 40$, the former 20 ton is charged as $T_c(20,d)$, and the left weight $(w-20)$ is charged by δ per kilometer per ton, where δ is a rate that linearly decreases on weight from β to 0.5β; if $w > 40$, the former 40 ton is charged as $T_c(40,d)$, and the left weight $(w-40)$ is charged by 0.5β per kilometer per ton.
Free Type $T_f(w,d)$***:*** $T_f(w,d) = 0$

Rate β for each highway was randomly generated from $U[0.02, 0.1]$, where $U[a,b]$ denotes the uniform distribution in the interval $[a,b]$. The number of customers were taken from the data set {100, 200, 300, 400, 500}, thereby we have five types of instances. Then, we randomly selected some nodes in highway network as customer locations and the customer demands were randomly selected from $U[0.5, 10]$. We generated six instances for each type and therefore 30 instances entirely.

We implemented our SA algorithm with Java and ran all experiments on a desktop with 2.2GHz AMD Duo and 2 GB RAM. As RI is the initial solution for SA, we take it as a baseline of improvement. Observed from the results in table 1, we

Table 1 Experimental results

		RI	SA		
n		Objective	Objective	Improvement	Time
100	1	742727	403248	45.71%	99
	2	670308	322205	51.93%	94
	3	680410	329755	51.54%	91
	4	805910	461138	42.78%	61
	5	853661	511263	40.11%	88
	6	727444	334218	54.06%	64
200	1	1615711	819416	49.28%	131
	2	1356250	578247	57.36%	108
	3	1484392	627643	57.72%	100
	4	1532161	796394	48.02%	96
	5	1611121	844256	47.60%	132
	6	1385114	601520	56.57%	116
300	1	2396134	1131241	52.79%	152
	2	2034065	855802	57.93%	157
	3	2233764	884571	60.40%	178
	4	2368512	1167541	50.71%	146
	5	2519571	1243520	50.65%	125
	6	2050262	791984	61.37%	224
400	1	3081097	1442774	53.17%	164
	2	2852013	1120581	60.71%	162
	3	3001713	1166162	61.15%	181
	4	3162125	1598037	49.46%	164
	5	3295919	1636382	50.35%	149
	6	2781891	1070926	61.50%	205
500	1	3795836	1764981	53.50%	217
	2	3625957	1384036	61.83%	220
	3	3782834	1435840	62.04%	233
	4	3992848	1961578	50.87%	195
	5	4213092	2030622	51.80%	226
	6	3399905	1283921	62.24%	209

Objective: best known objective value; Improvement = (RI Objective - SA Objective)/RI Objective;

can easily find that SA can improve the quality of solution from 40.11% to 62.24% compared with the solutions generated simply by RI, which shows that this SA algorithm can provide the practitioners with relative high quality solutions.

In this paper, we studied a capacitated vehicle routing problem with toll-by-weight (CVRPW) rule. This is a variation of VRP, which has great practical value in China. After analyzing this problem, we implemented a simulated annealing algorithm to solve to it. The experiments showed that the SA algorithm was able to produce relatively high-quality solutions.

References

1. Chen, X., Li, J.: Design and implementation of vehicle routing optimization system based on toll-by-weight. MIE of China 17, 69–72 (2007)
2. Toth, P., Vigo, D.: The Vehicle Routing Porblem. SIAM, Philadelphia (2001)
3. Osman, I.H.: Metastrategy simulated annealing and tabu search algorithms for the vehicle routing problem. Annals of Operations Research 41, 421–451 (1993)
4. China map data posted on webmap, `http://sms.webmap.cn/xq.asp?dbid=127`
5. Document of chinese communication ministry (Website, 2005), `http://www.wzjt.gov.cn/dzzf/jtfg/jtgf/06092001203803223.htm`

A Novel Multidimensional Scaling Technique for Mapping Word-Of-Mouth Discussions*

B. John Oommen, Ole-Christoffer Granmo, and Zuoyuan Liang

Abstract. The techniques which utilize Multidimensional Scaling (MDS) as a fundamental statistical tool have been well developed since the late 1970's. In this paper we show how an MDS scheme can be enhanced by incorporating into it a Stochastic Point Location (SPL) strategy (one which optimizes the former's gradient descent learning phase) and a new *Stress* function. The enhanced method, referred to as MDS_SPL, has been used in conjunction with a combination of the TF-IDF and Cosine Similarities on a very noisy Word-Of-Mouth (WoM) discussion set consisting of postings concerning mobile phones, yielding extremely satisfying results.

1 Introduction

The science and art of the fascinating field of Multidimensional Scaling (MDS) is fairly well established. Even though the field is quite mature, the challenges encountered with regard to application domains seem to constantly test the limits of

B. John Oommen
School of Computer Science, Carleton University, Ottawa, Canada
e-mail: oommen@scs.carleton.ca

B. John Oommen and Ole-Christoffer Granmo
Dept. of ICT, University of Agder, Grimstad, Norway
e-mail: ole.granmo@uia.no

Zuoyuan Liang
C/o. Dr. B. John Oommen, School of Computer Science,
Carleton University, Ottawa, Canada: K1S 5B6

* The first author was partially supported by NSERC, the Natural Sciences and Engineering Research Council of Canada. The work of the second author was done in Canada while visiting the first author with the support of a *Leiv Eriksson Grant* funded by the Research Council of Norway. Both authors acknowledge the support of *Integrasco*, a company providing WoM services.

B.-C. Chien and T.-P. Hong (Eds.): Opportunities and Challenges, SCI 214, pp. 317–322.
springerlink.com

the available techniques, especially because the data domains become prohibitively larger and more dynamic. In particular, processing, representing and accessing the ever-expanding body of digital text collections (digital libraries) using MDS is far from trivial. Indeed, the amount of digitally stored text has grown exponentially since the pioneering works of Salton *et al* [4].

Text collections in the form of online discussion fora, referred to as *Word-Of-Mouth* (WoM), introduces additional challenges. In this case, the purpose of applying MDS may be to support monitoring of ongoing discussion trends and significant events, for example within the mobile phone business domain. On-line discussion fora form a particularly challenging application domain because they are typically less well-formed than, for instance, a regular magazine article. The content of these discussions is more "noisy" and also more "chatty" (less formal), in the sense that discussions contain significantly more spelling errors, abbreviations, slang, and information conveyed by means of symbols (e.g, "smileys").

This paper describes how an MDS scheme can be enhanced by incorporating into it a Stochastic Point Location (SPL) [3] strategy (which optimizes the former's gradient descent learning phase) and a new *Stress* function. The enhanced method, referred to as MDS_SPL, has been used in conjunction with a combination of the TF-IDF and Cosine similarities on a very noisy WoM discussion set, consisting of postings concerning mobile phones, to achieve an aesthetically satisfying mapping.

2 A Novel MDS Technique for Mapping Word-Of-Mouth Discussions

MDS is a family of related statistical techniques often used in data visualization for exploring similarities or dissimilarities among pairs of objects, as distances between points of a low-dimensional multidimensional space. MDS is considered a special case of ordination that maps a set of points into a finite dimensional flat (Euclidean) domain, where the only given data are the corresponding distances between every pair of points. An MDS algorithm starts with a matrix of inter-item similarities, then assigns a location of each item in a low-dimensional space, suitable for graphing or 3D visualization. The graphical display of the correlations generated by MDS provides a view of the data and allows the user to visually explore the structure of the data.

Fundamental to the MDS problem is the model of the underlying space. The first MDS algorithm was proposed by Torgerson [7], and in his model, the dissimilarity estimates were assumed equal to distances in an Euclidean multidimensional space. In text mining, documents, represented by their VSMs, are compared using the so-called *Cosine Similarity*. The latter is a measure of the similarity between two vectors computed in terms of the angle between them. Salton, in his term importance theory, intended to develop a weighting scheme by associating a weight with every token in the document [4]. One measure, Frequency-Inverse Document Frequency (TF-IDF), assumes that the importance of a term is proportional to the number of times the term appears in the document, however, offset by the number of documents that it appears in.

With the above in mind, we now discuss the problem of utilizing the MDS for classifying and analyzing WoM on-line discussions. The fundamental hurdles which render this problem "orders" of magnitude more complex are the semantic (as opposed to the syntactic issues used in the earlier MDS schemes) and the computational aspects. We shall first explain these before proceeding to explain how they have been resolved.

Semantic Issues: Our goal is to prioritize accurate representation of close WoM relationships. It turns out that in WoM discussions, it is important to discern related topics from unrelated ones. However, the *degree* of "un-relatedness" is of comparably less importance, and seems to be more semantic than syntactic. This observation imposes two separate requirements. First of all, objects that are relatively close to each other in the multi-dimensional space should be correspondingly close to each other in the two-dimensional map. Secondly, objects that are far from each other in the multi-dimensional space should not be placed close to each other in the two dimensional space. In order to prioritize these two preferences, we introduce a new stress function, distinct from stress functions examined earlier, e.g., [7, 1]:

$$S_{\text{new}} = \sum_{i,j} \frac{(\delta_{ij} - d_{ij})^2}{\delta_{ij} \cdot d_{ij}} \tag{1}$$

As the reader will observe, the stress measure is "doubly-normalized" – normalized both with respect to the *real* distance as well as the approximate distance. Thus, the smaller the distance, the higher its accurate representation is prioritized, *both* with respect to the true distance and the approximate distance.

Computational Issues: In the case of WoM on-line discussions, it turns out that due to the intricate properties of the dissimilarity matrix, the convergence of a traditional MDS scheme is pathetically slow. Even with the above mentioned "doubly-normalized" criterion, the convergence criteria seem to be conflicting - if we use a fixed updating constant. While the new stress function succeeds in keeping items which are distant from each other also distant in the mapped space, and simultaneously keeping items which are close to each other also close in the mapped space, it does this at a tragic pace. Indeed, the underlying problem can be described as follows: Determining a *fixed* (i.e, constant) updating factor to explore the search space during the gradient descent is quite challenging. If the constant is too large, the search will diverge, over-passing the minimum. If, on the other hand, the constant is too small, the convergence is excessively sluggish. In fact, we have observed that exploring different parts of the search space with different tailored updating factors is beneficial. To rectify this, we have enhanced the basic MDS with a solution to the Stochastic Point Location problem (SPL) [3], which dynamically adapts the updating factor.

The formal algorithm is given in **Algorithm MDS_SPL**. In the algorithm, d_{ij} refers to the distance between the points i and j, as represented in a two-dimensional space. In other words, d_{ij} is an approximation of δ_{ij} — the *true* distance between point i and point j, measured in the original multidimensional space.

Algorithm 1. MDS_SPL

Input: The original set of points in high dimension.

Output: A mapping obtained by an MDS.

Assumption 1: The inter-item distances are computed using the Cosine dissimilarities obtained from the TF-IDF indices.

Assumption 2: The step-size of the gradient descent is a, initially initialized to some constant a_o. We have initialized it, in our experiments, to unity. It is updated at the end of each loop by either multiplying it by a constant λ_a, or dividing it by λ_a, where $\lambda_a > 1$. We have set λ_a to have the value 2. a_o and λ_a are user-defined parameters.

Method:

Step 1: Initialize step size : $a = a_o$.

Step 2: Let $I = \{1,\ldots,n\}$ be a set of indexes where each index $i \in I$ is associated with a point (x_i, y_i). Note that x_i and y_i are initialized by drawing a random point from the 2-dimensional square of size J. If $J = 4$ this is the square $[-2,2] \times [-2,2]$, i.e., $(x_i, y_i) \in [-2,2] \times [-2,2]$ initially.

Step 3 - Loop: Select an index i randomly from I.

Step 4: Calculate current stress $S_{\text{new}}(I)$ as per Eq. (1).

Step 5: Calculate the candidate adjustment in the x-direction, Δx_i, for point (x_i, y_i), based on the distance to the other points $j \in (I - \{i\})$ as:

$$\Delta x_i = a \cdot \sum_{j \in (I - \{i\})} \frac{\partial S_{\text{new}}(I)}{\partial x_i}.$$

Step 6: Calculate candidate adjustment in the y-direction, Δy_i, for point (x_i, y_i), based on the distance to the other points $j \in (I - \{i\})$ as:

$$\Delta y_i = a \cdot \sum_{j \in (I - \{i\})} \frac{\partial S_{\text{new}}(I)}{\partial y_i}.$$

Step 7: Calculate the stress of the adjusted set I^* as per Eq. (1), where the point (x_i, y_i) has been replaced with the point $(x_i + \Delta x_i, y_i + \Delta y_i)$.

Step 8: **If** $S_{\text{New}}(I^*) \leq S_{\text{New}}(I)$

$(x_i, y_i) \leftarrow (x_i + \Delta x_i, y_i + \Delta y_i)$

$a \leftarrow a * \lambda_a$

Else:

$a \leftarrow a / \lambda_a$

Step 9: Goto **Step 3 - Loop**

End Algorithm MDS_SPL

3 Empirical Results

In this section we report our results from using MDS analysis on an on-line *discussion fora* that contains discussions about mobile phones. As mentioned earlier, for this application, we have built a dedicated implementation of the MDS, **Algorithm MDS_SPL**, that was implemented with the specific goal of analyzing *on-line* discussion fora. The purpose of this latter on-line MDS-analysis system was to support monitoring of ongoing discussion trends and significant events within the mobile

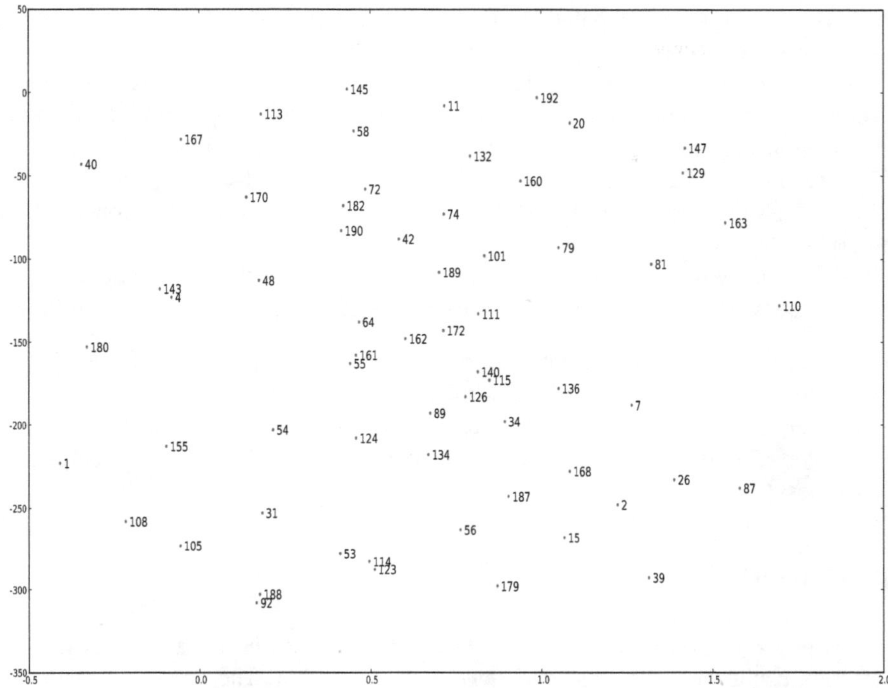

Fig. 1 The MDS map obtained for discussions within the fora titled "Map 1"

phone business domain. On-line discussion fora form a particularly challenging application domain because they are typically less well-formed than, for instance, a regular magazine article.

Typically, a discussion forum may consist of a collection of *Discussions*, where each discussion contains a sequence of *entries* written by *forum partakers*. The partakers normally use slang and abbreviations when writing discussion entries. Also, because of the often fragmentary nature of entries, the content of an entry may only have an unambiguous interpretation when the context of the entry is taken into account. That is, other entries may be needed to properly place a given entry into context. To make matters worse, the context of an entry may only be implicitly defined in the sense that it may lack syntactical or structural information relating it to its context. Consequently, only a semantic analysis will make the context apparent, and hence, the challenge!

Our discussion forum data set contained 199 discussions. Our aim was to visualize these discussions on a 2-dimensional topological map that brought forward and related the topics being discussed. As a starting point, we described each discussion using the TF-IDF approach, thus summarizing every discussion by means of a weighted term vector. In turn, these weighted term vectors formed the basis for calculating the Cosine similarity matrix, as discussed earlier. Then, in order to be able to fit the discussions on 2-dimensional maps of manageable size, we used a clustering method[1] to group the discussions into 4 partitions, using the Cosine matrix to measure similarity.

With regard to results, the 199 discussions have been visualized using four maps. Common to these maps is the fact that they partition the mobile phones into disjoint map sections. In Fig. 1, for instance, we observe that the discussions {132, 160, 72, 170, 182, 74, 190, 42, 189, 48, 4, 111, 4, 172, 162, 161, 55, 140, 115, 126, 34, 124, 134} discuss the phone "8801", while the discussions {53, 113, 123} concern the phone "6102".

In addition, within each formed cluster of products, finer details about the topics being discussed can be discerned. A few examples of such details from Fig. 1 follow:

- Cluster {72, 182, 190} is related to the keywords: {8801, compusa, battery}.
- Cluster {188, 192} covers the keywords: {scam, ip, n91, fraud}.
- Cluster {55, 64, 161} concerns the keywords: {firmware, 8801, update}.
- Cluster {53, 114, 123} associates with the keywords: {6101, 6102, cingular, wireless}.

4 Conclusion

In this paper we have described how MDS methods can be used for classifying data sets using their syntactic *and semantic* information. The enhanced method, referred to as MDS_SPL, has been used in conjunction with a combination of the TF-IDF and Cosine Similarities on very noisy WoM discussions consisting of postings concerning mobile phones, yielding extremely satisfying results.

References

1. Kruskal, J.B.: Nonmetric multidimensional scaling. Psychometrika 29, 1–27 (1964)
2. Oommen, B.J., Ma, D.C.Y.: Deterministic Learning Automata Solutions to the Equi-Partitioning Problem. IEEE Transactions on Computers 37, 2–14 (1988)
3. Oommen, B.J.: Stochastic searching on the line and its applications to parameter learning in nonlinear optimization. IEEE Transactions on Systems, Man and Cybernetics SMC-27B, 733–739 (1997)
4. Salton, G., McGill, M.J.: Introduction to Modern Information Retrieval. McGraw-Hill, Inc., New York (1983)
5. Salton, G., Wong, A., Yang, C.S.: A Vector Space Model for Automatic Indexing. Commun. of the ACM 18(11), 613–620 (1975)
6. Oommen, B.J., Granmo, O.C., Liang, Z.: Mapping Literature and Word-Of-Mouth Discussions Using Multidimensional Scaling, Unabridged version of this paper (submitted for publication) (2008)
7. Torgerson, W.S.: Multidimensional scaling: I. Theory and method. Psychometrika 17, 401–419 (1952)

[1] The clustering that we used was also based on a LA philosophy. The actual LA utilized was the Object Migrating Automaton (OMA) [2], the details of which are omitted here.

Comprehensible Knowledge Discovery Using Particle Swarm Optimization with Monotonicity Constraints

Chih-Chuan Chen, Chao-Chin Hsu, Yi-Chung Cheng,
Sheng-Tun Li, and Ying-Fang Chan

Abstract. Due to uncertain data quality, knowledge extracted by methods merely focusing on gaining high accuracy might result in contradiction to experts' knowledge or sometimes even common sense. In many application areas of data mining, taking into account the monotonic relations between the response variable and predictor variables could help extracting rules with better comprehensibility. This study incorporates Particle Swarm Optimization (PSO), which is a competitive heuristic technique for solving optimization tasks, with constraints of monotonicity for discovering accurate and comprehensible rules from databases. The results show that the proposed constraints-based PSO classifier can exploit rules with both comprehensibility and justifiability.

1 Introduction

Most data mining techniques adopted to discovering knowledge patterns from databases have been concentrated on obtaining high classification accuracy or

Chih-Chuan Chen and Sheng-Tun Li
Department of Industrial and Information Management, National Cheng Kung University, No.1, University Road, Tainan City 701, Taiwan, R.O.C.
e-mail: {r3895101,stli}@mail.ncku.edu.tw

Chih-Chuan Chen
Department of Information Management, Diwan College of Management, 87-1, anshih Li,Madou,Tainan 721,TaiwanTaiwan, R.O.C.

Chao-Chin Hsu
Department of Obstetrics and Gynecology, National Cheng Kung University, No.1, University Road, Tainan City 701, Taiwan, R.O.C.
e-mail: tube11@ms25.hinet.net

Yi-Chung Cheng
Department of International Business Management, Tainan University of Technology, 529 Jhongjheng Rd., Yongkang, Tainan 710, Taiwan, R.O.C.
e-mail: t20042@mail.tut.edu.tw

Sheng-Tun Li, and Ying-Fang Chan
Institute of Information Management, National Cheng Kung University, No.1, University Road, Tainan City 701, Taiwan, R.O.C.

B.-C. Chien and T.-P. Hong (Eds.): Opportunities and Challenges, SCI 214, pp. 323–328.
springerlink.com

precise prediction, however, in so doing knowledge induced from data could be conflictive to domain knowledge and very often becomes counter-intuitive. In medicine, for example, blood pressure is one of the general significant indices in diagnoses of diseases such as diseases of the circulatory system. The higher blood pressure is, the greater possibility that one would get ill. The aforementioned property can be characterized as monotonicity, which could be an important requirement for explaining and justifying model outcomes. Other examples of monotonicity constrains can be found in finance [1], law [2], and other areas.

In this paper, we present an algorithm for rule extraction which adapts the heuristic Particle Swarm Optimization (PSO) method with monotonicity constraints (PSO-MC) to extract rules from databases. The PSO algorithm has the advantage of particle expression, on account of each particle representing a position coordinates, and it is easy to calculate the distance between particles to evaluate monotonicity. Results show that the proposed PSO-MC can yield better performance as for comprehensibility and justifiability.

The paper is organized as follows. In Section 2, preliminaries of the PSO-MC are briefly described. In Section 3, we explain how the algorithm is developed. In Section 4, the algorithm is experimented with a real medical data set. The performance analysis is presented. Finally, we draw conclusions in Section 5.

2 Preliminaries

PSO is an heuristic method imitating natural evolution such as movements of a flock of birds or a school of fish searching for food [3]. In PSO, where and how fast a particle would move to its next position are based on its own experience and the experience of its most successful neighbor. By updating their previous best position and the best position in the neighborhood, the particles approach to the optimal position for feeding and the PSO algorithm reaches toward its optimal solution. Recent research showed that PSO is competitive with tree induction algorithms, such as J48, and Genetic Algorithm (GA) [4]. An evolution classifier adopting GA and Genetic Programming (GP) is used to produce accurate and comprehensible rules [5], and this two-phase evolutionary classifier was extended to solve dual objective classification rules mining with promising results [6].

Let $\vec{X_i}(t)$ and $\vec{V_i}(t)$ symbolize the previous position and velocity of particle P_i, respectively. Also, let the best experience position of P_i be denoted as $\vec{X}_{i\,Pbest}$ and the position of leader be denoted as $\vec{X}_{Leader}$. The new position and velocity respectively denoted as $\vec{X_i}(t+1)$ and $\vec{V_i}(t+1)$ can be updated as follows.

$$\begin{aligned}&\vec{V_i}(t+1)=\chi\left[\vec{V_i}(t)+c_1 rand_1\left(\vec{X}_{i\,Pbest}-\vec{X_i}(t)\right)+c_2 rand_2\left(\vec{X}_{Leader}-\vec{X_i}(t)\right)\right],\\&\vec{X_i}(t+1)=\vec{X_i}(t)+\vec{V_i}(t+1),\end{aligned} \tag{1}$$

where $rand_1$, $rand_2$ are random numbers between 0 and 1, and $c_1, c_2 > 0$; and χ is the constriction factor [7]. The fitness function used to evaluate fitness of the i_{th} particle at time t is shown as follows:

$$fitness\left(\overrightarrow{X_i}(t)\right)=\frac{tp}{tp+fn}\times\left(\frac{tp}{tp+fp}+\frac{tn}{tn+fp}\right), \quad (2)$$

where *tp*, *fp*, *tn*, *fn* are true positive, false positive, true negative and false negative, respectively.

In data mining, given a dataset $D=\{x^i,y^i\}_{i=1}^N$, with $x^i=\left(x_1^i,x_2^i,\ldots,x_n^i\right)\in X$ denoting the feature space, a partial ordering "$\leq$" defined over this input space X, and a linear ordering "$\leq$" defined over the space Y of class values y^i, a monotonicity constraint holds for the classifier if statement in (3) holds.

$$x^i \leq x^j \Rightarrow f\left(x^i\right)\leq f\left(x^j\right)\ or\ f\left(x^i\right)\geq f\left(x^j\right),\ \text{for}\ \forall i,j. \quad (3)$$

To incorporate monotonicity constraints in PSO algorithm, we modify the updating algorithm as follows:

$$\overrightarrow{V_{Mi}}(t+1)=\chi\left[\overrightarrow{V_i}(t)+c_{m1}rand_1\left(\overrightarrow{X_{Mi\ Pbest}}-\overrightarrow{X_i}(t)\right)+c_{m2}rand_2\left(\overrightarrow{X_{M\ Gbest}}-\overrightarrow{X_i}(t)\right)\right], \quad (4)$$

where acceleration constants c_{m1}, c_{m2} > 0; and $\overrightarrow{X_{Mi\ Pbest}}$ and $\overrightarrow{X}_{MGbest}$ are personal best position of the i_{th} particle and the global best position of the swarm with constraints. The fitness function of monotonicity is defined as follows:

$$M\_fitness\left(\overrightarrow{X_i}(t)\right)=w\times\frac{tp}{tp+fp}\times\left(1+\frac{tp}{tp+fn}\right), \quad (5)$$

where $w=1/\left(1+Nd(P_i,P_j)\right)$ and $Nd\left(P_i,P_j\right)=\sqrt{\sum_{l=1}^{n}(P_i^l-P_j^l)^2}/\sqrt{d}$, for any i, j = 1,2,…,N, l = 1,2,…, n.

Positions are updated according to the synthesis velocity defined as

$$\overrightarrow{V_{Si}}(t+1)=(1-u)\times\vec{V}(t+1)+u\times\overrightarrow{V_{Mi}}(t+1). \quad (6)$$

3 A Two-Phase Classifier Based on PSO-MC

In this study, a two-phase evolutionary classifier is designated to generate an interesting and comprehensible rule set. The evolution process is shown in Figure 1.

In the first phase, candidate classification rules with good quality are extracted . Michigan coding approach that each particle represents an individual rule is adopted. To enhance the diversity of candidate rules, token competition is used to generate candidate leaders in this phase.

In the second phase, the best rule set is extracted. Pittsburgh coding approach is adopted to take into account the interaction between rules, and the Gago and Bento's Distance Metric [8] is used to determine the optimal rule set with the highest coverage for the given data set in this phase. A rule set predicts the class of an instance by majority voting, and the classification accuracy of the rule set is computed as hit rate, the number of instances covered by the rule set divided by

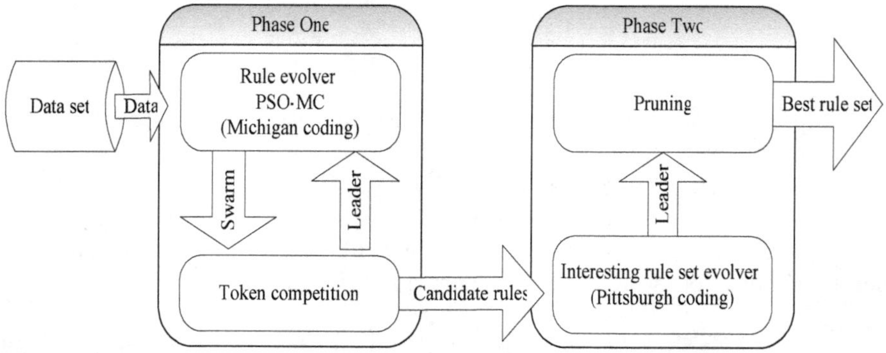

Fig. 1 A two-phase classifier based on PSO-MC

the number of training instances. In the final stage of the second phase, post-processing based on distance between rules is implemented to eliminate redundant rules and enhance comprehensibility.

4 Experiment and Results Analysis

We use the in-vitro fertilization (IVF) dataset collected from a women and infants clinic in Taiwan. After removing instances with missing values, there are 654 instances gathered from 2000 to 2006 in order to keep the same kind of medication. Each instance consists of 10 attributes, namely embryos transferred easy/difficult, ICSI/IVF, number of eggs, number of transferred embryos, transferred embryo grade, quality of follicle, number of mature eggs, embryo grade, age, dosage of Gonal_F/Puregon.

The PSO-MC algorithm was implemented in MATLAB 7.4.0 from Microsoft Vista operation system. In order to ensure the validity and get the best estimate of accuracy, 10-fold cross validation is implemented.

Classification results of the IVF showed in Table 1. We have made reference to the Waikato Environment for Knowledge Analysis (WEKA) system release 3.4 [9] which contains a large number of techniques. We have chosen some representatives such as decision tree J48, Naïve Bayes, Bayes network, and multi-layer perceptron to compare classification performance with PSO-MC. Parameter values used for above mentioned techniques are those set as default in WEKA.

Table 1 The comparison results of the IVF dataset

	PSO - MC	Decision tree J48	Naïve Bayes	Bayes Net	MLP ANN
Accuracy/ S.D. (%)	73.03/ 4.3	68.96	66.67	65.13	69.57
Times of win (test)	-	69	82	94	69
Times of win (train)	-	100	100	100	100
Rank	1	3	4	5	2

Table 2 A part of the best comprehensible rule set of IVF dataset

No.	Rule	Predicted class	Support	Confidence
1	IF transferred embryo grade <= 137 AND age >= 29	No	0.59	0.77
2	IF age >=38	No	0.18	0.85
3	IF ICSI/IVF = ICSI AND embryo transfer = difficult AND number of eggs <= 23 AND number of transferred embryos <= 4 AND transferred embryo grade <= 130 AND quality of follicle <=45 AND number of mature eggs <=11 AND embryo grade<=225 AND age >= 37 AND dosage of Gonal_F/Puregon >= 34	No	0.01	1.00
4	IF number of transferred embryos <= 1 AND transferred embryo grade <= 27 AND quality of follicle <= 9 AND embryo grade<= 47	No	0.03	1.00
5	IF transferred embryo grade <= 32 AND quality of follicle <= 12	No	0.07	0.95
Testing accuracy: 74.36%				

Table 1. shows that our proposed PSO-MC method performs better than the others. The results show that the proposed two-phase PSO-MC classifier considering monotonicity constraint would prevent unwanted rules and enhance justifiability. A part of the best comprehensible rule set of IVF dataset is listed at Table 2. Support factor and confidence factor are used to measure the performance of each rule.

5 Conclusions

A two-phase PSO-MC evolutionary classifier is adopted to generate an interesting and comprehensible rule set in medical diagnosis. The classifier considers not only classification accuracy but also monotonicity of the data in order to enhance comprehensibility. The second phase produces accurate and interesting rule set by optimizing the distance metric which quantifies the heterogeneity of rule set in order to provide the highest coverage for given data. The proposed two-phase PSO classifier has been validated upon datasets such as in-vitro fertilization database. The results obtained in this study indicate that when applied to IVF database, the proposed constraints-based PSO is competitive in classification works compared to many existing classifiers in literature. Moreover, the rules extracted with constraints takes account of hard constraints and monotonicity constraints to enhance comprehensibility and avoid extracting the whole range of attribute value tests.

References

1. Gamarnik, D.: Efficient learning of monotone concepts via quadratic optimization. In: The eleventh annual conference on computational learning theory, pp. 134–143. ACM Press, New York (1998)
2. Karpf, J.: Inductive modeling in law: example based expert systems in administrative law. In: The third international conference on artificial intelligence in law, pp. 297–306. ACM Press, New York (1991)
3. Kennedy, J., Eberhart, R.: Particle swarm optimization. In: Proceedings of the 1995 IEEE International Conference on Neural Networks, Piscataway, NJ (1995)
4. Sousa, T., Silva, A., Neves, A.: Particle swarm based data mining algorithms for classification tasks. Parallel Computing 30, 767–783 (2004)
5. Tan, K.C., Yu, Q., Heng, C.M., Lee, T.H.: Evolutionary computing for knowledge discovery in medical diagnosis. Artificial Intelligence in Medicine 27, 129–154 (2003)
6. Tan, K.C., Yu, Q., Ang, J.H.: A dual-objective evolutionary algorithm for rules extraction in data mining. Computational optimization and applications 34, 273–294 (2006)
7. Clerc, M., Kennedy, J.: The particle swarm-explosion, stability and convergence in a multi-dimentional complex space. IEEE Transactions on Evolutionary Computation 6, 58–73 (2002)
8. Gago, P., Bentos, C.: A metric for selection of the most promising rules. In: Principles of data mining and knowledge discovery, Nantes, France (1998)
9. Witten, I.H., Frank, E.: Data Mining: Practical Machine Learning Tools and Techniques. Morgan Kaufmann, San Francisco (2000)

Incorporation of Adapted Real World Tournament Selection into Artificial Immune Recognition System

Shahram Golzari, Shyamala Doraisamy, Md. Nasir Sulaiman, and Nur Izura Udzir

Abstract. The resource competition phase of the Artificial Immune Recognition System (AIRS) incorporates a selection mechanism with a high selective pressure and loss of diversity. This selection mechanism generates premature memory cells and decreases the classification accuracy. In this study, the Real World Tournament Selection (RWTS) method is incorporated in resource competition phase of AIRS to tackle this limitation. Some experiments are conducted to evaluate the accuracy of new algorithm, named RWTSAIRS. Algorithms were tested on benchmark datasets of UCI machine learning repository and RWTSAIRS achieved better classification accuracy in all cases.

1 Introduction

Natural computation is the study of computational systems that uses ideas based on natural systems, including biological, ecological and physical systems. One branch of natural computation is Artificial Immune System (AIS). AIS is a computational method inspired by the biological immune system. It is progressing slowly and steadily as a new branch of computational intelligence and soft computing [1, 2]. It has been used in several applications such as: machine learning, pattern recognition, computer virus detection, anomaly detection, optimization and robotics [2]. One of the AIS based algorithms is Artificial Immune Recognition System (AIRS). AIRS is a supervised immune-inspired classification system capable of assigning data items unseen during training to one of any number of classes based on previous training experience. AIRS is probably the first and best known AIS for classification, having been developed in 2001 [3].

AIRS has four main steps: Initialization, ARB generation, Competition for resources and nomination of candidate memory cell, and finally promotion of candidate memory cell into memory pool.

Shahram Golzari, Shyamala Doraisamy, Md. Nasir Sulaiman, and Nur Izura Udzir
Faculty of Computer Science and Information Technology,
Universiti Putra Malaysia, 43400, Serdang, Selangor, Malaysia
e-mail: golzari@ieee.org, {shyamala,nasir,izura}@fsktm.upm.edu.my

B.-C. Chien and T.-P. Hong (Eds.): Opportunities and Challenges, SCI 214, pp. 329–334.
springerlink.com

The resource competition phase plays an important role in AIRS [4]. The goal of resource competition is the development of fittest individuals. The resource competition phase improves the quality of high affinity ARB (Artificial Recognition Ball). Resource competition phase removes weakest ARBs and selects strongest (seemly good) ARBs. This type of selection has high selective pressure and also a loss of diversity. It may generate premature memory cells. The stochastic selection methods in evolutionary computation attempts to balance between selection pressure and diversity to prevent the algorithms from converging towards wrong solutions [5]. One of the recently proposed selection methods is Real Word Tournament Selection (RWTS) [6]. RWTS is derived from the real world tournament competition method, and statistical analysis shows that RWTS has a higher selection pressure with a relatively small level of diversity and a higher sampling accuracy than other conventional selection methods [7].

In this study, we adapt the RWTS in order to incorporate it in resource competition phase of AIRS and increase the accuracy of AIRS. To evaluate the performance of this new algorithm, named RWTSAIRS, both algorithms are run on some benchmark datasets from the UCI machine learning repository and ten-fold cross validation method is used to estimate the accuracy of algorithms.

The rest of paper is organized as follows; the following section introduces the AIRS algorithm in briefly. Section 3 describes the resource competition phase and RWTS method. Section 4 illustrates the evaluation method, experiments and results.

2 AIRS

Artificial Immune Recognition System (AIRS) is investigated by Watkins [3]. AIRS can be applied to classification problems, which is a very common real world data mining task. To show the capability of AIS for performing classification was the initial objective of developing AIRS, but results shown that AIRS is comparable with famous classifiers. Before AIRS, most artificial immune system researches focused on unsupervised learning and clustering. The only other attempt to use immune systems for supervised learning was the work of Carter [8]. AIRS uses the several concepts of immune system including resource competition, clonal selection, affinity maturation, memory cell production and also used the resource limited artificial immune system concept investigated by [9]. In fact, AIRS is a hybrid algorithm that uses the concepts of different immune system theories.

Feature vectors (labeled data) presented for training and testing are named as antigens and the system units are called as ARBs (Artificial Recognition Balls) or B cells. In theory, similar B cells are represented as Artificial Recognition Balls (ARBs) and ARBs compete with each other for a fixed maximum number of B cells. AIRS adapts these concepts. In AIRS, ARB and B cells are the same and ARBs compete for a fixed number of resources. The algorithm generates new instances as memory cells that are used in classification task finally. Memory cells are best ARBs. These ARBs have highest affinities to training antigens and generated based on the immune metaphors.

AIRS has four stages: The first stage is performed once at the beginning of the algorithm. This stage includes normalization and initialization. Other stages are performed for each antigen in the training set. These stages are ARB generation, resource competition and insertion candidate memory cell into memory pool. Finally kNN classifier is used to predict the class of unseen data. Comprehensive description of AIRS has been presented in [3, 4].

AIRS has been shown to be comparable with famous classifiers [3]. In [4] some updates, such as modification in ARB pool structure, mutation routine, cloning routine and stopping criteria, were applied on AIRS. These modifications didn't put significant positive effect on accuracy of AIRS. Also some researches have been done to improve and evaluate the performance of AIRS [10-13].

3 Resource Competition

AIRS uses the adapted version of resource limited artificial immune system introduced in [9]. In AIRS, ARB and B-cell concepts are same and ARBs of each class compete for resources. Each class could have the maximum number of resources. This amount is a portion of resources of system. The resources of system are also limited. Each ARB claims the number of resources depend on its affinity to antigen and its class. If the total number of resources allocated for one class is greater than the maximum number of resources allowed for it, then the number of resources make up the difference must be removed. To remove the extra resources, the weakest ARB (less rewarded ARB) is selected and a sufficient number of its resources are removed. If the number of its resources falls to zero then the ARB is removed from the system. This process is repeated until the allocated resources and maximum amount allowed resources for class be equal. Therefore, the resource competition phase selects the strongest ARBs to stay in system and removes weakest ARBs from system (survival the fittest individuals). The resource competition then is a selection process and generating memory cells is an evolutionary process.

There are two important issues in the evolutionary process: diversity and selective pressure. These factors are strongly related: an increase in the selective pressure decreases the diversity and vice versa. In other words strong selective pressure supports the premature convergence of process; a weak selective pressure can make the process ineffective [5]. Thus it is important to strike a balance between these two factors. Stochastic selection will maintain the diversity in the population by occasionally choosing not-so-good solutions. The resource competition phase of AIRS supports high loss of diversity, because the weakest ARBs don't have even a little chance to be selected for next step.

RWTS is a tournament-based selection method that introduced in [6]. Statistical analysis show that RWTS has a higher selection pressure with a relatively small less of diversity and higher sampling accuracy than conventional tournament selection method [6,7]; In addition, tournament selection has been shown to be stable and reliable when under pressure, in contrast to other stochastic methods [5]. Therefore, we use the RWTS method in this study.

The idea of RWTS is derived from real world tournament competitions. The participants compete in elimination races, and if one wins a competition, then one survives; if not, one is eliminated. In RWTS, each individual in the population is sequentially paired with a neighbor. The competition factor is fitness. When all competitions in the present tournament level are completed, only the winners are inserted into the mating pool and go on to the next tournament level. The process is repeated until the suitable numbers of individuals are selected [6].

To use RWTS in the resource competition phase of AIRS, some adaptation must be done to RWTS. In resource competition, individuals are the ARBs of each class. These ARBs participate in competition. Each ARB competes with a neighbor (ARB B is neighbor of ARB A, and if ARB B is the first ARB among the ARBs that is generated in the system after ARB A, then its class of them is similar to the class of ARB A). The basis of competition is the amount of allocated resources of ARBs. The amount of allocated resources for each ARB is calculated in the resource allocation process. The ARB with more allocated resources remains in the system and another ARB leaves the system. This competition repeated for pairs of ARBs until the allowed resources for class will be equal the sum of allocated resources of remained ARBs, or all ARBs participate in competition. If all ARBs participate in competition and the allowed resources for class is greater that sum of allocated resources yet, the next tournament level of competition will be started. The competition rules are same for all levels of tournament. The competition is repeated until the equality of allowed resources with sum of allocated resources would be satisfied finally.

4 Experiments and Results

Experiments were carried out in order to determine how RWTSAIRS performed compared to AIRS. The WEKA programme codes of AIRS [14] was used to incorporate the adapted RWTS in resource competition phase of AIRS. To have as fair as possible comparison between the two algorithms, both the AIRS and RWTSAIRS were run with the default parameters of the code. AIRS is a self-determinant classifier [12]; therefore the used parameters in experiments have the little effect on system performance. The values of used parameters are shown in Table 1.

Table 1 Algorithm Parameters

Used Parameter	Value
Clonal rate	10
Mutation rate	0.1
ATS	0.2
Stimulation threshold	0.9
Resources	150
Hypermutation rate	2
K value in KNN classifier	3
Seed	1

Twelve benchmark datasets were retrieved from the well-known UCI machine learning repository [15]. We selected datasets with varying number of attributes, instances and classes, from simple toy datasets to difficult real world learning problems to cover the complete characteristics of data. The N-fold cross validation method was used to estimate the classification accuracy. Several theoretical ideas, and also several tests on numerous different data sets by using different classification algorithms have shown that ten fold cross validation gets best estimate of accuracy [16]; therefore we used ten fold cross validation in experiments. Table 2 shows the accuracies obtained by algorithms. Based on the results RWTSAIRS obtains the higher accuracy in all cases.

Table 2 Comparison of classification accuracy

Dataset	**AIRS (%)**	**RWTSAIRS (%)**
Balance-Scale	83.6892	84.32412
Breast-Cancer	96.6347	97.80051
Credit-Crx	82.89855	83.91304
German	66.3	72.3
Glass	60.75758	63.50649
Hepatitis	83.125	83.16667
Image-Segment	82.38095	83.80952
Ionosphere	86.06349	86.3254
Iris	94.66667	95.33333
Pima-Diabetes	70.58271	73.44668
Wine	94.93464	94.93464
Zoo	94.96364	96

5 Conclusions

In this study, we incorporated RWTS method in resource competition phase of AIRS in order to prevent the possible generation of premature memory cells and increase the accuracy of AIRS. Ten fold cross validation was used as evaluation method to compare the accuracy of AIRS and proposed algorithm, RWTSAIRS. The results of experiments showed that RWTSAIRS increases the accuracy of AIRS in all used datasets.

References

1. de Castro, L.N., Timmis, J.: Artificial Immune Systems as a novel Soft Computing Paradigm. Soft Computing Journal 7(8), 526–544 (2003)
2. de Castro, L.N., Timmis, J.: Artificial Immune Systems: A New Computational Intelligence Approach. Springer, Heidelberg (2002)

3. Watkins, A.: AIRS: A Resource Limited Artificial Immune Classifier. M.S. thesis, Department of Computer Science, Mississippi State University (2001)
4. Watkins, A., Timmis, J., Boggess, L.: Artificial Immune Recognition System (AIRS): An Immune-Inspired Supervised Learning Algorithm. Genetic Programming and Evolvable Machines 5(3), 291–317 (2004)
5. Bäck, T., Fogel, D.B., Michalewicz, Z. (eds.): Evolutionary Computation 1: Basic Algorithms and Operators. IOP, Bristol (2000)
6. Soak, S., Corne, D., Ahn, B.: A Powerful New Encoding for Tree-Based Combinatorial Optimization Problems. In: Yao, X., Burke, E.K., Lozano, J.A., Smith, J., Merelo-Guervós, J.J., Bullinaria, J.A., Rowe, J.E., Ti o, P., Kabán, A., Schwefel, H.-P. (eds.) PPSN 2004. LNCS, vol. 3242, pp. 430–439. Springer, Heidelberg (2004)
7. Lee, S., Soak, S., Kim, K., Park, H.: Statistical properties analysis of real world tournament selection in genetic algorithms. Applied Intelligence 28(2), 195–205 (2008)
8. Carter, J.H.: The immune systems as a model for pattern recognition and classification. Journal of the American Medical Informatics Association 7(1), 28–41 (2000)
9. Timmis, J., Neal, M.: A Resource Limited Artificial Immune System. Knowledge Based Systems 14(3), 121–130 (2001)
10. Marwah, G., Boggess, L.: Artificial immune systems for classification: Some issues. In: Proceedings of the first international conference on artificial immune systems, University of Kent Canterbury, England, pp. 149–153 (2002)
11. Goodman, D.E., Boggess, L., Watkins, A.: Artificial immune system classification of multiple-class problems. In: Proceedings of the artificial neural networks in engineering (ANNIE 2002), pp. 179–183 (2002)
12. Goodman, D.E., Boggess, L., Watkins, A.: An investigation into the source of power for AIRS, an artificial immune classification system. In: Proceedings of the international joint conference on neural networks (IJCNN 2003), Portland, Oregon, pp. 1678–1683 (2003)
13. Boggess, L., Hamaker, J.: The Effect of Irrelevant Features on AIRS, an Artificial Immune-Based Classifier. In: Proceedings of the intelligent engineering system through artificial neural networks (AINNIE 2003), pp. 219–224 (2003)
14. Brownlee, J.: Artificial Immune Recognition System (AIRS) - A Review and Analysis. Tech. Rep., Centre for Intelligent Systems and Complex Processes (CISCP), Faculty of Information and Communication Technologies (ICT), Swinburne University of Technology (2005)
15. Newman, D.J., et al.: CI Repository of machine learning databases (1998), `http://www.ics.uci.edu/~mlearn/MLRepository.html` (retrieved, July 2008)
16. Witten, H., Frank, E.: Data Mining: Practical Machine Learning Tools and Techniques, 2nd edn. Morgan Kaufmann, San Francisco (2005)

Competition State Visualization for Sales Record Mining

Yoshihiro Hayashi, Ryosuke Saga, and Hiroshi Tsuji

Abstract. To determine the competition state among products, this paper is used to present a visualization technique for mining interesting knowledge. First, the technique determines the relations among products from volumes of sales records. Then, it identifies the preference transition as a direction for the identified relations. According to the type of direction and the number of connected links, the technique assigns the size and color of a product a node and generates a graph. Finally, this paper describes the semantics of the competition state among the products in a graph. A numerical example of the sales records of 172 kinds of products and their 90,000 sales records is also illustrated.

1 Introduction

Data mining [1] has been a major topic for knowledge engineering. In particular, there are many works on mining recommendation knowledge from volumes of sale records, such as for deciding the similarities among customers, identifying customers' preferences [2], and identifying the preferable products that are selected instead of the others [3]. However, these techniques have actually been developed for customers not for the providers. The providers are generally more interested in their products state of competition, basically wanting to know which are their rival products and what are their strong and/or weak points?

However, techniques for visualizing knowledge embedded in volumes of data have also been remarkable topics in the field of knowledge engineering [4]. Some examples of these techniques include the Key-graph [5] and FACT-graph [6]. While Key-graph has paid much attention on co-occurrence among topics,

Yoshihiro Hayashi and Hiroshi Tsuji
Graduate School of Engineering, Osaka Prefecture University,
1-1, Gakuencho, Naka-ku, Sakai, 599-8531 Osaka, Japan
e-mail: {hayashi@mis., tsuji@}cs.osakafu-u.ac.jp

Ryosuke Saga
Department of Information and Computer Sciences
Kanagawa Institute of Technology 1030 Shimo-ogino, Atsugi,
243-0292 Kanagawa, Japan
e-mail: saga@ic.kanagawa-it.ac.jp

B.-C. Chien and T.-P. Hong (Eds.): Opportunities and Challenges, SCI 214, pp. 335–340.
springerlink.com

FACT-graph has paid attention on recency as well as frequency for expressing trend. Such visualizing techniques are owed to such drawing tools as Graphviz [7].

A method for visualizing product competition states from volumes of sales records based on this background is proposed in this paper. The target products are commodities, such as beverages and toothbrushes. Widely reservation of business hotel is also included. This means: 1) there are [volume kinds of products, 2) a customer does not always decide the brand of products, and 3) he frequently purchase in his life.

The organization of the paper is as follows. First, there the problem is described. Next, is a proposal on how to visualize the competition state among products. Then, to demonstrate the proposed method, the results from a numerical experimentation are presented.

2 Problem

The method of visualizing the competition state of products for market analysts not for consumers is proposed in this paper. Developing general methods for every product is unrealistic because there too many product properties,. Therefore, this section starts by describing a targeted class of products and their available sales records.

- There are a lot of product instances (items) in a specific class. 'A lot of kinds' is assumed to be more than about 100.
- The users frequently buy the items. 'Frequently' is several times a year at least.
- There is a possibility that the users' preference moves to another item from the current one. On the other hand, even after the preference seems to move to new items from the previous ones, the users may again buy the previous items.

An overview of our approach is shown in Fig. 1 and will be discussed later.

- The method of visualizing the entire competition state.
- The method of visualizing a partial competition state.

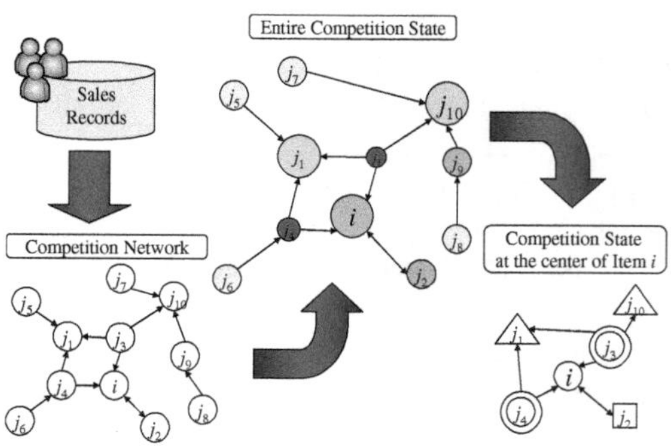

Fig. 1 Target System Image

3 Visualizing Product Competition State

This section describes how to make a graph for visualizing the product competition state. First, let us introduce a competition network, and then explain the semantics for visualizing the entire and partial competition states.

3.1 Creation of Competition Network

A competition network that expresses the movement of a user's preference between items is created to visualize the competition state [3]. The network is generated from the sales records as a directed graph.

The competition network is studied to visualize the competition state. The process to create a competition network consists of the following steps: 1) expressing the relations between items, 2) calculating the co-occurrence (cross-purchasing) and filtering the relations by using the Simpson coefficient, and 3) identifying a user's preference transition by using a Mann-Whitney U test [8].

3.2 Method for Visualizing Entire Competition State

An entire competition state is determined by changing the sizes and colors of the nodes based on the proposed competition network. First, in order to classify the nodes by their strengths, $InDegree_i$ and $OutDegree_i$ are defined. $InDegree_i$ which is a products degree of strength shows how many items transfer to item *i*. However, $OutDegree_i$ which is a products degree of weakness degree shows how many items are transferred from item *i*.

1) Deciding node size

Then formula (1) is used to evaluate item *i* by using $InDegree_i$ and $OutDegree_i$. The sizes of the nodes are classified by formula (1). When all the nodes are compared, the size of the nodes whose P_i is big are expanded and those whose P_i are small are reduced.

$$P_i = InDegree_i - OutDegree_i \tag{1}$$

2) Changing color of nodes

To change the color of the nodes, the average of *InDegree* of all nodes (*Av.In*) and the average of *OutDegree* of all the nodes (*Av.Out*) are used. An example method for changing color of the nodes is listed in Table 1. When item *i* belongs to 1 in Table 1, $Indegree_i$ is also more than *Av.In* but $OutDegree_i$ is more than *Av.Out*, so item *i* is regarded as an item that should be noted. In the case of 2, item *i* is regarded as a good item. In the case of 3, item *i* is regarded as a bad condition. In the case of 4, item *i* is regarded as an isolated item whose relation to the others is thin.

Table 1 Classification color of nodes

	$Av.In > InDegree_i$	$Av.In < InDegree_i$
$Av.Out < OutDegree_i$	3 (red)	1 (yellow)
$Av.Out > OutDegree_i$	4 (gray)	2 (blue)

3.3 Method for Visualizing Partial Competition States

When we look at one item *i*, determining what the competition state of item *i* is with the items around item *i* is shown by changing the shape of the nodes based on the directions of the links in the competition network (Table 2). The partial competition state is made visible up to distance *L* from item *i*. The visualizing method up to $L = 2$ is discussed in this paper. First, in regard to item *j*, as distance *L* between items *i* and *j* is a 1, and in regard to item k, as distance *L* between items *i* and *k* is a 2, the shape of the nodes is classified based on the following rules.

Table 2 Classification of nodes

The shape of nodes	Domination (◎)	Threat (△)	Apposition (□)	Uncertainty (◇)
Mean	This item give a chance to item *i*	This item deprive item *i* of a chance	This item is no difference with item *i*	The relation between item *i* and this is uncertainty

- Rule for L=1
  1. If node $i \leftarrow j$ then node *j* is (◎)
  2. If node $i \rightarrow j$ then node *j* is (△)
  3. If node $i \leftrightarrow j$ then node *j* is (□)
- Rule for L=2
  1. If node $i \leftarrow j \leftarrow k$ then node *k* is (◎)
  2. If node $i \leftarrow j \rightarrow k$ or $i \leftarrow j \leftrightarrow k$ then node *k* is (□)
  3. If node $i \rightarrow j \rightarrow k$ then node *k* is (△)
  4. If node $i \rightarrow j \leftarrow k$ or $i \rightarrow j \leftrightarrow k$ then node *k* is (◇)
  5. If node $i \leftrightarrow j \leftarrow k$ then node *k* is (◎)
  6. If node $i \leftrightarrow j \rightarrow k$ then node *k* is (◇)
  7. If node $i \leftrightarrow j \leftrightarrow k$ then node *k* is (□)
  8. If some rules overlaps and a shape of node *k* is different then node *k* is (◇)

4 Example

To visualize the competition state for real sales records, we used data containing 2,227 users who stayed in several of the 172 business hotels in Tokyo for more than twenty times. Note that the Simpson coefficient is 0.5.

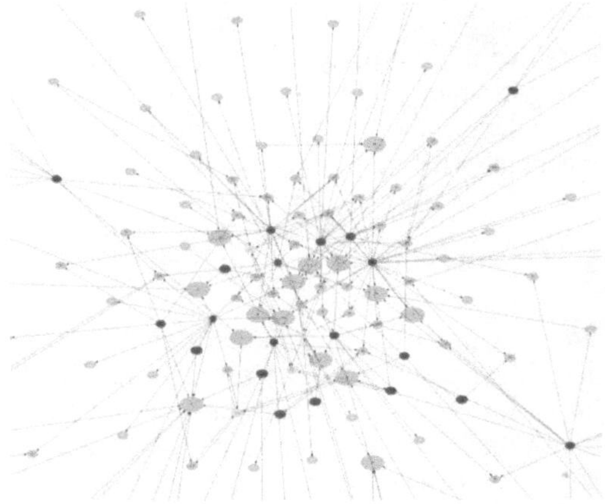

Fig. 2 Visualized entire competition state

4.1 Visualizing Entire Competition State

First, the nodes whose values P in formula (1) are ranked at the top and the lower 10 % have either strengths or weaknesses. Next, the color of the nodes is assigned by using the InDegree and OutDegree. As a result, Fig. 2 shows which items are favorable. This figure shows that the blue nodes are dominant in a sense and the red nodes may have defects. On the other hand, an item whose size is small and whose color is yellow can also be found in this figure. This indicates that the item might be used temporarily instead of the other items. Then, items with a large yellow node are assumed to be excellent and have few competitors.

4.2 Visualizing Partial Competition State

To demonstrate the power of the proposed method, let us illustrate the competition state for item 16 in Fig. 2. Figure 3 shows that item 7 offers a new chance to item 16 via item 52, and item 16 is superior to item 7. Then, since the preference transitions between items 16 and 147 are bi-directional, items 52 and 147 are as good as item 16. The relation between item 16 and the other items is uncertainty. In this way, we can see the partial competition state. Here, if items 16 and 147 are regarded as rivals, items 11, 75, 78, and 109 seem to deprive item 147 of a chance,

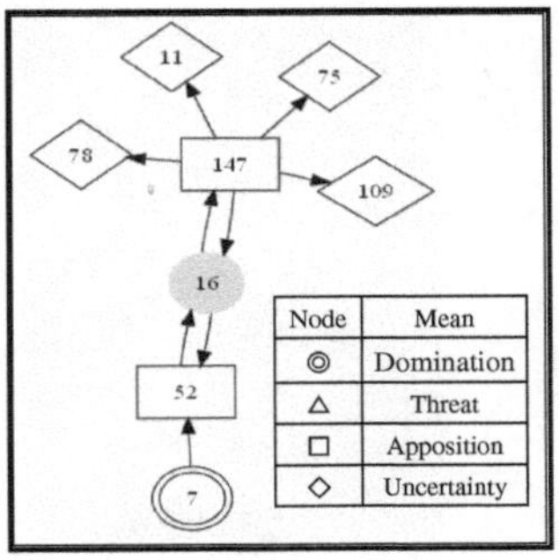

Fig. 3 Visualized competition state at the center of item 16

and thus, help item 16. Then, item 7 is set as domination, but if item 52 is regarded as a rival, item 7 gives a chance to item 52 too, so item 7 seems to interfere with item 16. In this way, how the nodes whose distance L is a 1 are classified may change the meaning of the nodes whose distance L is a 2 and give new meaning to the nodes regarded as uncertainty.

5 Conclusion

We have described a method for visualizing the competition state among products and illustrated it using an example graph. We have shown the power of finding interesting knowledge that was unclear in original sales records by describing how to use the visualized graph for analyzing the competition state. Although the technique was developed for special classes of products like beverages and toothbrushes, the coming research should widen the class of products for use with the visualization technique.

References

1. Adriaans, P., Zantinge, D.: Data Mining. Addison-Wesley, Reading (1996)
2. Saga, R., Tsuji, H., Onoda, J.: Agent System for Notifying Hotel Room Reservation Alterna-tives. In: 11th International Conference on Human Computer Interaction (HCII 2005), vol. 5, Emergent Application Domains in HCI, pp. 1–10 (2005) (CD-Rom)
3. Saga, R., Hayashi, Y., Tsuji, H.: Hotel Recommender System based on User's Preference Transition. In: IEEE International Conference on Systems, Man & Cybernetics (IEEE/SMC 2008), pp. 2437–2442 (2008)
4. Yoshikawa, T.: Visualization Techniques of Multi-Dimensional Data. Systems, Control and Information, Vol 52(7), 232–238
5. Ohsawa, Y., Benson, N.E., Yachida, M.: KeyGraph: Automatic Indexing by Segmenting and Unifing Co-occurrence Graphs. IEICE D-I, J82-D-I(2), 391–400 (1999)
6. Terachi, M., Saga, R., Sheng, Z., Tsuji, H.: Visualized Technique for Trend Analysis of News Articles. In: Nguyen, N.T., Borzemski, L., Grzech, A., Ali, M. (eds.) IEA/AIE 2008. LNCS (LNAI), vol. 5027, pp. 659–668. Springer, Heidelberg (2008)
7. Graphviz, `http://www.graphviz.org/`
8. Lehmann, E.L.: Nonparametric Statistical Methods Based on Ranks. McGraw-Hill, New York (1975)

Author Index

Abreu, Rui 233

Bosse, Tibor 285

Calot, Enrique 19
Chalabi, Nida 193
Chan, Chien-Chung 241
Chan, Ying-Fang 323
Chang, Fengming M. 147
Chang, Hsueh-Wei 13
Chang, Tsair-Yuan 271
Chang, Ye-In 133
Chen, Austin H. 1
Chen, Chen-Tung 85
Chen, Chih-Chuan 323
Chen, Ching-Fa 127
Chen, I-Fei 99
Chen, Jyun-Da 175
Chen, Peter P. 297
Chen, Yun 119
Cheng, Yi-Chung 323
Cheng, Yu-Huei 13
Chiang, Chang Yu 265
Chiang, Chia Hsun 303
Chou, Wenya 139
Chou, Yo-Shing 303
Chouza, Mariano 153
Chuang, Li-Yeh 13
Chung, Chia-Hsien 187
Clua, Osvaldo 153

de Vos, Michael 285
Doraisamy, Shyamala 329
Duell, Rob 285

Faghihi, Usef 279
Fan, Chin-Yuan 213
Fournier-Viger, Philippe 279

Gao, Tao 39
García-Martínez, Ramón 19, 153
Gnanvo, Kondo 225
Golzari, Shahram 329
Granmo, Ole-Christoffer 317
Griffin, Terry 167

Hao, Pei-Yi 7
Hayashi, Yoshihiro 335
Hoch, Richard 225
Hohlmann, Marcus 225
Hong, Tzung-Pei 175
Hoogendoorn, Mark 285
Hsieh, Fu-Shiung 105
Hsu, Chao-Chin 323
Hsu, Chi-Chun 147
Hua, Cheng Chung 105
Huang, Hung-Hsuan 219
Huang, Jen-Peng 59
Huang, Lee-Wen 133
Huang, Yan 167
Hung, Jui-Chung 207
Hung, Wei-Zhan 85

Ju, Jiun-Huang 79
Ju, Ming-Yi 25

Kimes, Sheryl E. 53
Klein, Michel 285

Komatani, Kazunori 111
Ku, Chung-Chao 147
Kuan-Liang, Chen 31
Kubota, Hidekazu 219
Kubota, Yuji 111
Kuo, Huang-Cheng 59

Lambalgen, Rianne van 285
Lee, Chung-Ta 99
Lee, Huey-Ming 187, 259
Lee, Tian-Shyug 99
Lee, Tsang-Yean 187
Lee, Wei-Chi 207
Li, Chunping 73
Li, Guo 139
Li, Nan 73
Li, Sheng-Tun 323
Li, Xinhui 45
Liang, Zuoyuan 317
Lim, Andrew 291, 311
Lin, Cha-Hwa 247
Lin, Ching-Heng 1
Lin, Huan-Yu 147
Lin, Lily 259
Lin, Pei-Chun 93
Lin, Shouxun 45
Lin, Yuan-Horng 199
Liu, Alan 297
Liu, Chen-Hao 213
Liu, Hung-Ze 133
Liu, Jeng-Fu 271
Liu, Jing 159
Liu, Jing-Sin 303
Liu, Yuncai 181
Liu, Zheng-guang 39
Lompo, Salomon 167
Lu, Chi-Jie 99

Ma, Shihua 139
Merlino, Hernan 19
Mephu Nguifo, Engelbert 279
Mitra, Debasis 225
Mohammad, Yasser 253

Nishida, Toyoaki 219, 253
Nkambou, Roger 279

Ogata, Tetsuya 111
Okuno, Hiroshi G. 111
Oommen, B. John 317
Oorburg, Rogier 285
Ouyang, Chen-Sen 25

Pan, Shing-Tai 127
Pham, Huy Nguyen Anh 297

Qian, Yuelian 45
Qin, Hu 291, 311

Rancan, Claudio 19, 153

Saga, Ryosuke 335
Shao, Yuehjen E. 271
Sharpanskykh, Alexei 285
Shen, Chenghao 311
Shihab, Khalil 193
Shiramatsu, Shun 111
Shr, Arthur 297
Shue, Li-Yen 93
Stirpe, Paul A. 175
Su, Chuan Jun 265
Su, Jin-Shieh 187
Su, Jun-Ming 147
Sulaiman, Md. Nasir 329

Takahashi, Toru 111
Treur, Jan 285
Tsai, Cheng-Fa 79
Tsai, Chia-Yin 93
Tsai, Chih-Hung 247
Tsai, Jui-Pin 147
Tsai, Lung-Biao 7
Tsai, Tsung-Hsien 53
Tsan-Ying, Yu 31
Tseng, Shian-Shyong 147
Tsuji, Hiroshi 335

Udzir, Nur Izura 329

van der Mee, Andy 285
van Gemund, Arjan J.C. 233

Wang, Chien-Chih 247
Wang, Kesheng 119
Wang, Shyue-Liang 175
Wang, Xiangdong 45
Wang, Yen-Wen 213

Wang, Yi 65, 119
Wang, Yong 65
Wei-Chih, Hsu 31
Wu, Chien-Ho 271
Wu, Qi 159

Xu, Dongsheng 291

Yang, Cheng-Hong 13
Yang, Han-Lin 25
Yang, Ying 45

Zhang, Jun 39
Zhang, Yang 65, 181
Zhu, Daqi 159
Zoeteweij, Peter 233

Kolibri 360 GmbH
Ferdinand-Jühlke-Straße 7
99095 Erfurt, Deutschland
produktsicherheit@kolibri360.de